Ecological Statistics: Contemporary Theory and Application

Ecological Statistics: Contemporary Theory and Application

Edited By

GORDON A. FOX
University of South Florida

SIMONETA NEGRETE-YANKELEVICH
Instituto de Ecología A. C.

VINICIO J. SOSA
Instituto de Ecología A. C.

OXFORD
UNIVERSITY PRESS

OXFORD
UNIVERSITY PRESS

Great Clarendon Street, Oxford, OX2 6DP,
United Kingdom

Oxford University Press is a department of the University of Oxford.
It furthers the University's objective of excellence in research, scholarship,
and education by publishing worldwide. Oxford is a registered trade mark of
Oxford University Press in the UK and in certain other countries

© Oxford University Press 2015

The moral rights of the authors have been asserted

First published 2015

Published in the United States of America by Oxford University Press
198 Madison Avenue, New York, NY 10016, United States of America

British Library Cataloguing in Publication Data

Data available

Library of Congress Control Number: 2014956959

ISBN 978–0–19–967255–4

Acknowledgments

The contributors did much more than write their chapters; they provided invaluable help in critiquing other chapters and in helping to think out many questions about the book as a whole. We would like to especially thank Ben Bolker for his thinking on many of these questions. Graciela Sánchez Ríos provided much-needed help with the bibliography. Fox was supported by grant number DEB-1120330 from the U.S. National Science Foundation. The Instituto de Ecología A.C. (INECOL) encouraged this project from beginning to end, and gracefully allocated needed funding (through the *Programa de Fomento a las Publicaciones de Alto Impacto/Avances Conceptuales y Patentes* 2012) to allow several crucial work meetings of the editors; without this help this book would probably not have seen the light of day.

Contents

4 Missing data: mechanisms, methods, and messages

Shinichi Nakagawa

5 What you don't know can hurt you: censored and truncated data in ecological research   *Gordon A. Fox*

9 Research synthesis methods in ecology   *Jessica Gurevitch and Shinichi Nakagawa*

List of contributors

Benjamin M. Bolker
Departments of Mathematics & Statistics and Biology
McMaster University
1280 Main Street West
Hamilton, Ontario L8S 4K1
Canada
bolker@mcmaster.ca

Yvonne M. Buckley
School of Natural Sciences
Trinity College , University of Dublin
Dublin 2
Ireland
buckleyy@tcd.ie
and
The University of Queensland
School of Biological Sciences
Queensland 4072
Australia

Gordon A. Fox
Department of Integrative Biology (SCA 110)
University of South Florida
4202 E. Fowler Ave.
Tampa, FL 33620
USA
gfox@usf.edu

James B. Grace
US Geological Survey
700 Cajundome Blvd.
Lafayette, LA 70506
USA
gracej@usgs.gov

Jessica Gurevitch
Department of Ecology and Evolution
Stony Brook University
Stony Brook, NY 11794-5245
USA
Jessica.Gurevitch@stonybrook.edu

Bruce E. Kendall
Bren School of Environmental Science & Management
University of California, Santa Barbara
Santa Barbara CA 93106-5131
USA
kendall@bren.ucsb.edu

Marc J. Lajeunesse
Department of Integrative Biology (SCA 110)
University of South Florida
4202 E. Fowler Ave.
Tampa, FL 33620
USA
lajeunesse@usf.edu

Michael A. McCarthy
School of BioSciences
The University of Melbourne
Parkville VIC 3010
Australia
mamcca@unimelb.edu.au

Earl D. McCoy
Department of Integrative Biology (SCA 110)
University of South Florida
4202 E. Fowler Ave.
Tampa, FL 33620
USA
edm@mail.usf.edu

Shinichi Nakagawa
Department of Zoology
University of Otago
340 Great King Street
P.O. Box 56
Dunedin
New Zealand
shinichi.nakagawa@otago.ac.nz
and
School of Biological, Earth and Environmental Sciences
University of New South Wales
Sydney
NSW 2052
Australia

Simoneta Negrete-Yankelevich
Instituto de Ecología A. C. (INECOL)
Carretera Antigua a Coatepec 351
El Haya Xalapa 91070

Veracruz
México
simoneta.negrete@inecol.mx

Jonathan R. Rhodes
The University of Queensland
School of Geography, Planning, and Environmental Management
Brisbane
Queensland 4072
Australia
jrhodes@uq.edu.au

Shane A. Richards
School of Biological & Biomedical Sciences
Durham University
South Road
Durham, DH1 3LE
UK
s.a.richards@durham.ac.uk

Samuel M. Scheiner
Division of Environmental Biology
National Science Foundation
Arlington, VA 22230
USA
sscheine@nsf.gov

Donald R. Schoolmaster Jr.
U. S. Geological Survey
700 Cajundome Blvd.
Lafayette, LA 70506
USA
schoolmasterd@usgs.gov

Vinicio J. Sosa
Instituto de Ecología A. C. (INECOL)
Carretera Antigua a Coatepec 351
El Haya Xalapa 91070
Veracruz
México
vinicio.sosa@inecol.mx

Introduction

*Vinicio J. Sosa, Simoneta Negrete-Yankelevich,
and Gordon A. Fox*

Why another book on statistics for ecologists?

This is a fair question, given the number of available volumes on the subject. The reason is deceptively simple: our use and understanding of statistics has changed substantially over the last decade or so. Many contemporary papers in major ecological journals use statistical techniques that were little known (or not yet invented) a decade or two ago. This book aims at synthesizing a number of the major changes in our understanding and practice of ecological statistics.

There are several reasons for this change in statistical practice. The most obvious cause is the continued growth of computing power and the availability of software that can make use of that power (including, but by no means restricted to, the R language). Certainly, the notebook and desktop computers of today are vastly more powerful than the mainframe computers that many ecologists (still alive and working today) once had to use. Both hardware and software can still impose limits on the questions we ask, but the constraints are less severe than in the past.

The ability to ask new questions, together with a growing body of practical experience and a growing cadre of ecological statisticians, has led to an increased level of statistical sophistication among ecologists. Today, many ecologists recognize that the questions we ask should be dictated by the scientific questions we would like to address, and not by the limitations of our statistical toolkit. You may be surprised to hear that this has ever been an issue, but letting our statistical toolkit determine the questions we address was a dominant practice in the past and is still quite common. However, increasingly today we see ecologists adapting procedures from other disciplines, or developing their own, to answer the questions that arise from their research. This change in statistical practice is what we mean by "deceptively simple" in the first paragraph: the difference between ecologists' statistical practice today and a decade or two ago is not just that we can compute quantities more quickly, or crunch more (complex) data. We are using our data to consider problems that are more complex. For example, a growing number of studies use statistical methods to estimate parameters (say, the probability that the seed of an invasive pest will disperse X meters) for use in models that consider questions like rates of population growth or spread, risks of extinction, or changes to species' ranges; fundamental questions, but ones that were previously divorced from statistics. Meaningful estimates of these quantities require careful choice of statistical approaches, and sometimes these approaches cannot be

Ecological Statistics: Contemporary Theory and Application. First Edition. Edited by Gordon A. Fox, Simoneta Negrete-Yankelevich, and Vinicio J. Sosa. © Oxford University Press 2015. Published in 2015 by Oxford University Press.

limited to the contents of traditional statistics courses. This is of course only a point in a continuum; future techniques will continue to extend our repertoire of tractable questions and new books like this will continue to appear.

There is nothing wrong with using basic or old statistical techniques. Techniques like linear regression and analysis of variance (ANOVA) are powerful, and we continue to use them. But using techniques because we know them (rather than because they are appropriate) amounts to fitting things into a Procrustean bed–it does not necessarily ask the question we want to ask. We encountered recently a small but illustrative example in one of our labs: identifying environmental characteristics predicting presence of a lily, *Lilium catesbaei* (Sommers et al. 2011). It seemed reasonable to approach this problem with logistic regression (GLM with a binomial link; chapter 6), using site characteristics as the predictors and probability of presence/absence as the outcome. In reviewing literature on prediction of site occupancy, we found that a very large fraction of studies used a very different approach: ANOVA to compare the mean site characteristics of occupied with unoccupied sites. These might seem like comparable approaches, but they are quite different: logistic regression models probability of occupancy as a function of site characteristics, while ANOVA considers occupancy to be like an experimental treatment that somehow causes site characteristics! Yet many studies had used just this approach. To explore the problem, we analyzed the data using both approaches. The set of explanatory variables that we found predicted lily presence (using logistic regression) was not the same as the set of predictors for which occupied and unoccupied sites differed significantly (using ANOVA). The difference is not because the two approaches differ in power, or because we strongly violated underlying assumptions using one of the methods; the different results occur because the questions asked by the two approaches are quite different. This underlines a point that is often not obvious to beginners: the same data processed with different methods leads to different answers. By choosing a statistical method because it is convenient, we run the risk of answering questions we do not intend to ask. Worse still, we may not even realize that we have answered the wrong question.

The idea for this book emerged during a couple of occasions on which Fox came to Mexico to teach a survival module in the Sosa–Negrete statistics course for ecology graduate students. Dinner conversations often converged on the conclusion that, despite considerable efforts, learning statistics continues to be boring for many ecologists and more often than not, it feels a bit like having dental work done: frightening and painful but necessary for survival.

However, nothing could be further from the truth. Statistics is at the core of our science, because it provides us with tools that help us interpret our complex (and noisy) picture of the natural world (figure I.1). Ecologists today are leading in the development of a number of areas of statistics, and potentially we have a lot more to contribute. Many techniques used by ecologists are thoughtful, efficient, powerful, and diverse. For young ecologists to be able to keep up with this phenomenal advance, old ways of teaching statistics (based on memorizing which ready-made test to use for each data type) no longer suffice; ecologists today need to learn concepts enabling them to understand overarching themes. This is especially clear in the contribution that ecologists and ecological problems have made to the development of roll-your-own models (Hilborn and Mangel, 1997; Bolker, 2008).

The chapters of this book are by experienced ecologists who are actively working to upgrade ecologists' statistical toolkit. This upgrade involves developing models and statistical techniques, as well as testing the utility, usability, and power of these techniques in real ecological problems. Some of the techniques highlighted in the book are not new, but are underused in ecology, and can be a great aid in data analysis.

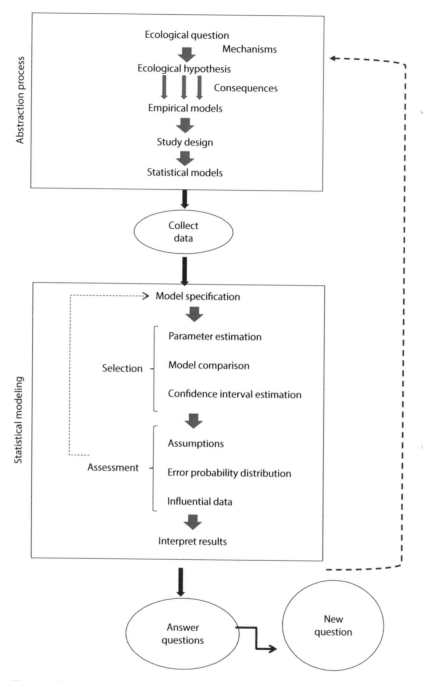

Fig. I.1 The cycle of ecological research and the role of statistical modeling.

One reason statistics is forbidding for some ecologists is its sometimes complex mathematical core. Most ecologists lack strong mathematical training. We have made a strong effort in this book to keep the mathematical explanations to the minimum necessary. At the same time, we have concentrated many of the overarching conceptual tools in the first three chapters (e.g., approaches to inference, probability distributions, statistical models, likelihood functions, model parameter estimation using maximum likelihood, model selection, and diagnostics). This should give you a stronger foundation in the ideas that underlie the methods used, and thus tie those methods together. Most methods in the book, in fact, are applications of these concepts (figure I.2).

There is another reason it makes sense to have a strong conceptual understanding of statistics: in the end, it is a lot easier than trying to memorize! For example, understanding the difference between fixed- and random-effects renders memorizing all the different traditional ANOVA designs (split plots, Latin square, etc.) unnecessary, because this knowledge will let you develop the linear model to analyze correctly any experimental design accordingly with more modern fitting techniques. You can, of course, use parts of this book as a cookbook, and simply use particular analyses for your own problems. Nevertheless, knowing how to follow a few recipes doesn't make you a chef, or even someone able to understand different techniques in cooking. We think you will be better off using the book to master the concepts and tools that underlie many different applications.

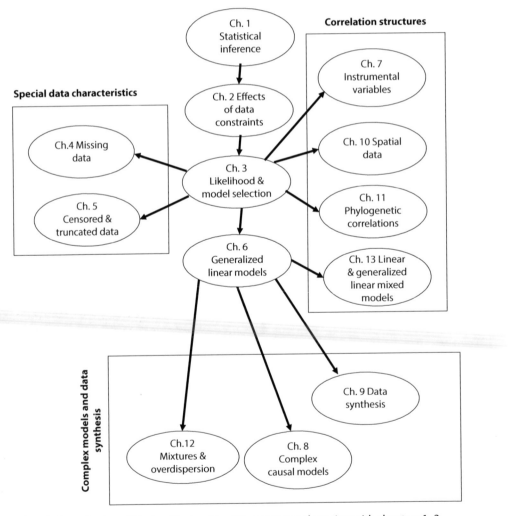

Fig. I.2 Organization of the book's chapters. We recommend starting with chapters 1–3.

Relating ecological questions to statistics

In doing science, we are acquiring knowledge about nature. In this effort, we are constantly proposing models of how the world works based on our observations, intuition, and previous knowledge. These models can be pictorial, verbal, or mathematical representations of a process. Mathematical models are of great utility because they are formal statements that focus our thinking, forcing us to be explicit about our assumptions and about the ways we envision that variables measured in our study objects are related (Motulsky and Chistopoulus 2003). Statistical models are mathematical statements in which at least one variable is assumed to be a *random variable* (compare with a deterministic model; see Bolker (2008) and Cox and Donnelly (2011) for a discussion of different kinds of models used in ecology and other life sciences).

While there is no unique way to approach a research problem, most modern ecologists follow an iterative cycle (figure I.1) in which they formulate the research questions and hypotheses; develop empirical models; design a plan to collect data; and develop statistical models to analyze the data in the light of their empirical models. The latter step includes plans for parameter and confidence interval estimation, model evaluation in terms of assumptions, and selection among alternative models or reformulation of models. Having made these plans, one can then collect the data, implement the planned data analyses, interpret the results (both statistically and ecologically), draw conclusions about what the data reveal, and then refine old or formulate new research hypotheses. Preliminary data analysis, model evaluation and criticism, and the theory behind your question usually guide the modeling process (Hilborn and Mangel 1997; Bolker 2008; Cox and Donnelly 2011).

The ability to go from an ecological question to a statistical model is a skill that involves an abstraction process and can only be acquired with practice and guidance from experienced researchers. In this sense, science is still a trade learned in the master's workshop. There are many types of ecological questions, such as: How is survival of a fish population affected by the concentration of a pollutant? Does this depend on the sex or genetic background of the fish? On the density of its population, or of its predators? What is the difference in the dispersal dynamics of a native and an exotic shrub species? How do their population growth rates compare, and what are the causes of the differences? Do earthworms distribute randomly across a maize field? If they aggregate, what are the causes? How does distance to the shore affect predation on shrimp larvae? Is pollen competition acting as a selection pressure on stigma morphology? Does the quality of nurse trees depend on phylogeny? When you read these questions, you probably automatically thought of the possible mechanisms that could drive the answers. For example, earthworms may aggregate in patches where the organic matter they feed on accumulates, or an exotic shrub species may spread farther than a native one because it is mammal-dispersed and not wind-dispersed. These are ecological hypotheses.

If the link between the ecological question/hypothesis and the statistical models is inadequate, the ecological question will remain unaddressed, and a lot of work and resources wasted. In our experience, young researchers (often students under great pressure) frequently rush into the field or lab to gather whatever data they can, because they fear running out of time—perhaps because the breeding, flowering, migrating, planting (or you name it) season has started, or their advisor is breathing down their neck. A common result is that they end up being able to use only a very small fraction of the data they gathered and, on many occasions, the original ecological question ends up being replaced by one (no matter how boring) that is tractable, given the data in hand.

The step between the ecological question and the empirical model requires concentration, thinking, drawing graphs, or scribbling equations, erasing, rethinking, letting it rest, and thinking again. You may have to do this many times. This is not surprising if you consider that the ecological hypothesis often embodies the proposal of an ecological mechanism, while the empirical models are usually mathematical representations of the measurable consequences of that mechanism occurring. Once combined with the concept of sampling, they become statistical models.

In summary, a study needs to rest on a plan, and you need to stick to the plan. Advanced readers will know that (of course) plans sometimes change; one should be able to learn from the data. But the key thing is to have a plan. This plan should include the way you are going to collect and analyze your data before going to the field or the lab to actually collect them. Although it sounds quite logical, this happens surprisingly rarely, so we invite you to (at least) think about it, and to consult a statistician early in the planning of the study. Because considerations about sampling/experimental design are crucial for selecting an approach to data analysis, many of the chapters in this book discuss how to plan your sampling or research design prior to collecting data. We emphasize that, if your study is reasonably well-designed (proper replication, randomization, control, blocking, etc.) you will be able to dig some useful information out of the data, even if you struggle with finding a convenient statistical method; if it's poorly designed, you'll certainly be in trouble, no matter how fancy your statistical method is.

How you record data can also be very specific for certain techniques like spatial or survival statistics. For example, how do you control for possible confounding factors (chapter 7)? How do you account for aggregation of experimental or sampling units (chapters 12 and 13)? How do you deal with spatial or phylogenetic correlation of your data (chapters 10 and 11)? How do you record data points with missing measurements (chapter 4), or values that are only known to be greater/less than some value (chapter 5)? The way you plan for these sorts of peculiarities—and therefore the way you collect your data—will determine the model you need to build, the extent to which your data are an appropriate sample for studying that model, and the degree to which your study coheres with the ecological theory underlying that model. Chapter authors have also made efforts to highlight methods (such as graphical analyses) that can help you understand the nature of your data and whether they conform to the requirements of the statistical techniques you plan to use.

Going from ecological thinking (hypothesis) to statistical thinking (model) is a skill that you cannot easily learn from books; but we can offer some guidelines (they are not rules—see Crome 1997; Karban and Huntzinger 2006 for more thorough treatment):

(1) Privilege *your* ecological problem, questions, and hypotheses. These should not be bent to accommodate the statistical tools you know.
(2) Define the type of study. Is it a description of a pattern or a causal explanation of a phenomenon? Or do you aim to predict the outcome of that phenomenon? For reviews of the main types of studies, consult Gotelli and Ellison (2004), Lindsey (2004), and Cox and Donnelly (2011).
(3) Define clearly the study units. Are you comparing individual organisms, plots, or some other group of objects (statistical populations)? Study units are those individuals, things, quadrats, populations, etc., for which different conditions or treatments can occur.

(4) Define clearly the changing characteristics (variables) that you are measuring in the study units: Is it a label for some kind of category? Is it a percentage? Does your variable only take count values? Is it an index calculated with a few variables? Are you obtaining a proportion? Is it a measure on a continuous or a discrete scale (Dobson 2002)? Is it a binary state (dead or alive; visited or not visited; germinated or not germinated; present or absent)?

(5) Draw a diagram of, or write down equations for, your predictions. If your hypotheses are about possible relations or associations between variables, you probably can propose, at least initially, some sort of cause–effect relationship represented by

$$A \rightarrow B.$$

This can be understood as meaning that A causes B, B is a consequence of A, or the values of A determine the values of B. A and B are characteristics of all study units. More complicated versions of your hypothesis may suggest that $A, B \rightarrow C$ (A and B together cause C) or that $A \rightarrow B, C$ (A causes or affects B and C) or something more complex (chapter 8). Draw boxes and arrows, or flow diagrams that help represent the relationships you are envisioning. Sketch what a graph of the data is expected to look like. Now revisit your ecological hypothesis, and ask whether the empirical models that you wrote down provide a clear set of consequences for your ecological hypothesis. Iterate through 2–5 until the answer is unequivocally yes. You are now ready to incorporate the sampling scheme into your model and, therefore, construct the statistical model.

A conceptual foundation: the statistical linear model

The cause–effect relationship in an empirical model is often represented as a linear statistical model. The abstraction of a relationship or association between variables into a linear statistical model may be difficult to many beginners, but once you get the idea it will be difficult to avoid thinking in linear model terms. Linear models are the basis for many methods for analyzing data, including several included in this book. In its simplest form, the linear model says that the *expected value* (the true, underlying mean of a random variable) of a variable Y in a study unit is proportionally and deterministically a function of the value of the variable X (predictor or explanatory or independent variable). For any individual data point, there will also be a random component (or error) e caused by the effect of many unidentified but relatively unimportant variables and measurement error:

$$Y_i = F(X_i) + e_i,$$

Where $F(X_i) = \beta_0 + \beta_1 X_i$. Here $e_i = Y_i - \hat{Y}$, i.e., the difference between the observed value and the predicted value, and is called the residual or error. In the simplest case, the distribution of the residuals is $e \sim N(0, \sigma_e^2)$ —that is, they have a Normal distribution with mean zero and variance σ_e^2. $\hat{Y}$ is the predicted value of the response variable, and Y_i is the observed value of the response variable. β_0 and β_1 are parameters we need to estimate, in this case the intercept and the slope of a straight line. If this relationship exists, the researcher generally wants the quantity e to be as small as possible, because that increases the predictive power of the model. You may have recognized in this model the simple linear regression equation.

Box I.1 LINEAR REGRESSION AND LEAST SQUARES

The statistical linear model for multiple regression and ANOVA analyses is:

$$Y_i = \beta_0 + \beta_1 X_{i1} + \beta_2 X_{i2} + \cdots + \beta_n X_{in} + e_i.$$

In matrix notation, the linear model in this equation is

$$\mathbf{Y} = \mathbf{X}\boldsymbol{\beta} + \mathbf{e},$$

where $\mathbf{Y}$ is a vector of responses and $\mathbf{e}$ is the vector of residuals, both of length N (the number of observations). $\boldsymbol{\beta}$ is a vector of parameters (to be estimated) with length p (the number of parameters). $\mathbf{X}$ is an $N \times p$ matrix of known constants; $\mathbf{X}$ is called the design matrix whose elements are values of the predictor variables or zeros and ones.

As an example of a model with both response and explanatory continuous variables, suppose you have measured volume, height, and diameter at breast height (DBH, in m) of 20 trees of the same species cut down by chainsaw; you want to predict the volume of other standing trees with a rapid method based on measuring height and DBH. The data and R scripts for this analysis are in the online appendices. A plausible model would be

$$\text{Volume}_i = \beta_0 + \beta_1 \, \text{Diameter}_i + \beta_2 \, \text{Height}_i + e_i.$$

Given the model, we need to estimate the parameters–i.e., we have to find some b_i as best estimators of β for equation (I.2). Historically, the method used for estimation consists of minimizing the sum of squares of the errors. Writing this out in matrix form, we get

$$L = \sum_{i=1}^{n} e_i^2 = \mathbf{e}'\mathbf{e} = (\mathbf{y} - \mathbf{Xb})'(\mathbf{y} - \mathbf{Xb}).$$

Using calculus and some algebra

$$\frac{\delta L}{\delta \mathbf{b}} \Big| = -2\mathbf{X}'\mathbf{y} + 2\mathbf{X}'\mathbf{Xb} = \mathbf{0}$$

$$\mathbf{X}'\mathbf{Xb} = \mathbf{X}'\mathbf{y}.$$

This allows us to solve for $\mathbf{b}$:

$$\mathbf{b} = (\mathbf{X}'\mathbf{X})^{-1}\mathbf{X}'\mathbf{y}.$$

This is the ordinary least squares (OLS) estimate. Under the assumption that $Y_i \sim N(\bar{Y}, \sigma_Y^2)$, it turns out that the $\mathbf{b}$ are maximum likelihood estimators (MLE). Parameter estimation by MLE is a more general, convenient, and modern way of fitting models (chapter 3) and we use it extensively throughout the book.

To see how an example looks, download the R script and data file from the companion website (http://www.oup.co.uk/companion/ecologicalstatistics), and run the script in R. Make sure you read the comments in the script! For the tree data, the estimated parameters are $\beta_0 = 35.81$, $\beta_1 = 34.10$, $\beta_2 = 0.32$, and the model for predicting the volumes of trees is

$$V_i = -35.81 + 34.1 \, \text{Diameter}_i + 0.32 \, \text{Height}_i.$$

We can extend this model to cases where there is more than one predictor variable; it is then called multiple regression (see the worked example in Box I.1) and is given by:

$$Y_i = \beta_0 + \beta_1 X_{i1} + \beta_2 X_{i2} + \cdots + \beta_n X_{in} + e_i$$

$$= \beta_0 + \sum_{j=1}^{n} \beta_j X_{ij} + e_i. \tag{I.1}$$

This can also be written in terms of the expected value of Y:

$$\mathbf{E}[Y] = \beta_0 + \sum_{j=1}^{n} \beta_j X_j.$$

Many advanced readers fail to recognize that this is also the model for ANOVA. This is in part because basic texts on statistics often keep it as a secret. The only difference between the regression and ANOVA models is that the predictor variables are categorical in ANOVA (for instance, levels of a treatment), while they are continuous in linear regression; for a more detailed explanation see Grafen and Hails (2002) and Crawley (2007).

For quantitative explanatory variables, the model contains terms of the form $\beta_j X_{ij}$, where the parameter β_0 represents $\hat{Y}$ when there is no influence of X on Y. β_j represents the rate of change in the response corresponding to changes in the jth predictor variable. For qualitative (categorical) variables, there is one parameter per level of the factor. The corresponding elements of $\mathbf{X}$ either include or exclude the appropriate parameters for each observation, usually by taking the values of 1 or 0, depending on whether the

---

Box I.2 MATRICES IN LINEAR MODELS

We can write linear models in matrix form:

$$\mathbf{Y} = \mathbf{X}\boldsymbol{\beta} + \mathbf{e}.$$

How do we interpret these matrices? Here we give two examples.

First, let's consider the model for tree volume, considered in Box I.1:

Volume		Intercept	Diameter	Height			e
10.2		1	0.6917	70			e_1
10.3		1	0.7167	65			e_2
10.2		1	0.7333	63			e_3
16.4		1	0.875	72			e_4
18.8		1	0.8917	81			e_5
19.7		1	0.9	83			e_6
15.6		1	0.9167	66			e_7
18.2		1	0.9167	75			e_8
22.6		1	0.925	80	β_1		e_9
19.9	=	1	0.9333	75	β_2	+	e_{10}
24.2		1	0.9417	79	β_3		e_{11}
21.0		1	0.95	76			e_{12}
21.4		1	0.95	76			e_{13}
21.3		1	0.975	69			e_{14}
19.1		1	1	75			e_{15}
22.2		1	1.075	74			e_{16}
33.8		1	1.075	85			e_{17}
27.4		1	1.1083	86			e_{18}
25.7		1	1.1417	71			e_{19}
24.9		1	1.15	64			e_{20}

Box I.2 (*continued*)

The first column of the design matrix (**X**) of independent variables contains only 1s. This is the general convention to be used for any regression model containing a constant term β_0. To see why this is so, imagine the β_0 term to be of the form $\beta_0 X_0$, where X_0 is a dummy variable always taking the value 1 (Kleinbaum and Kupper 1978).

As a second example, consider an experiment consisting of four different diets (**D**) applied randomly to 19 pigs of the same sex, gender, and age (Zar 2010). The response variable Y_i is the pig body weight (**W**) in kg, after being raised on these diets. The question is whether the mean pig weights are the same for all four diets. The data and R script are on the companion website (http://www.oup.co.uk/companion/ecologicalstatistics).

In this case, the independent variables **X** are categorical or dummy variables that label each level of the diet treatment. The model, with the full matrices, is:

$$
\begin{pmatrix} \text{Weight} \\ 60.8 \\ 57 \\ 65 \\ 58.6 \\ 61.7 \\ 68.7 \\ 67.7 \\ 74 \\ 66.3 \\ 69.8 \\ 102.6 \\ 102.1 \\ 100.2 \\ 96.5 \\ 87.9 \\ 84.2 \\ 83.1 \\ 85.7 \\ 90.3 \end{pmatrix}
=
\begin{pmatrix} \text{DietA} & \text{DietB} & \text{DietC} & \text{DietD} \\ 1 & 0 & 0 & 0 \\ 1 & 0 & 0 & 0 \\ 1 & 0 & 0 & 0 \\ 1 & 0 & 0 & 0 \\ 1 & 0 & 0 & 0 \\ 0 & 1 & 0 & 0 \\ 0 & 1 & 0 & 0 \\ 0 & 1 & 0 & 0 \\ 0 & 1 & 0 & 0 \\ 0 & 1 & 0 & 0 \\ 0 & 0 & 1 & 0 \\ 0 & 0 & 1 & 0 \\ 0 & 0 & 1 & 0 \\ 0 & 0 & 1 & 0 \\ 0 & 0 & 0 & 1 \\ 0 & 0 & 0 & 1 \\ 0 & 0 & 0 & 1 \\ 0 & 0 & 0 & 1 \\ 0 & 0 & 0 & 1 \end{pmatrix}
\begin{pmatrix} \beta_0 \\ \beta_1 \\ \beta_2 \\ \beta_3 \end{pmatrix}
+
\begin{pmatrix} e \\ e_1 \\ e_2 \\ e_3 \\ e_4 \\ e_5 \\ e_6 \\ e_7 \\ e_8 \\ e_9 \\ e_{10} \\ e_{11} \\ e_{12} \\ e_{13} \\ e_{14} \\ e_{15} \\ e_{16} \\ e_{17} \\ e_{18} \\ e_{19} \end{pmatrix}
$$

observation is or is not in that level. See Box I.2 for an example; consult Grafen and Hails (2002) or more basic texts if you need background information. Therefore, there is a difference in the meaning of the estimated parameters: in regression models, they express a proportionality constant, but in ANOVA they are (depending on how one chooses to parameterize the model) either mean differences between the treatments, or they are treatment level means (Crawley 2007).

Finally, equation (I.1) can also be written as

$$Y_i \sim N\left(\beta_0 + \sum_{i=1}^{n} \beta_i X_i, \sigma_Y^2\right),$$

which can be read as saying that each observation Y_i is drawn from a Normal distribution with mean $= \beta_0 + \beta_1 X_1 + \beta_2 X_2 + \cdots + \beta_n X_n$ and variance σ_Y^2 (figure 6.2a). In the traditional view, the β_i are assumed to be fixed values, but in Bayesian analysis (chapter 1) they are themselves random variables with estimated b_i and corresponding variances s_i^2.

Many common statistical methods taught in basic statistics courses—in addition to linear regression and ANOVA—rely on models in the form of equation (I.1). These are called linear models because the deterministic part of the model (also known as *signal*) is a linear combination of the parameters, and the noise part, **e** is added to it; see Crawley (2007) for examples. A useful feature of this type of model is that if the parameters β_1 to β_n are not different from zero, then the best description of the data set is its overall mean ($\beta_0 = \mu$)

$$Y_{ij} = \beta_0 + e_{ij}.$$

This is often called a *null model* and provides an important reference when examining the explanatory power of candidate models.

Having estimated the model parameters, we can ask whether they improve our understanding, as compared with a simpler model. Often one compares a series of hypothetical *nested models*, in which one or more of the proposed predictor variables are deleted from the *saturated model*. In the model for tree volume considered in Box I.1, we retain both diameter and height in the model, since they have explanatory power. This method is explained in chapters 1, 3, and 6.

But we are not yet finished. The conclusions drawn from the tree model depend on some important assumptions: (1) the relationships between **Y** and **X** are linear; (2) the data are a random sample of the population, i.e., the errors are statistically independent of one another; (3) the expected value of the errors is always zero; (4) the independent variables are not correlated with one another; (5) the independent variables are measured without error; (6) the residuals have constant variance; and (7) the residuals are Normally distributed. These are obviously strong assumptions, and most ecologists know that, very often, they do not hold for our data; or that assuming a linear relationship among variables is naïve or not supported by theory. This renders many of the statistical methods learned in basic courses limited and a source of great disappointment among beginners and graduate students!

But you don't have to ditch your basic stats textbook. Consider an analogy for a moment: imagine that you are writing a novel. You would need to know spelling and some rules of grammar and syntax at the outset. Once you knew those, you would discover that you still couldn't write a readable (we won't even ask for interesting!) novel, for two reasons. First, good writers often violate some of these rules, but they do so knowingly and deliberately. One can violate some assumptions in statistics too—but doing so requires having a strong understanding of the underlying methods. This does not include blanket claims (which you have probably heard) that "this method is robust," so violation of the assumptions is fine. Second, no matter how great your knowledge of the rules of spelling, grammar, and syntax, you have only begun to know how to use the language. The analogy holds for statistics: if a simple method (like linear regression) is poorly justified, for your question, use the conceptual basis of statistical linear models with Normal distribution of errors, and go on to other methods that allow you to deal more realistically with ecological data.

Many techniques in this book elaborate on the conceptual basis of the statistical linear model to model ecological problems. For example, by modeling the random variation of residuals with other probability distributions than the Normal, we can model the response of binary variables or variables measured as counts or proportions (chapter 6). In addition, you will be able to consider explicitly cases where errors are likely to be correlated due to spatial (chapter 10) or phylogenetic (chapter 11) correlation of the study units.

The assumption of a linear relationship with $\beta\mathbf{X}$ can be relaxed by working with the exponential family of distributions (chapters 5, 6, and 13). You may have to incorporate into your linear model the fact that the data include missing values (you have no information about the value), censored values (you have partial information, e.g., it is less than the smallest value you can measure) or that they come from a truncated sample (certain values are never sampled). Failing to do so can cause your estimates to differ systematically from their true value (chapters 4 and 5). Many ecological questions involve complex causal relationships; many problems are best addressed by considering results from a large number of research projects. These require special statistical models that often use linear models as building blocks (chapters 8 and 9).

Historically, the term *error*, used to refer to the residuals, comes from the assumption made in physics that any deviation from the true value was the consequence of measurement error. However, in ecology and other fields (including geology, meteorology, economics, psychology, and medicine) the variability around predicted values is very often the result of many small influences on the study units, and often more important than measurement error. Modeling this variability appropriately contributes to our understanding of the phenomena under investigation. In this sense, a proper statistical model has a heuristic component. Variability is not just a nuisance, but actually tells us something about the ecological processes (Bolker 2008). You will find discussion and several approaches to this end in several chapters (2, 3, 6, 10, 12, and 13) of this book.

Thus, the methods discussed in this book will broaden our capabilities for analyzing data and addressing interesting questions about our study subjects. The details of each model—and the scientific questions from which it comes—matter a lot; they determine how we make inferences about nature. Once you have analyzed your numerical results with a statistical technique, you have to interpret the results in light of the theory behind your question/hypothesis, and draw ecological conclusions. Put differently, you need to return from statistics to ecology.

Where does statistics end? There is no general answer, but in each chapter authors will point out some of the places where you must exercise your biological judgment. Keep your focus on the *ecological* questions you set out to answer (Bolker 2008). You should ask constantly "Does this make sense?" and "What does this answer really mean?" Statistics does not tell you the answer; at a basic level, it helps provide methods for evaluating the internal and external validity of your conclusions (Shadish et al. 2002). You need to be a scientist, not someone performing an obscure procedure.

What we need readers to know

Our intended audience is composed of graduate students and professionals in ecology and related fields, including evolutionary biology and environmental sciences. We assume that readers have an introductory background in statistics, covering most of the topics in Gotelli and Ellison (2004). This includes probability, distributions, descriptive statistics, hypothesis testing, confidence intervals, correlation, simple regression, and ANOVA. Other sources you may find useful for filling in your background as needed are Underwood (1997), Scheiner and Gurevitch (2001), Quinn and Keough (2002), Crawley (2007), Zar (2010), and Cox and Donnelly (2011).

Having read the previous paragraph, don't panic! We assume familiarity with these methods, not encyclopedic command. We generally assume that you have taken no more mathematics than an introductory calculus course. We do not expect that you remember the computations used in that course, but recalling the basic concepts involved in log-arithms, exponentials, derivatives, and integrals will be a big help to you. Almost any introductory calculus text is adequate for a refresher, but if you want a text with a biological motivation, use Adler (2004) or Neuhauser (2010). Most chapters in this book are still readable if you choose to skip mathematical sections.

Readers can get a fair amount out of this book even if they have no knowledge of the R language (R Core Team 2014), but most examples are analyzed using this software environment. Sorry: we won't try to teach you how to use R, but there are many books, handbooks, and web pages that already do this very well. We emphasize use of R because it is free, open source, used around the world, and works for all major operating systems. R has become the dominant system for statistical analysis in much of ecology and many other scientific fields, and is the most popular package among statistical researchers; many new methods are available first in R. For most examples in the book, usable R code is provided on our companion web site (http://www.oup.co.uk/companion/ecologicalstatistics). Thoroughly commented R scripts are provided in downloadable files; the book itself contains code snippets to explain the process of data analysis. The code is not optimized for speed or elegance, but rather for transparency and understanding. We are aware that there are multiple ways of writing code to do the same job and you will see differences in style and sophistication throughout the book.

How to get the most out of this book

One virtue of this book (compared with others at a comparable level) is that it begins not with the description and application of specific techniques but rather with the discussion, in the first three chapters, of concepts that constitute the fundamental building blocks of the rest of the book. The chapters in the remainder of the book can be understood as developments of ideas discussed in the first chapters, tied together around unifying concepts. These include data with special characteristics (chapters 4 and 5), complications of the statistical linear model (chapters 6, 11–13), combining results from different studies (chapter 9), and combining models in more complex causal models (chapters 8), or correlation structure of data (chapters 10–13; figure I.2). While it is logical to begin with chapters 1–3 or 1–6 before any of the other chapters, not all of us learn logically; it is certainly possible to start elsewhere in the book, but we do recommend starting with chapters 1–3, in that order. In figure I.2, we show the relationship between chapters. Chapters from the top to the bottom are linked logically and loosely. Horizontally, chapters deal with different conceptual aspects: correlation structures (confounded effects, spatial, and phylogenetic correlation); complex variance structures; complex models and data synthesis; and data special characteristics (missingness, censorship, and truncation).

Most of us, at some point, act as though we believe in the following hypothesis of learning: if I get a book and just read it casually, I will learn something from it. We have tested it extensively and it is literally true: if you read bits of this book casually, you will probably learn something. But you won't learn very much, and you won't learn it very quickly. To get much from it, you need to read the chapters carefully enough that you can explain the main ideas to someone else (always an excellent way to see whether you understand something). Moreover, you need to *work examples*—either the examples the chapter contributors provide, or examples of your own choice. We strongly encourage you to read the chapters and go through the code carefully. Just running the R code provided by the contributors won't teach you much; we recommend users to hack it, take it apart and put it back together again in new ways, because this is the best way of learning R, statistics, and formal hypothesis/model formulation.

Inevitably, there are important topics the book does not cover. These include multivariate techniques such as classification and ordination (McCune and Grace 2002), regression trees, etc.; time series (Diggle 1990); compositional analysis; diversity and similitude analysis; non-linear models; and several techniques related to data mining (Hastie et al. 2009). These useful techniques are explained well in many books and some are widely used in ecological research. This does not imply any slight to those methods and, in many cases, the concepts covered here will still be useful in those other specific contexts.

In sum, we hope that after using this book, readers will have learned how to apply some new statistical approaches to ecological problems, but also that they will gain the ability to dissect new or emerging techniques that they encounter in the literature. Sometimes statistical theory (or our data) is not up to the complexity we confront. For example, spatial and genetic structure in data usually cannot be assessed simultaneously (and they are typically confounded). Therefore, we often have to use a good bit of thought and creativity. Ultimately, we hope that readers will be able to use the methods discussed in this book to tailor their own statistical analyses. Last but not least, we hope that after reading this book, you will be confident enough to consult a statistician for more complex problems than those presented here.

So, take the road, good luck, and enjoy your trip.

CHAPTER 1

Approaches to statistical inference

Michael A. McCarthy

1.1 Introduction to statistical inference

Statistical inference is needed in ecology because the natural world is variable. Ernest Rutherford, one of the world's greatest scientists, is supposed to have said "If your experiment needs statistics, you ought to have done a better experiment." Such a quote applies to deterministic systems or easily replicated experiments. In contrast, ecology faces variable data and replication constrained by ethics, costs, and logistics.

Ecology—often defined as the study of the distribution and abundance of organisms and their causes and consequences—requires that quantities are measured and relationships analyzed. However, data are imperfect. Species fluctuate unpredictably over time and space. Fates of individuals, even in the same location, differ due to different genetic composition, individual history or chance encounters with resources, diseases, and predators.

Further to these intrinsic sources of uncertainty, observation error makes the true state of the environment uncertain. The composition of communities and the abundance of species are rarely known exactly because of imperfect detection of species and individuals (Pollock et al. 1990; Parris et al. 1999; Kéry 2002; Tyre et al. 2003). Measured variables do not describe all aspects of the environment, are observed with error, and are often only indirect drivers of distribution and abundance.

The various sources of error and the complexity of ecological systems mean that statistical inference is required to distinguish between the signal and the noise. Statistical inference uses logical and repeatable methods to extract information from noisy data, so it plays a central role in ecological sciences.

While statistical inference is important, the choice of statistical method can seem controversial (e.g., Dennis 1996; Anderson et al. 2000; Burnham and Anderson 2002; Stephens et al. 2005). This chapter outlines the range of approaches to statistical inference that are used in ecology. I take a pluralistic view; if a logical method is applied and interpreted appropriately, then it should be acceptable. However, I also identify common major errors in the application of the various statistical methods in ecology, and note some strategies to avoid them.

Mathematics underpins statistical inference. While anxiety about mathematics is painful (Lyons and Beilock 2012), without mathematics, ecologists need to follow Rutherford's

Ecological Statistics: Contemporary Theory and Application. First Edition. Edited by Gordon A. Fox, Simoneta Negrete-Yankelevich, and Vinicio J. Sosa. © Oxford University Press 2015. Published in 2015 by Oxford University Press.

advice, and do a better experiment. However, designing a better experiment is often prohibitively expensive or otherwise impossible, so statistics, and mathematics more generally, are critical to ecology. I limit the complexity of mathematics in this chapter. However, some mathematics is critical to understanding statistical inference. I am asking you, the reader, to meet me halfway. If you can put aside any mathematical anxiety, you might find it less painful. Please work at any mathematics that you find difficult; it is important for a proper understanding of your science.

1.2 A short overview of some probability and sampling theory

Ecological data are variable—that is the crux of why we need to use statistics in ecology. Probability is a powerful way to describe unexplained variability (Jaynes 2003). One of probability's chief benefits is its logical consistency. That logic is underpinned by mathematics, which is a great strength because it imparts repeatability and precise definition. Here I introduce, as briefly as I can, some of the key concepts and terms used in probability that are most relevant to statistical inference.

All ecologists will have encountered the Normal distribution, which also goes by the name of the Gaussian distribution, named for Carl Friedrich Gauss who first described it (figure 1.1). Excluding the constant of proportionality, the probability density function (box 1.1) of the Normal distribution is:

$$f(x) \propto e^{\frac{-(x-\mu)^2}{2\sigma^2}}.$$

Fig. 1.1 The Normal distribution was first described by Carl Friedrich Gauss (left; by G. Biermann; reproduced with permission of Georg-August-Universität Göttingen). Pierre-Simon Laplace (right; by P. Guérin; © RMN-Grand Palais (Château de Versailles) / Franck Raux) was the first to determine the constant of proportionality $\left(\frac{1}{\sqrt{2\pi\sigma^2}}\right)$, and hence was able to write the full probability density function.

Box 1.1 PROBABILITY DENSITY AND PROBABILITY MASS

Consider a discrete *random variable* (appendix 1.A) that takes values of non-negative integers (0, 1, 2, . . .), perhaps being the number of individuals of a species within a field site. We could use a distribution to define the probability that the number of individuals is 0, 1, 2, etc. Let X be the random variable, then for any x in the set of numbers {0, 1, 2, . . .}, we could define the probability that X takes that number; $Pr(X = x)$. This is known as the *probability mass function*, with the sum of $Pr(X = x)$ over all possible values of x being equal to 1.

For example, consider a probability distribution for a random variable X that can take only values of 1, 2 or 3, with $Pr(X = 1) = 0.1$, $Pr(X = 2) = 0.6$, and $Pr(X = 3) = 0.3$ (see the figure, *a*). In this case, X would take a value of 2 twice as frequently as a value of 3 because $Pr(X = 2) = 2 \times Pr(X = 3)$. The sum of probabilities is 1, which is necessary for a probability distribution.

Probability mass functions cannot be used for continuous probability distributions, such as the Normal distribution, because the random variable can take any one of infinitely many possible values. Instead, continuous random variables can be defined in terms of probability density.

Let $f(x)$ be the *probability density function* of a continuous random variable X, which describes how the probability density of the random variable changes across its range. The probability that X will occur in the interval $[x, x + dx]$ approaches $dx \times f(x)$ as dx becomes small. More precisely, the probability that X will fall in the interval $[x, x + dx]$ is given by the integral of the probability density function $\int_x^{x+dx} f(u)du$. This integral is the area under the probability density function between the values x and $x + dx$.

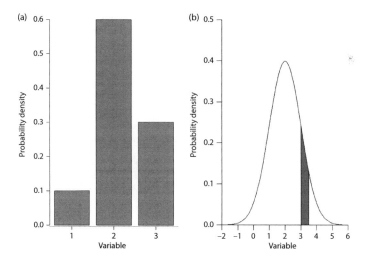

An example (using the Normal distribution) is shown in part *b* of the figure. A continuous random variable can take any value in its domain; in this case of a Normal distribution, any real number. The shaded area equals the probability that the random variable will take a value between 3.0 and 3.5. It is equal to the definite integral of the probability function: $\int_{3.0}^{3.5} f(u)du$. The entire area under the probability density function equals 1.

The *cumulative distribution function* $F(x)$ is the probability that the random variable X is less than x. Hence, $F(x) = \int_{-\infty}^{x} f(u)du$, and $f(x) = \frac{dF(x)}{dx}$. Thus, probability density is the rate at which the cumulative distribution function changes.

The probability density at x is defined by two parameters μ and σ. In this formulation of the Normal distribution, the mean is equal to μ and the standard deviation is equal to σ.

Many of the examples in this chapter will be based on the assumption that data are drawn from a Normal distribution. This is primarily for the sake of consistency, and because of its prevalence in ecological statistics. However, the same basic concepts apply when considering data generated by other distributions.

The behavior of random variables can be explored through simulation. Consider a Normal distribution with mean 2 and standard deviation 1 (figure 1.2a). If we take 10 draws from this distribution, the mean and standard deviation of the data will not equal 2 and 1 exactly. The mean of the sample of 10 draws is named the sample mean. If we repeated this procedure multiple times, the sample mean will sometimes be greater than the true mean, and sometimes less (figure 1.2b). Similarly, the standard deviation of the data in each sample will vary around the true standard deviation. These statistics such as the sample mean and sample standard deviation are referred to as sample statistics.

The 10 different sample means have their own distribution; they vary around the mean of the distribution that generated them (figure 1.2b). These sample means are much less variable than the data; that is the nature of averages. The standard deviation of a sampling

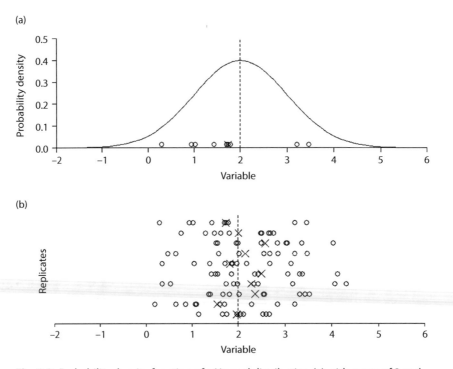

Fig. 1.2 Probability density function of a Normal distribution (a) with mean of 2 and standard deviation of 1. The circles represent a random sample of 10 values from the distribution, the mean of which (cross) is different from the mean of the Normal distribution (dashed line). In (b), this sample, and nine other replicate samples from the Normal distribution, each with a sample size of $n = 10$, are shown. The means of each sample (crosses) are different from the mean of the Normal distribution that generated them (dashed line), but these sample means are less variable than the data. The standard deviation of the distribution of sample means is the standard error ($se = \sigma/\sqrt{n}$), where σ is the standard deviation of the data.

statistic, such as a sample mean, is usually called a standard error. Using the property that, for any distribution, the variance of the sum of independent variables is equal to the sum of their variances, it can be shown that the standard error of the mean is given by

$$se = \sigma/\sqrt{n},$$

where n is the sample size.

While standard errors are often used to measure uncertainty about sample means, they can be calculated for other sampling statistics such as variances, regression coefficients, correlations, or any other value that is derived from a sample of data.

Wainer (2007) describes the equation for the standard error of the mean as "the most dangerous equation." Why? Not because it is dangerous to use, but because ignorance of it causes waste and misunderstanding. The standard error of the mean indicates how different the true population mean might be from the sample mean. This makes the standard error very useful for determining how reliably the sample mean estimates the population mean. The equation for the standard error indicates that uncertainty declines with the square root of the sample size; to halve the standard error one needs to quadruple the sample size. This provides a simple but useful rule of thumb about how much data would be required to achieve a particular level of precision in an estimate.

These aspects of probability (the meaning of probability density, the concept of sampling statistics, and precision of estimates changing with sample size) are key concepts underpinning statistical inference. With this introduction complete, I now describe different approaches to statistical inference.

1.3 Approaches to statistical inference

The two main approaches to statistical inference are frequentist methods and Bayesian methods. Frequentist methods are based on determining the probability of obtaining the observed data (or in the case of null hypothesis significance testing, the probability of data more extreme than that observed), given that particular conditions exist. I regard likelihood-based methods, such as maximum-likelihood estimation, as a form of frequentist analysis because inference is based on the probability of obtaining the data. Bayesian methods are based on determining the probability that particular conditions exist given the data that have been collected. They both offer powerful approaches for estimation and considering the strength of evidence in favor of hypotheses.

There is some controversy about the legitimacy of these two approaches. In my opinion, the importance of the controversy has sometimes been overstated. The controversy has also seemingly distracted attention from, or completely overlooked, more important issues such as the misinterpretation and misreporting of statistical methods, regardless of whether Bayesian or frequentist methods are used. I address the misuse of different statistical methods at the end of the chapter, although I touch on aspects earlier. First, I introduce the range of approaches that are used.

Frequentist methods are so named because they are based on thinking about the frequency with which an outcome (e.g., the data, or the mean of the data, or a parameter estimate) would be observed if a particular model had truly generated those data. It uses the notion of hypothetical replicates of the data collection and method of analysis. Probability is defined as the proportion of these hypothetical replicates that generate the observed data. That probability can be used in several different ways, which define the type of frequentist method.

1.3.1 *Sample statistics and confidence intervals*

I previously noted the equation for the standard error, which defines the relationship between the standard error of the mean and the standard deviation of the data. Therefore, if we knew the standard deviation of the data, we would know how variable the sample means from replicate samples would be. With this knowledge, and assuming that the sample means from replicate samples have a particular probability distribution, it is possible to calculate a confidence interval for the mean.

A confidence interval is calculated such that if we collected many replicate sets of data and built a $Z\%$ confidence interval for each case, those intervals would encompass the true value of the parameter $Z\%$ of the time (assuming the assumptions of the statistical model are true). Thus, a confidence interval for a sample mean indicates the reliability of a sample statistic.

Note that the limits to the interval are usually chosen such that the confidence interval is symmetric around the sample mean $\bar{x}$, especially when assuming a Normal distribution, so the confidence interval would be $[\bar{x}-\varepsilon, \bar{x}+\varepsilon]$. When data are assumed to be drawn from a Normal distribution, the value of ε is given by $\varepsilon = z\sigma/\sqrt{n}$, where σ is the standard deviation of the data, and n is the sample size. The value of z is determined by the cumulative distribution function for a Normal distribution (box 1.1) with mean of 0 and standard deviation of 1. For example, for a 95% confidence interval, $z = 1.96$, while for a 70% confidence interval, $z = 1.04$.

Of course, we rarely will know σ exactly, but will have the sample standard deviation s as an estimate. Typically, any estimate of σ will be uncertain. Uncertainty about the standard deviation increases uncertainty about the variability in the sample mean. When assuming a Normal distribution for the sample means, this inflated uncertainty can be incorporated by using a t-distribution to describe the variation. The degree of extra variation due to uncertainty about the standard deviation is controlled by an extra parameter known as "the degrees of freedom" (box 1.2). For this example of estimating the mean, the degrees of freedom equals $n-1$.

When the standard deviation is estimated, the difference between the mean and the limits of the confidence interval is $\varepsilon = t_{n-1}s/\sqrt{n}$, where the value of t_{n-1} is derived from the t distribution. The value of t_{n-1} approaches the corresponding value of z as the sample size increases. This makes sense; if we have a large sample, the sample standard deviation s will provide a reliable estimate of σ so the value of t_{n-1} should approach a value that is based on assuming σ is known.

Box 1.2 DEGREES OF FREEDOM

The degrees of freedom parameter reflects the number of data points in an estimate that are free to vary. For calculating a sample standard deviation, this is $n-1$ where n is the sample size (number of data points).

The "−1" term arises because the standard deviation relies on a particular mean; the standard deviation is a measure of deviation from this mean. Usually this mean is the sample mean of the same data used to calculate the standard deviation; if this is the case, once $n-1$ data points take their particular values, then the nth (final) data point is defined by the mean. Thus, this nth data point is not free to vary, so the degrees of freedom is $n-1$.

However, in general for a particular percentage confidence interval, $t_{n-1} > z$, which inflates the confidence interval. For example, when $n = 10$, as for the data in figure 1.2, we require $t_9 = 2.262$ for a 95% confidence interval. The resulting 95% confidence intervals for each of the data sets in figure 1.2 differ from one another, but they are somewhat similar (figure 1.3). As well as indicating the likely true mean, each interval is quite good at indicating how different one confidence interval is from another. Thus, confidence intervals are valuable for communicating the likely value of a parameter, but they can also foreshadow how replicable the results of a particular study might be. Confidence intervals are also critical for meta-analysis (see chapter 9).

1.3.2 *Null hypothesis significance testing*

Null hypothesis significance testing is another type of frequentist analysis. It is commonly an amalgam of Fisher's significance testing and Neyman–Pearson's hypothesis testing (Hurlbert and Lombardi 2009). It has close relationships with confidence intervals, and is used in a clear majority of ecological manuscripts, yet it is rarely used well (Fidler et al. 2006). It works in the following steps:

(1) define a null hypothesis (and often a complementary alternative hypothesis; Fisher's original approach did not use an explicit alternative);
(2) collect some data that are related to the null hypothesis;

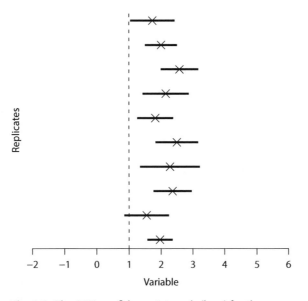

Fig. 1.3 The 95% confidence intervals (bars) for the sample means (crosses) for each of the replicate samples in figure 1.2, assuming a Normal distribution with an estimated standard deviation. The confidence intervals tend to encompass the true mean of 2. The dashed line indicates an effect that might be used as a null hypothesis (figures 1.4 and 1.5). Bayesian credible intervals constructed using a flat prior are essentially identical to these confidence intervals.

(3) use a statistical model to determine the probability of obtaining those data or more extreme data when assuming the null hypothesis is true (this is the p-value); and

(4) if those data are unusual given the null hypothesis (if the p-value is sufficiently small), then reject the null hypothesis and accept the alternative hypothesis.

Note that Fisher's original approach merely used the p-value to assess whether the data were inconsistent with the null hypothesis, without considering rejection of hypotheses for particular p-values.

There is no "else" statement here, particularly in Fisher's original formulation. If the data are not unusual (i.e., if the p-value is large), then we do not "accept" the null hypothesis; we simply fail to reject it. Null hypothesis significance testing is confined to rejecting null hypotheses, so a reasonable null hypothesis is needed in the first place.

Unfortunately, generating reasonable and useful hypotheses in ecology is difficult, because null hypothesis significance testing requires a precise prediction. Let me illustrate this point by using the species–area relationship that defines species richness S as a power function of the area of vegetation (A) such that $S = cA^z$. The parameter c is the constant of proportionality and z is the scaling coefficient. The latter is typically in the approximate range 0.15–0.4 (Durrett and Levin 1996). Taking logarithms, we have $\log(S) = \log(c) + z\log(A)$, which might be analyzed by linear regression, in which the linear relationship between a response variable ($\log(S)$ in this case) and an explanatory variable ($\log(A)$ in this case) is estimated.

A null hypothesis cannot be simply "we expect a positive relationship between the logarithm of species richness and the logarithm of area." The null hypothesis would need to be precise about that relationship, for example, specifying that the coefficient z is equal to a particular value. We could choose $z = 0$ as our null hypothesis, but logic and the wealth of previous studies tell us that z must be greater than zero. The null hypothesis $z = 0$ would be a nil null, which are relatively common in ecology. In some case, null hypotheses of "no effect" (e.g., $z = 0$ in this case) might be informative because a nil null is actually plausible. However, rejecting a nil null of $z = 0$ is, in this case, uninformative, because we already know it to be false.

A much more useful null hypothesis would be one derived from a specific theory. There are ecological examples where theory can make specific predictions about particular parameters, including for species–area relationships (Durrett and Levin 1996). For example, models in metabolic ecology predict how various traits, such as metabolic rate, scale with body mass (West et al. 1997; Kooijman 2010). Rejecting a null hypothesis based on these models is informative, at least to some extent, because it would demonstrate that the model made predictions that were inconsistent with data. Subsequently, we might investigate the nature of that mismatch and its generality (other systems might conform more closely to the model), and seek to understand the failure of the model (or the data).

Of course, there are degrees by which the data will depart from the prediction of the null hypothesis. In null hypothesis testing, the probability of generating the data or more extreme data is calculated assuming the null hypothesis is true. This is the p-value, which measures departure from the null hypothesis. A small p-value suggests the data are unusual given the null hypothesis.

How is a p-value calculated? Look at the data in figure 1.2a, and assume we have a null hypothesis that the mean is 1.0 and that the data are drawn from a Normal distribution. The sample mean is 1.73, marked by the cross. We then ask "What is the probability of obtaining, just by chance alone, a sample mean from 10 data points that is 0.73 units (or further) away from the true mean?" That probability depends on the variation in the data. If the standard deviation were known to be 1.0, we would know that the sample mean would have a Normal distribution with a standard deviation (the standard error of the mean) equal to $1/\sqrt{10}$. We could then calculate the probability of obtaining a deviation larger than that observed.

But, as noted above, we rarely know the true standard deviation of the distribution that generated the data. When we don't know, the p-value is calculated by assuming the distribution of the sample mean around the null hypothesis is defined by a t-distribution, which accounts for uncertainty in the standard deviation. We then determine the probability that a deviation as large as the sample mean would occur by chance alone, which is the area under the relevant tails of the distribution (sum of the two gray areas in figure 1.4). In this case, the area is 0.04, which is the p-value.

Note that we have done a "two-tailed test." This implies the alternative hypothesis is "the true mean is greater than or less than 1.0"; more extreme data are defined as deviations in either direction from the null hypothesis. If the alternative hypothesis was that the mean is greater than the null hypothesis, then only the area under the right-hand tail is relevant. In this case, more extreme data are defined only by deviations that exceed the sample mean, and the p-value would be 0.02 (the area of the right-hand tail). The other one-sided alternative hypothesis, that the mean is less than 1.0, would only consider deviations that are less than the sample mean, and the p-value would be $1 - 0.02 = 0.98$. The point here is that the definition of "more extreme data" needs to be considered carefully by clearly defining the alternative hypothesis when the null hypothesis is defined.

We can think of the p-value as the degree of evidence against the null hypothesis; the evidence mounts as the p-value declines. However, it is important to note that p-values are typically variable. Cumming (2011) describes this variability as "the dance of the p-values." Consider the 10 data sets in figure 1.2. Testing a null hypothesis that the mean is

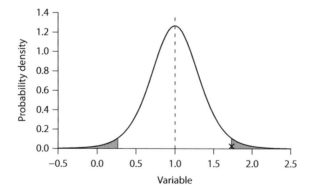

Fig. 1.4 The distribution of the sample mean under the null hypothesis that the mean is 1.0, derived from a t-distribution and the sample standard deviation (0.97) of the values in figure 1.2a. The p-value for a two sided null hypothesis test is the probability of obtaining a sample mean (cross) that deviates from the null hypothesis more than that observed. This probability is equal to the sum of the gray areas in the two tails.

1.0 leads to p-values that vary from 0.00018 to 0.11—almost three orders of magnitude—even though the process generating the data is identical in all cases (figure 1.5). Further, the magnitude of any one p-value does not indicate how different other p-values, generated by the same process, might be. How much more variable might p-values be when data are collected from real systems?

Rather than simply focusing on the p-value as a measure of evidence (variable as it is), many ecologists seem to perceive a need to make a dichotomous decision about whether the null hypothesis can be rejected or not. Whether a dichotomous decision is needed is often debatable. Some cases require a choice, such as when a manager must determine whether a particular intervention is required. Even in that case, alternatives such as decision theory exist. More broadly in ecology, the need for a dichotomous decision is less clear (Hurlbert and Lombardi 2009), but if we assume we must decide whether or not to reject a null hypothesis, a threshold p-value is required. If the p-value is less than this particular threshold, which is known as the type I error rate, then the null hypothesis is rejected. The type I error rate is the probability of falsely rejecting the null hypothesis when it is true. The type I error rate is almost universally set at 0.05, although this is a matter of convention and is rarely based on logic (chapter 2).

Null hypothesis significance tests with a type 1 error rate of α are closely related to $100(1-\alpha)\%$ confidence intervals. Note that the one case where the 95% confidence interval overlaps the null hypothesis of 1.0 (figure 1.3) is the one case in which the p-value is greater than 0.05. More generally, a p-value for a two-sided null hypothesis significance test will be less than α when the $100(1-\alpha)\%$ confidence interval does not overlap the

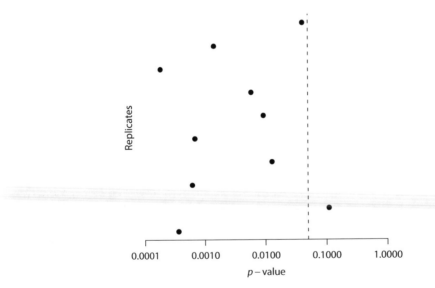

Fig. 1.5 p–values for the hypothesis that the mean is equal to 1.0 (figure 1.4). Cumming (2011) describes variation in p-values as "the dance of the p-values." The p-values (axis has a logarithmic scale) were calculated from the replicate samples in figure 1.2b. The dashed line is the conventional type I error rate of 0.05, with the p-value less than this in 9 of the 10 replicate samples.

null hypothesis. Thus, null hypothesis significance testing is equivalent to comparing the range of a confidence interval to the null hypothesis.

While assessing overlap of a confidence interval with a null hypothesis is equivalent to significance testing, statistical significance when comparing two estimates is not simply a case of considering whether the two confidence intervals overlap. When comparing two independent means with confidence intervals of similar width, a p-value of 0.05 occurs when their confidence intervals overlap by approximately one half the length of their arms (Cumming, 2011). Such "rules of eye" have not been determined for all statistical tests, but the one half overlap rule is useful, and more accurate than assuming statistical significance only occurs when two intervals do not overlap.

While the type I error rate specifies the probability of falsely rejecting a true null hypothesis, such a dichotomous decision also entails the risk of failing to reject a false null hypothesis. The probability of this occurring is known as the type II error rate. For example, in the 10 data sets shown in figure 1.2, only 9 of them lead to p-values that that are less than the conventional type I error rate of 0.05. Thus, we would reject the null hypothesis in only 9 of the 10 cases, despite it being false.

The type I and type II error rates are related, such that one increases as the other declines. For example, if the type I error rate in figure 1.5 were set at 0.01, then the null hypothesis would be rejected for only 7 of the 10 data sets.

The type II error rate also depends on the difference between the null hypothesis and the true value. If the null hypothesis was that the mean is 0 (a difference of 2 units from the truth), then all the data sets in figure 1.2 would generate p-values less than 0.05. In contrast, a null hypothesis of 1.5 (0.5 units from the truth) would be rejected (with a type I error rate of 0.05) in only 4 of the 10 data sets. The type II error rate also changes with variation in the data and the sample size. Less variable data and larger sample sizes both decrease the type II error rate; they increase the chance of rejecting the null hypothesis when it is false.

In summary, the type II error rate depends on the type of statistical analysis being conducted (e.g., a comparison of means, a linear regression, etc.), the difference between the truth and the null hypothesis, the chosen type I error rate, the variation in the data, and the sample size. Because the truth is not known, the type II error rate is usually calculated for different possible truths. These calculations indicate the size of the deviation from the null hypothesis that might be reliably detected with a given analysis and sample size.

Calculating type II error rates is not straightforward to do by hand, but software for the task exists (e.g., G*power, or R libraries like `pwr`). However, analytical *power* calculations are available only for particular forms of analysis. Power can, more generally, be calculated using simulation. Data are generated from a particular model, these data are analyzed, the statistical significance is recorded, and the process is iterated many times. The proportion of iterations that lead to statistically significant results measures the statistical power (box 1.3).

While the analysis here is relatively simple (figure 1.6 reports power for a basic logistic regression), more detail can be added, such as a null hypothesis other than zero, extra variation among years, temporal correlation in reporting rate, imperfect detection, etc. (e.g., Guillera-Arroita and Lahoz-Monfort 2012).

I estimated statistical power for an initial reporting rate of 50%, for different rates of decline and number of survey sites, a type I error rate of 0.05, and using a null hypothesis of no decline (figure 1.7). When there is no decline (when the null is true), the probability of obtaining a statistically significant decline is 0.05 (the type I error rate),

---

Box 1.3 SIMULATION TO CALCULATE POWER

Assume we plan to monitor a bird species over a 10-year period, recording the proportion of sites at which the species is detected each year (reporting rate). The survey design requires us to choose the number of sites to be sampled. If the reporting rate declines at a particular rate over time, then we wish to determine how the probability of observing a statistically significant decline increases with the number of sites sampled per year.

This form of data lends itself to logistic regression, with the reporting rate being a function of time. We sample over $T + 1$ years from time 0 to time T, with n sites surveyed per year.

Thus, the reporting rate in year t (p_t) is given by:

$$\text{logit}(p_t) = \text{logit}(p_0) - bt,$$

where p_0 is the initial reporting rate, b controls the rate of decline, and the logit function is the natural logarithm of the odds ratio: $\text{logit}(p_t) = \ln(p_t/[1-p_t])$. Then, in any given year, the number of sites (y_t) at which the species is detected (out of the n surveyed) will have a binomial distribution given by:

$$y_t \sim \text{dbin}(p_t, n),$$

where $\text{dbin}(p_t, n)$ defines a random sample from a binomial distribution derived from n independent trials, each with a probability of success of p_t.

If we assume particular values for p_0 and b, which reflect the initial reporting rate and rate of decline of the species, we can simulate the data y_t in each year. We then apply a logistic regression analysis to these data.

Note, the trend in y_t will not perfectly match that in p_t because the data have a random element, so the regression will not perfectly estimate the underlying trend (see code, appendix 1.A). We are interested in determining how well the proposed sampling can estimate the trend in p_t, and in particular the probability of detecting a statistically significant decline.

If we record whether a statistically significant decline is detected, and then iterate the data generation and analysis multiple times, then the proportion of iterations in which a statistically significant decline occurs will estimate the statistical power.

---

as expected (figure 1.6). In this case, a statistically significant decline is only evident when the rate of decline and the number of survey sites per year are sufficiently large (figure 1.6).

The analyses described here are based on an *a priori* power analysis. The effect sizes are not those that are measured, but those deemed important. Some statistical packages report retrospective power analyses that are based on the effect sizes as estimated by the analysis. Such power analyses should usually be avoided (Steidl and Thomas 2001) because they do not help design a study or understand the power of the study to detect important effects (as opposed to those observed).

The type II error rate (β), or its complement power ($1-\beta$), is clearly important in ecology. What is the point of designing an expensive experiment to test a theory if that experiment has little chance of identifying a false null hypothesis? Ecologists who practice null hypothesis testing should routinely calculate type II error rates, but the evidence is that they do not. In fact, they almost *never* calculate it (Fidler et al. 2006). The focus of ecologists

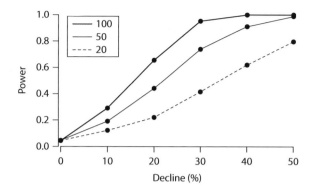

Fig. 1.6 The probability of detecting a statistically significant decline (power) for a program that monitors reporting rate over 10 years. Results are shown for various rates of decline from an initial reporting rate of 50% and for 20, 50, or 100 sites monitored each year. A null hypothesis of no decline, and a type I error rate of 0.05 was assumed.

on the type I error rate and failure to account for the type II error rate might reflect the greater effort required to calculate the latter. This is possibly compounded by practice, which seems to accept ignorance of type II error rates. If type II error rates are hard to calculate, and people can publish papers without them, why would one bother? The answer about why one *should* bother is discussed later.

1.3.3 *Likelihood*

A third approach to frequentist statistical methods is based on the concept of *likelihood* (see also chapter 3). Assume we have collected sample data of size n. Further, we will assume these data were generated according to a statistical model. For example, we might assume that the sample data are random draws from a Normal distribution with two parameters (the mean and standard deviation). In this case, a likelihood analysis would proceed by determining the likelihood that the available data would be observed if the true mean were μ and the true standard deviation were σ. Maximum likelihood estimation finds the parameter values (μ and σ in this case) that were most likely to have generated the observed data (i.e., the parameter values that maximize the likelihood).

The likelihood of observing each data point can simply equal the probability density, $f(x)$, for each; likelihood need only be proportional to probability. The likelihood of observing the first data point x_1 is $f(x_1)$, the likelihood of observing the second data point x_2 is $f(x_2)$, and so forth. In general, the likelihood of observing the ith data point x_i is $f(x_i)$. If we assume that each data point is generated independently of each other, the likelihood of observing all n data points is simply the product of the n different values of $f(x_i)$. This is expressed mathematically using the product operator (Π):

$$L = \prod_{i=1}^{n} f(x_i).$$

For various reasons, it is often simpler to use the logarithm of the likelihood, with the product of the likelihoods then becoming the sum of the logarithms of $f(x_i)$. Thus,

$$\ln L = \sum_{i=1}^{n} \ln f(x_i).$$

By expressing the equation in terms of the log-likelihood, a sum (which can be much easier to manipulate mathematically) has replaced the product. Further, because $\ln L$ is a monotonic function of L, maximizing $\ln L$ is equivalent to maximizing L. Thus, maximum likelihood estimation usually involves finding the parameter values (μ and σ in the case of a Normal distribution) that maximize $\ln L$.

While it is possible to derive the maximum likelihood estimators for the Normal model (box 1.4) and some other statistical models, for many other statistical models such expressions do not exist. In these cases, the likelihood needs to be maximized numerically.

Maximum likelihood estimation can also be used to place confidence intervals on the estimates. A $Z\%$ confidence interval is defined by the values of the parameters for which values of $\ln L$ are within $\chi^2_{1-Z/100}/2$ units of the maximum, where $\chi^2_{1-Z/100}$ is the chi-squared value with 1 degree of freedom corresponding to a p-value of $1 - Z/100$.

For example, in the case of the Normal distribution, the 95% confidence interval based on the likelihood method reduces to the expression $\bar{x} \pm 1.96\frac{\sigma}{\sqrt{n}}$, which is the standard frequentist confidence interval.

Box 1.4 MAXIMUM LIKELIHOOD ESTIMATION AND THE NORMAL DISTRIBUTION

For the case of the Normal distribution, the log-likelihood function is given by:

$$\ln L = \sum_{i=1}^{n} [\ln(\frac{1}{\sigma\sqrt{2\pi}}) - \frac{(x_i - \mu)^2}{2\sigma^2}] = -n\ln(\sigma) - n\ln(\sqrt{2\pi}) - \frac{1}{2\sigma^2}\sum_{i=1}^{n}(x_i - \mu)^2. \quad (1.4A)$$

Note that by expressing the equation in terms of the log-likelihood, we have avoided the exponential terms for the Normal probability density function, simplifying the expression for the likelihood substantially.

For the case of a Normal distribution, it is possible to obtain mathematical expressions for the values of μ and σ (known as the maximum likelihood estimators) that maximize $\ln L$. Inspecting equation (1.4A) reveals that the value of μ that maximizes $\ln L$ is the value that minimizes $S_x = \sum_{i=1}^{n}(x_i - \mu)^2$, because μ does not appear in the other terms. This term is the sum of squares, so the value of μ that maximizes the likelihood is the same as the value that minimizes the sum of squares. Thus, the maximum likelihood estimate of μ is the same as the least-squares estimate in the case of a Normal distribution. Differentiating S_x with respect to μ, setting the derivative to zero, and solving for μ gives the value of μ that minimizes S_x. This procedure shows that the maximum likelihood estimator for μ is the sample mean $\bar{x} = \sum_{i=1}^{n} x_i/n$ because this maximizes $\ln L$.

The maximum likelihood estimate of σ can be obtained similarly. Note that $\mu = \bar{x}$ when $\ln L$ is maximized, so at this point $\sum_{i=1}^{n}(x_i - \mu)^2 = s^2(n - 1)$, where s^2 is the sample variance. Thus, the value of σ that maximizes $\ln L$ is the one that maximizes $-n\ln(\sigma) - \frac{s^2(n-1)}{2\sigma^2}$. Taking the derivative of this expression with respect to σ, setting it to zero, and solving for σ yields its maximum likelihood estimate. This procedure reveals that the maximum likelihood estimator of σ is the "population" standard deviation, $s\sqrt{(n - 1)/n}$.

Maximum likelihood estimation might appear a convoluted way of estimating the mean, standard deviation and confidence intervals that could be obtained using conventional methods when data are generated by a Normal distribution. However, the power of maximum likelihood estimation is that it can be used to estimate parameters for probability distributions other than the Normal, using the same procedure of finding the parameter values under which the likelihood of generating the data is maximized (box 1.5).

Maximum likelihood estimation also extends generally to other statistical models. If we think of the data as being generated by a particular probability distribution, and relate the parameters of that distribution to explanatory variables, we have various forms of regression analysis. For example, if we assume the mean of a Normal distribution is a linear function of explanatory variables, while the standard deviation is constant, we have standard linear regression (chapter 3). In this case, maximum likelihood methods would be used to estimate the regression coefficients of the relationship between the mean and the explanatory variables. Assuming a non-linear relationship leads to non-linear regression; change the assumed probability distribution, and we have a generalized linear model (McCullagh and Nelder 1989; chapter 6). Include both stochastic and deterministic components in the relationships between the parameters and the explanatory variables and we have mixed models (Gelman and Hill 2007; chapter 13). Thus, maximum likelihood estimation provides a powerful general approach to statistical inference.

Just as null hypothesis significance testing can be related to overlap of confidence intervals with the null hypotheses, intervals defined using likelihood methods can also be used for null hypothesis significance testing. Indeed, null hypothesis significance testing can be performed via the likelihood ratio—the likelihood of the analysis based on the

---

Box 1.5 MAXIMUM LIKELIHOOD ESTIMATION OF A PROPORTION

Assume that we wish to estimate the probability (p) that a species occurs within study quadrats. With the species observed in y of n surveyed quadrats (and ignoring imperfect detectability), the likelihood of observing the data is proportional to $p^y(1-p)^{n-y}$. That is, the species occurred in y quadrats, an outcome that has likelihood p for each, and it was absent from $n-y$ quadrats, an outcome that has likelihood $(1-p)$ for each.

The log-likelihood in this case is $y\ln(p)+(n-y)\ln(1-p)$. The derivative of this with respect to p is $y/p-(n-y)/(1-p)$, which equals zero at the maximum likelihood estimate of p. Some simple algebra yields the maximum likelihood estimator for p as y/n.

A $Z\%$ confidence interval can be obtained by finding the values of p such that the log-likelihood is within $\chi^2_{1-Z/100}/2$ units of the maximum. The maximum log-likelihood is $y\ln(y/n)+(n-y)\ln(1-y/n)$, so the limits of the confidence interval are obtained by solving

$$y\ln(y/n)+(n-y)\ln(1-y/n)-y\ln(p)-(n-y)\ln(1-p)=\chi^2_{1-Z}/2.$$

When $y=0$ or $y=n$, the terms beginning with y or $(n-y)$ are zero, respectively, so analytical solutions are possible. In the former case, the confidence interval is $[0, 1-\exp(-\chi^2_{1-Z/100}/2n)]$, while in the latter case it is $[\exp(-\chi^2_{1-Z/100}/2n), 1]$. In other cases, a numerical solution is required. For example, for $y=1$ and $n=10$, the 95% confidence interval is $[0.006, 0.37]$. For $y=2$ and $n=4$, the 95% confidence interval is $[0.107, 0.893]$.

null hypothesis model, relative to the likelihood of a model when using maximum likelihood estimates. The log of this ratio multiplied by –2 can be compared to a chi-squared distribution to determine statistical significance.

1.3.4 *Information-theoretic methods*

In general, adding parameters to a statistical model will improve its fit. Inspecting figure 1.7 might suggest that a 3- or 4-parameter function is sufficient to describe the relationship in the data. While the fit of the 10-parameter function is "perfect" in the sense that it intersects every point, it fails to capture what might be the main elements of the relationship (figure 1.7). In this case, using 10 parameters leads to over-fitting.

As well as failing to capture the apparent essence of the relationship, the 10-parameter function might make poor predictions. For example, the prediction when the dependent variable equals 1.5 might be wildly inaccurate (figure 1.7). So while providing a very good fit to one particular set of data, an over-fitted model might both complicate understanding and predict poorly. In contrast, the two parameter function might under-fit the data, failing to capture a non-linear relationship. Information theoretic methods address the trade-off between over-fitting and under-fitting.

Information theoretic methods (see also chapter 3) use information theory, which measures uncertainty in a random variable by its entropy (Kullback 1959; Burnham and Anderson 2002; Jaynes 2003). Over the range of a random variable with probability density function $f(x)$, entropy is measured by $-\sum_{x \in X} f(x) \ln f(x)$. Note that this is simply the expected value of the log-likelihood. If we think of $f(x)$ as the true probability density function for the random variable, and we have an estimate ($g(x)$) of that density function, then the difference between the information content of the estimate and the truth is the Kullback–Leibler divergence, or the relative entropy (Kullback 1959):

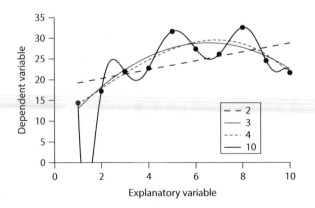

Fig. 1.7 The relationship between a dependent variable and an explanatory variable based on hypothetical data (dots). Polynomial functions with 2, 3, 4, and 10 estimated parameters were fit to the data using least-squares estimation. An *n*th order polynomial is a function with $n + 1$ parameters.

$$KL = -\sum_{x \in X} f(x) \ln g(x) + \sum_{x \in X} f(x) \ln f(x) = \sum_{x \in X} f(x) \ln \frac{f(x)}{g(x)}.$$

The Kullback–Leibler divergence can measure the relative distance of different possible models from the truth. When comparing two estimates of $f(x)$, we can determine which departs least from the true density function, and use that as the best model because it minimizes the information lost relative to the truth.

Of course, we rarely know $f(x)$. Indeed, an estimate of $f(x)$ is often the purpose of the statistical analysis. Overcoming this issue is the key contribution of Akaike (1973), who derived an estimate of the relative amount of information lost or gained by using one model to represent the truth, compared with another model, when only a sample of data is available to estimate $f(x)$. This relative measure of information loss, known as *Akaike's Information Criteria (AIC)*, is (asymptotically for large sample sizes)

$$\text{AIC} = -2\ln L_{\max} + 2k,$$

where $\ln L_{\max}$ is the value of the log-likelihood at its maximized value and k is the number of estimated parameters in the model. Thus, there is a close correspondence between maximum likelihood estimation and information theoretic methods based on AIC.

AIC is a biased estimate of the relative information loss when the sample size (n) is small, in which case a bias-corrected approximation can be used (Hurvich and Tsai 1989):

$$\text{AIC}_c = -2\ln L_{\max} + 2k + \frac{2k(k+1)}{n-k+1}.$$

Without this bias correction, more complicated models will tend to be selected too frequently, although the correction might not be reliable for models with non-linear terms or non-Normal errors (chapter 3).

AIC is based on an estimate of information loss, so a model with the lowest AIC is predicted to lose the least amount of information relative to the unknown truth. The surety with which AIC selects the best model (best in the sense of losing the least information) depends on the difference in AIC between the models. The symbol ΔAIC is used to represent the difference in AIC between one model and another, usually expressed relative to the model with the smallest AIC for a particular data set. Burnham and Anderson (2002) suggest rules of thumb to compare the relative support for the different models using ΔAIC.

For example, the ΔAIC$_c$ values indicate that the 3-parameter (quadratic) function has most support relative to those in figure 1.7 (table 1.1). This is perhaps reassuring given that these data were actually generated using a quadratic function with an error term added.

The term $-2\ln L_{\max}$ is known as the *deviance*, which increases as the likelihood L declines. Thus, AIC increases with the number of parameters and declines with the fit to the data, capturing the trade-off between under-fitting and over-fitting the data. While the formula for AIC is simple and implies a direct trade-off between $\ln L_{\max}$ and k, it is important to note that this trade-off is not arbitrary. Akaike (1973) did not simply decide to weight $\ln L_{\max}$ and k equally in the trade-off. Instead, the trade-off arises from an estimate of the information lost when using a model to approximate an unknown truth.

Table 1.1 The ΔAIC_c values for the functions shown in figure 1.7, assuming Normal distributions of the residuals. The clearly over-fitted 10-parameter function is excluded; in this case it fits the data so closely that the deviance $-2 \ln L$ approaches negative infinity. Akaike weights (w_i) are also shown

Number of parameters	ΔAIC_c	w_i
2	8.47	0.0002
3	0	0.977
4	3.74	0.023

Information theoretic methods provide a valuable framework for determining an appropriate choice of statistical models when aiming to parsimoniously describe variation in a particular data set. In this sense, a model with a lower AIC is likely to predict a replicate set of data better, as measured by relative entropy, than a model with a higher AIC. However, other factors, such as improved discrimination ability, greater simplicity for the sake of rough approximation, extra complexity to allow further model development, or predictive ability for particular times or places, might not be reflected in AIC values. In these cases, AIC will not necessarily be the best criteria to select models (see chapter 3 and examples in chapters 5 and 10).

AIC measures the relative support of pairs of models. The model with the best AIC might make poor predictions for a particular purpose. Instead of using information theoretic methods, the predictive accuracy of a model needs to be evaluated using methods such as cross-validation or comparison with independent data (see also chapter 3).

Use of AIC extends to weighting the support for different models. For example, with a set of m candidate models, the *Akaike weight* assigned to model i is (Burnham and Anderson 2002):

$$w_i = \frac{\exp(-AIC_i/2)}{\sum_{j=1}^{m} \exp(-AIC_j/2)}.$$

Standardizing by the sum in the denominator, the weights sum to 1 across the m models. In addition to assessing the relative support for individual models, the support for including different parameters can be evaluated by summing the weights of those models that contain the parameter.

Relative support as measured by AIC is relevant to the particular data set being analyzed. In fact, it is meaningless to compare the AICs of models fit to different data sets. A variable is not demonstrated to be unimportant simply because a set of models might hold little support for a variable as measured by AIC. Instead, a focus on estimated effects is important. Consider the case of a sample of data that is used to compare one model in which the mean is allowed to differ from zero (and the mean is estimated from the data) and another model in which the mean is assumed equal to zero (figure 1.8). An information theoretic approach might conclude, in this case, that there is at most only modest support for a model in which the mean can differ from zero ($\Delta AIC = 1.34$ in this case).

Values in the second data set are much more tightly clustered around the value of zero (figure 1.8). One might expect that the second data set would provide much greater support for the model in which the mean is zero. Yet the relative support for this model,

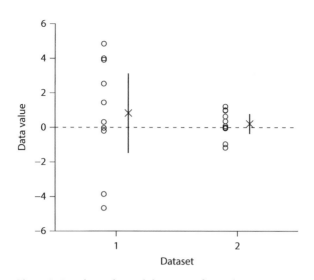

Fig. 1.8 Two hypothetical data sets of sample size 10 showing the means (crosses) and 95% confidence intervals for the means (bars). For each data set, the AIC of a model in which the mean is allowed to differ from zero is 1.34 units larger than a model in which the mean is set equal to zero (dashed line). While the AIC values do not indicate that data set 2 provides greater support for the mean being close to zero, this greater support is well represented by the confidence intervals. The values in data set 2 are the values in data set 1, divided by 4.

as measured by AIC, is the same for both. The possible value of the parameter is better reflected in the confidence interval for each data set (figure 1.8), which suggests that the estimate of the mean in data set 2 is much more clearly close to zero than in data set 1.

This is a critical point when interpreting results using information theoretic methods. The possible importance of a parameter, as measured by the width of its confidence, is not necessarily reflected in the AIC value of the model that contains it as an estimated parameter. For example, if a mean of 2 or more was deemed a biologically important effect, then data set 2 provides good evidence that the effect is not biologically important, while data set 1 is somewhat equivocal with regard to this question. Unless referenced directly to biologically important effect sizes, AIC does not indicate biological importance.

1.3.5 *Bayesian methods*

If a set of data estimated the annual adult survival rate of a population of bears to be 0.5, but with a wide 95% confidence interval of [0.11, 0.89] (e.g., two survivors from four individuals monitored for a year; box 1.5) what should I conclude? Clearly, more data would be helpful, but what if waiting for more data and a better estimate were undesirable?

Being Australian, I have little personal knowledge of bears, even drop bears (Janssen 2012), but theory and data (e.g., Haroldson et al. 2006; Taylor et al. 2005; McCarthy et al. 2008) suggest that mammals with large body masses are likely to have high survival rates. Using relationships between annual survival and body mass of mammals, and

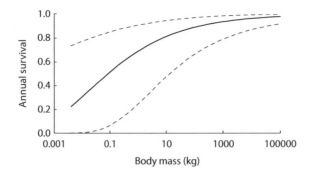

Fig. 1.9 Predicted annual survival of carnivores (solid line is the mean, dashed line is 95% credible intervals) versus annual survival based on a regression model of mammals that accounts for differences among species, studies, and taxonomic orders. Redrawn from McCarthy et al. 2008; Copyright © 2008, The University of Chicago.

accounting for variation among species, among studies and among taxonomic orders, the survival rate for carnivores can be predicted (figure 1.9). For a large bear of 245 kg (the approximate average body mass of male grizzly bears, Nagy and Haroldson 1990), the 95% *prediction interval* is [0.72, 0.98].

This prediction interval can be thought of as my expectation of the survival rate of a large bear. Against this a priori prediction, I would think that the relatively low estimate of 0.5 from the data (with 95% confidence interval of [0.11, 0.89]) might be due to (bad) luck. But now I have two estimates, one based on limited data from a population in which I am particularly interested, and another based on global data for all mammal species.

Bayesian methods can combine these two pieces of information to form a coherent estimate of the annual survival rate (McCarthy 2007). Bayesian inference is derived from a simple rearrangement of conditional probability. Bayes' rule states that the probability of a parameter value (e.g., the annual survival rate of the bear, s) given a set of new data (D) is

$$\Pr(s|D) = \frac{\Pr(D|s)\Pr(s)}{\Pr(D)}.$$

Here $\Pr(D \mid s)$ is the probability of the new data given a particular value for survival; this is simply the likelihood, so $\Pr(D \mid s) = L(D \mid s)$. $\Pr(s)$ is the unconditional probability of the parameter value, and $\Pr(D)$ is the unconditional probability of the new data. By unconditional probability, I mean that these values do not depend on the present data.

$\Pr(s)$, being independent of the data, represents the prior understanding about the values of the parameter s. A probability density function $f(s)$ can represent this prior understanding. A narrow density function indicates that the parameter is already estimated quite precisely, while a wide interval indicates that there is little prior information.

To make $\Pr(D)$ independent of a particular value of s, it is necessary to integrate over the possible values of s. Thus, for continuous values of s, $\Pr(D) = \int_{-\infty}^{\infty} L(D|u)f(u)du$, and Bayes' rule becomes

$$\Pr(s|D) = \frac{L(D|s)f(s)}{\int_{-\infty}^{\infty} L(D|u)f(u)du}.$$

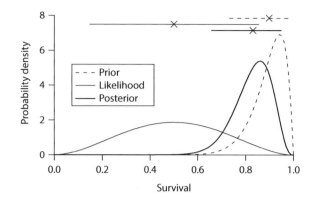

Fig. 1.10 Estimated annual survival of a large carnivore, showing a prior derived from mammalian data (dashed; derived from McCarthy et al. 2008), the likelihood of two of four individuals surviving for a year (thin line), and the posterior distribution that combines these two sources of information (thick line). The 95% credible intervals (horizontal lines) and means (crosses) based on each of the prior, likelihood and posterior are also shown.

When the parameter values are discrete, the integral in the denominator is replaced by a summation, but it is otherwise identical. The probability distribution $f(s)$ is known as the *prior distribution* or simply the "prior." The *posterior distribution* $\Pr(s \mid D)$ describes the estimate of s that includes information from the prior, the data, and the statistical model.

In the case of the bear example, the prior (from figure 1.9) combines with the data and statistical model to give the posterior (figure 1.10). The posterior is a weighted average of the prior and the likelihood, and is weighted more toward whichever of the two is more precise (in figure 1.10, the prior is more precise). The 95% credible interval of the posterior is [0.655, 0.949].

Difficulties of calculating the denominator of Bayes' rule partly explain why Bayesian methods, despite being first described 250 years ago (Bayes 1763), are only now becoming more widely used. Computational methods to calculate the posterior distribution, particularly *Markov chain Monte Carlo (MCMC)* methods, coupled with sufficiently fast computers and available software are making Bayesian analysis of realistically complicated methods feasible. Indeed, the methods are sufficiently advanced that arbitrarily complicated statistical models can be analyzed.

Previously, statistical models were limited to those provided in computer packages. Bayesian MCMC methods mean that ecologists can now easily develop and analyze their own statistical models. For example, linear regression is based on four assumptions: a linear relationship for the mean, residuals being drawn from a Normal distribution, equal variance of the residuals along the regression line, and no dependence among those residuals. Bayesian MCMC methods allow you to relax any number of those assumptions in your statistical model.

Posterior distributions contain all the information about parameter estimates from Bayesian analyses, and are often summarized by calculating various statistics. The mean or median of a posterior distribution can indicate its central tendency. Its standard deviation indicates the uncertainty of the estimate; it is analogous to the standard error of a statistic in frequentist analysis. Inner percentile ranges are used to calculate credible intervals. For example, the range of values bounded by the 2.5 percentile and the 97.5 percentile of the posterior distribution is commonly reported as a 95% credible interval.

Credible intervals of Bayesian analyses are analogous to confidence intervals of frequentist analyses, but they differ. Because credible intervals are based on posterior distributions, we can say that the probability is 0.95 that the true value of a parameter occurs within its 95% credible interval (conditional on the prior, data, and the statistical model). In contrast, confidence intervals are based on the notion of replicate sampling and analysis; if we conducted this study a large number of times, the true value of a parameter would be contained in a Z% confidence interval constructed in this particular way Z% of the time (conditional on the data and the statistical model). In most cases, the practical distinction between the two definitions of intervals is inconsequential because they are similar (see below).

The relative influence of the prior and the posterior is well illustrated by estimates of annual survival of female European dippers based on mark–recapture analysis. As for the mammals, a relationship between annual survival and body mass of European passerines can be used to generate a prior for dippers (McCarthy and Masters 2005).

Three years of data (Marzolin 1988) are required to estimate survival rate in mark–recapture models that require joint estimation of survival and recapture probabilities. If only the first three years of data were available, the estimate of annual survival is very imprecise. In the relatively short time it takes to compile and analyze the data (about half a day with ready access to a library), a prior estimate can be generated that is noticeably more precise (left-most interval in figure 1.11).

Three years of data might be the limit of what could be collected during a PhD project. If you are a PhD student at this point, you might be a bit depressed that a more precise

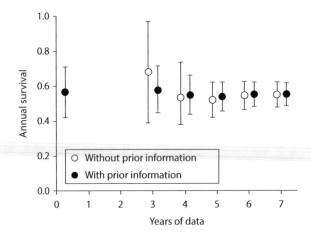

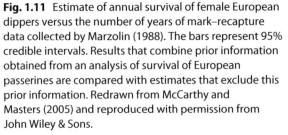

Fig. 1.11 Estimate of annual survival of female European dippers versus the number of years of mark–recapture data collected by Marzolin (1988). The bars represent 95% credible intervals. Results that combine prior information obtained from an analysis of survival of European passerines are compared with estimates that exclude this prior information. Redrawn from McCarthy and Masters (2005) and reproduced with permission from John Wiley & Sons.

estimate can be obtained by simply analyzing existing data compared with enduring the trials (and pleasures) of field work for three years.

However, since you are reading this, you clearly have an interest in Bayesian statistics. And hopefully you have already realized that you can use my analysis of previous data as a prior, combine it with the data, and obtain an estimate that is even more precise. The resulting posterior is shown by the credible interval at year 3 (figure 1.11). Note that because the estimate based only on the data is much less precise than the prior, the posterior is very similar to the estimate based only on the prior. In fact, five years of data are required before the estimate based only on the data is more precise than the prior. Thus, the prior is initially worth approximately 4–5 years of data, as measured by the precision of the resulting estimate.

In contrast, the estimate based only on seven years of data has approximately the same precision as the estimate using both the prior and six years of data (figure 1.11). Thus, with this much data, the prior is worth about one year of data. The influence of the prior on the posterior in this case is reduced because the estimate based on the data is more precise than the prior. Still, half a day of data compilation and analysis seems a valuable investment when it is worth another year of data collection in the field.

In Bayes' rule, probability is being used as measure of how much a rational person should "believe" that a particular value is the true value of the parameter, given the information at hand. In this case, the information consists of the prior knowledge of the parameter as represented by $f(s)$, and the likelihood of the data for the different possible values of the parameter. As for any statistical model, the likelihood is conditional on the model being analyzed, so it is relatively uncontroversial. Nevertheless, uncertainty about the best choice of model remains, so this question also needs to be addressed in Bayesian analyses.

The priors for annual survival of mammals and European passerines are derived from an explicit statistical model of available data. In this sense, the priors are no more controversial than the choice of statistical model for data analysis; it is simply a judgment about whether the statistical model is appropriate. Controversy arises, however, because I extrapolated from previous data, different species, and different study areas to generate a prior for a unique situation. I attempted to account for various factors in the analysis by including random effects such as those for studies, species, taxonomic orders and particular cases within studies. However, a lingering doubt will persist; is this new situation somehow unique such that it lies outside the bounds of what has been recorded previously? This doubt is equivalent to questions about whether a particular data point in a sample is representative of the population that is the intended focus of sampling. However, the stakes with Bayesian priors can be higher when the prior contains significant amounts of information relative to the data. There is little if any empirical evidence on the degree to which biases occur when using priors derived from different sources (e.g., different species, different times, etc.).

Controversy in the choice of the prior essentially reflects a concern that the prior will bias the estimates if it is unrepresentative. Partly in response to this concern, and partly because prior information might have little influence on the results (consider using seven years of data in figure 1.11), most ecologists use Bayesian methods with what are known as "uninformative," "vague," or "flat" priors.

In Bayes' rule, the numerator is the prior multiplied by the likelihood. The denominator of Bayes' rule recalibrates this product so the posterior conforms to probability (i.e., the area under the probability density function equals 1). Therefore, the posterior is simply proportional to the product of the prior and the likelihood. If the prior is flat across the range of the likelihood function, then the posterior will have the same shape as the likelihood. Consequently, parameter estimates based on uninformative priors are very similar to parameter estimates based only on the likelihood function (i.e., a frequentist analysis). For example, a Bayesian analysis of the data in figure 1.2 with uninformative priors produces 95% credible intervals that are so similar to the confidence intervals in figure 1.3 that it is not worth reproducing them.

In this example, I know the Bayesian prior is uninformative because the resulting credible intervals and confidence intervals are the same. In essence, close correspondence between the posterior and the likelihood is the only surety that the prior is indeed uninformative. However, if the likelihood function can be calculated directly, why bother with the Bayesian approach? In practice, ecologists using Bayesian methods tend to assume that particular priors are uninformative, or use a range of different reasonable priors. The former is relatively safe for experienced users of standard statistical models, who might compare the prior and the posterior to be sure the prior has little influence. The latter is a form of robust Bayesian analysis, whereby a robust result is one that is insensitive to the often arbitrary choice of prior (Berger 1985).

Why would an ecologist bother to use Bayesian methods when informative priors are rarely used in practice, when uninformative priors provide answers that are essentially the same as those based on likelihood analysis, and when priors are only surely noninformative when the posterior can be compared with the likelihood? The answer is the convenience of fitting statistical models that conform to the data. Hierarchical models represent one class of such models.

While frequentist methods can also be used, hierarchical models in ecology are especially well suited to Bayesian analyses (Clark 2005; Gelman and Hill 2007). Hierarchical models consider responses at more than one level in the analysis. For example, they can accommodate nested data (e.g., one level modeling variation among groups, and another modeling variation within groups), random coefficient models (regression coefficients themselves being modeled as a function of other attributes), or state–space models. State–space models include, for example, a model of the underlying (but unobserved) ecological process overlaid by a model of the data collection, but which then allows inference about the underlying process not just the generated data (McCarthy 2011).

Because prior and posterior distributions represent the degree of belief in the true value of a parameter, an analyst can base priors on subjective judgments. The advantage of using Bayesian methods in this case is that these subjective judgments are updated logically as data are analyzed. Use of subjective priors with Bayesian analyses might, therefore, be useful for personal judgments. However, such subjective judgments of an individual might be of little interest to others, and might have little role in wider decisions or scientific consensus (unless that individual were particularly influential, but even then such influence might be undesirable).

In contrast, when priors reflect the combined judgments of a broad range of relevant people and are compiled in a repeatable and unbiased manner (Martin et al. 2005a), combining them with data via Bayes' rule can be extremely useful. In this case, Bayesian methods provide a means to combine a large body of expert knowledge with new data. While the expert knowledge might be wrong (Burgman 2005), the important aspect of Bayesian analysis is that its integration with data is logical and repeatable.

Additionally, priors that are based on compilation and analysis of existing data are also valuable. Such compilation and analysis is essentially a form of meta-analysis (chapter 9). Indeed, Bayesian methods are often used for meta-analysis. Discussion sections of publications often compare and seek to integrate the new results with existing knowledge.

Bayesian methods do this formally using coherent and logical methods, moving that integration into the methods and results of the paper, rather than confining the integration to subjective assessment in the discussion. If ecology aims to have predictive capacity beyond particular case studies, then Bayesian methods with informative priors will be used more frequently.

1.3.6 *Non-parametric methods*

This chapter emphasizes statistical analyses that are founded on probabilistic models. These require an assumption that the data are generated according to a specified probability distribution. Non-parametric methods have been developed to avoid the need to pre-specify a probability distribution. Instead, the distribution of the collected data is used to define the sampling distribution. So while non-parametric methods are sometimes described as being "distribution-free," this simply means that the analyst does not choose a distribution; rather the data are used to define the distribution.

A wide range of non-parametric methods exist (Conover 1998). Instead of describing them all here, I will focus on only one method as an example. Non-parametric methods often work by iterative resampling of the data, calculating relevant statistics of each subsample, and then defining the distribution of the sample statistics by the distribution of the statistics of the subsamples.

Bootstrapping is one such resampling method (DiCiccio and Efron 1996). Assume that we have a sample of size n, for which we want to calculate a 95% confidence interval but are unable or unwilling to assume a particular probability distribution for the data. We can use bootstrapping to calculate a confidence interval by randomly resampling (with replacement) the values of the original sample, and generate a new sample of size n. We then calculate the relevant sample statistic (e.g., the mean), and record the value. This procedure is repeated many times. Percentiles of the resulting distribution of sample statistics can be used to define a confidence interval. For example, the 2.5 percentile and 97.5 percentile of the distribution of resampled statistics define a 95% confidence interval. For the data in figure 1.2, the resulting bootstrapped confidence intervals for the mean, while narrower than those derived assuming a Normal distribution, are largely similar (figure 1.12).

Non-parametric methods tend to be used when analysts are unwilling to assume a particular probabilistic model for their data. This reluctance was greatest when statistical models based on the Normal distribution were most common. With greater use of statistical models that use other distributions (e.g., generalized linear models, McCullagh and Nelder 1989), the impetus to use non-parametric methods is reduced, although not eliminated. Indeed, methods like the bootstrap are likely to remain important for a long time.

1.4 Appropriate use of statistical methods

With such a broad array of approaches to statistical inference in ecology, which approach should you choose? The literature contains debates about this (e.g., Dennis 1996; Anderson et al. 2000; Burnham and Anderson 2002; Stephens et al. 2005). To a small extent, I have contributed to those debates. For example, my book on Bayesian methods (McCarthy 2007) was partly motivated by misuses of statistics. I thought greater use of Bayesian methods would reduce that misuse. Now, I am less convinced. And the debates seem to distract from more important issues.

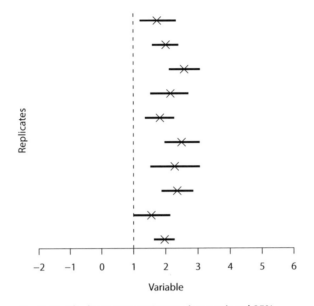

Fig. 1.12 The bootstrap estimates (crosses) and 95% confidence intervals (bars) for the mean using the data in figure 1.2.

The key problem with statistical inference in ecology is not resolving which statistical framework to choose, but appropriate reporting of the analyses. Consider the confidence and credible intervals for the data in figure 1.2. The intervals, representing estimates of the mean, are very similar regardless of the method of statistical inference (figures 1.3 and 1.12). In these cases, the choice of statistical philosophy to estimate parameters is not very important. Yes, the formal interpretation and meaning of a confidence interval and a credible interval differ. However, assume that I constructed a confidence interval using likelihood methods, and interpreted that confidence interval as if it were a Bayesian credible interval formed with a flat prior. Strictly, this is not correct. Practically, it makes no difference because I would have obtained the same numbers however I constructed the intervals.

Understanding the relatively infrequent cases when credible intervals differ from confidence intervals (Jaynes 2003) is valuable. For example, there is a difference between the probability of recording a species as being present at a site, and the probability that the species is present at a site given it is recorded (or not). The latter, quite rightly, requires a prior probability and Bayesian analysis (Wintle et al. 2012). However, the choice of statistical model and appropriate reporting and interpretation of the results are much more important matters. Here I note and briefly discuss some of the most important issues with the practice of statistical inference in ecology, and conclude with how to help overcome these problems.

Avoid nil nulls. Null hypothesis significance testing is frequently based on nil nulls, which usually leads to trivial inference (Anderson et al. 2000; Fidler et al. 2006). Nil nulls are often hypotheses that we know, a priori, have no hope of being true. Some might argue that null hypothesis significance testing conforms with Popperian logic based on falsification. But Popper requires bold conjectures, so the null hypothesis needs to be plausibly true. Rejecting a nil null, that is already known to be false, is unhelpful regardless of whether or not Popperian falsification is relevant in the particular circumstance. Ecologists should avoid nil nulls unless they are plausibly true. Null hypotheses should be based as much as possible on sound theory or empirical evidence of important effects. If a sensible null hypothesis cannot be constructed, which will be frequent in ecology, then null hypothesis significance testing should be abandoned and the analysis limited to estimation of effect sizes.

Failure to reject a null does not mean the null is true. Null hypothesis significance testing aims to reject the null hypothesis. Failure to reject a null hypothesis is often incorrectly used as evidence that the null hypothesis is true (Fidler et al. 2006). This is especially important because power is often low (Jennions and Møller 2003), it is almost never calculated in ecology (Fidler et al. 2006), and ecologists tend to overestimate statistical power when they judge it subjectively (Burgman 2005). Low statistical power means that the null hypothesis is unlikely to be rejected even if it were false. Failure to reject the null should never be reported as evidence in favor of the null unless power is known to be high.

Statistical significance is not biological importance. A confidence interval or credible interval for a parameter that overlaps zero is often used incorrectly as evidence that the associated effect is biologically unimportant. This is analogous to equating failure to reject a null hypothesis with a biologically unimportant effect. Users of all statistical methods are vulnerable to this fallacy. For example, low Akaike weights or high ΔAIC values are sometimes used to infer that a parameter is biologically unimportant. Yet AIC values are not necessarily sensitive to effect sizes (figure 1.8).

P-values do not indicate replicability. P-values are often viewed as being highly replicable, when in fact they are typically variable (Cumming 2011). Further, the size of the p-value does not necessarily indicate how different the p-value from a new replicate might be. In contrast, confidence intervals are less variable, and also indicate the magnitude of possible variation that might occur in a replicate of the experiment (Cumming 2011). They should be used and interpreted much more frequently.

Report and interpret confidence intervals. Effect sizes, and associated measures of precision such as confidence intervals, are often not reported. This is problematic for several reasons. Firstly, the size of the effect is often very informative. While statistical power is rarely calculated in ecology, the precision of an estimate conveys information about power (Cumming 2011). Many ecologists might not have the technical skills to calculate power, but all ecologists should be able to estimate and report effect sizes with confidence intervals. Further, failure to report effect sizes hampers meta-analysis because the most informative meta-analyses are based on them. Reporting effect sizes only for statistically significant effects can lead to reporting biases. Meta-analysis is extremely valuable for synthesizing and advancing scientific research (chapter 9), so failure to report effect sizes for all analyses directly hampers science.

Wide confidence intervals that encompass both important and trivial effects indicate that the data are insufficient to determine the size of effects. In such cases, firm inference about importance would require more data. However, when confidence intervals are limited to either trivial or important effects, we can ascribe importance with some reliability.

I addressed the failure to report effect sizes last because addressing it is relatively easy, and doing so overcomes many of the other problems (although in some complex statistical models, the meaning of particular parameters needs to be carefully considered). Reporting effect sizes with intervals invites an interpretation of biological importance. If variables are scaled by the magnitude of variation in the data, then effect sizes reflect

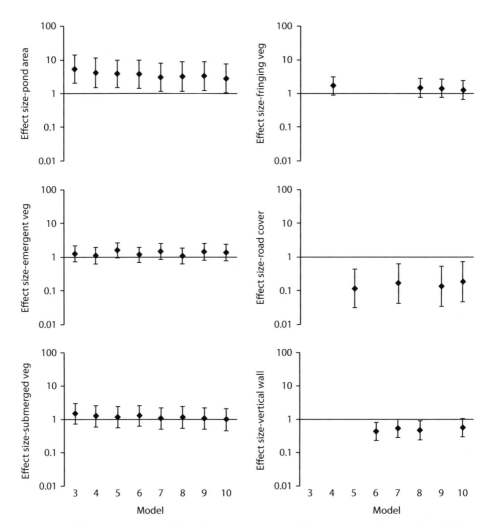

Fig. 1.13 Effect sizes estimated from Poisson regression of frog species richness of ponds for six different explanatory variables, showing means (dots) and 95% credible intervals. Results are shown for eight models that included different combinations of the explanatory variables. The effect size is the predicted proportional change in species richness from the lowest to the highest value of each explanatory variable recorded in the data set (i.e., a value of 10 is a tenfold increase while 0.1 is a tenfold decline). Reproduced from Parris (2006) with permission from John Wiley & Sons.

the predicted range of responses in that data set. For example, Parris (2006) reported regression coefficients in terms of how much the predicted species richness changed across the observed range of the explanatory variables in Poisson regression models (figure 1.13). This illustrates that more than tenfold changes in expected species richness are possible across the range of some variables (e.g., road cover) but such large effects are unlikely for other variables (e.g., fringing vegetation). Nevertheless, all the intervals encompass possible effects that are larger than a doubling of expected species richness regardless of the particular statistical model. These results quantify how precisely the parameters are estimated in this particular study. They also permit direct comparison with effect sizes in similar studies, either informally, or by using meta-analysis.

Even if confidence intervals are interpreted poorly by an author (for example, authors might incorrectly interpret a confidence interval that encompasses zero as evidence that the associated variable is unimportant), reporting them properly is critical, because they can still be interpreted appropriately by readers. Important effects might be currently unknown, so interpreting effect sizes might not always be possible. However, importance might be determined in the future. At that time, researchers can only interpret importance of reported results if effect sizes are presented with confidence intervals.

These problems in the use of statistical inference are not unique to ecology. It is valuable to look beyond ecology to understand how use of statistics has improved in some other disciplines. Some disciplines have largely overcome these problems (Fidler et al. 2006), while others are making progress by recommending reporting of effect sizes (Cumming 2011). The key to improvement is concerted effort across a discipline. This needs to involve authors and reviewers, but as the final arbiters of what constitutes acceptable scientific practice, editors are particularly influential.

Statistical inference is critical in ecology because data are variable and replication is often difficult. While statistical methods are becoming more complex, it is important that statistical practices are founded on sound principles of interpretation and reporting. A greater emphasis in ecology on basic estimation, reporting, and interpretation of effect sizes is critical for the discipline.

Acknowledgments

I'd like to thank the many people who provided comments on early drafts of this chapter including Ben Bolker, Barry Brook, Fiona Fidler, Gordon Fox, Eli Gurarie, Jessica Gurevitch and students in her graduate course, Jenny Koenig, Simoneta Negrete, Daniel Noble, Paco Rodriguez-Sanchez, and Vinicio Sosa.

CHAPTER 2

Having the right stuff: the effects of data constraints on ecological data analysis

Earl D. McCoy

2.1 Introduction to data constraints

Statistical analysis can be daunting for the novice, and, sometimes, even for the seasoned practitioner. The proliferation of techniques, in fact entire systems of analysis, over the past couple of decades truly is staggering. Accompanying this proliferation of statistical techniques has been a proliferation of statistical textbooks, many of which are written as "how-to" manuals. These textbooks typically are of the form: here is a kind of problem, here are some relevant techniques, and here are some things to watch out for (repeat). To reinforce the "how-to" approach, a textbook may come with flow charts to help one choose among the seemingly endless possibilities. Perhaps the extreme is to offer a dichotomous key, as some textbooks do, to make the difficult choices seem effortless. While I appreciate what the authors of these textbooks are trying to accomplish, I regret also their implicit suggestion that the application of statistics is pro forma, like using a hammer rather than a pliers to drive a nail. On the contrary, the successful application of statistics requires that a great deal of thought be given to research design, particularly to *data gathering*—the poor stepchild of statistical analysis. Without adequate forethought about data gathering, a researcher risks wasting time and effort in an endeavor that is doomed from the start, no matter how sophisticated the subsequent analysis.

How should one approach data gathering? Most knowledgeable persons likely would answer that the data should satisfy the requirements and assumptions of the statistical analysis that ultimately will be used to test the relevant hypothesis or to compare models. Planning data gathering for this purpose is a good idea: much grief can be avoided by asking oneself a few basic questions beforehand. Do I have a reasonable estimate of effect size? Will I have enough power to reject a hypothesis that actually is false? When one of my incubators, or cages, or enclosures fails—as it invariably will—will my blocking and replication scheme still allow testing of important effects? Although the potential problems reflected in questions such as these clearly are important, they usually can be eliminated, or at least eased, with proper planning. Another, very much underappreciated, aspect of data gathering is the suite of choices that have to be made: about which method

Ecological Statistics: Contemporary Theory and Application. First Edition. Edited by Gordon A. Fox, Simoneta Negrete-Yankelevich, and Vinicio J. Sosa. © Oxford University Press 2015. Published in 2015 by Oxford University Press.

to employ, about how to match data gathering to the available time and labor, and about how to decide which factors are likely to be explanatory and which are likely to be confounding, for example. It would seem at first blush that these choices should result more-or-less automatically from careful logical analysis; but, in reality, they are shaped by forces such as *presuppositions*, *values*, and *biases*. For instance, if the results of the statistical analysis of a data set were to be judged at an α-level of 0.05 and data gathering was based on this *Type I error* rate, then the researcher may have presupposed that the Type I error is more important to avoid than the *Type II error*. I shall refer to the forces that determine the data set used to address a particular question as *data constraints*. Potential problems arising from data constraints often are difficult even to recognize.

In this chapter, I first introduce some ecological data constraints. Then, I show how these constraints may affect the conclusions a researcher reaches about ecological patterns and processes. The topic of this chapter is closely related to the topic of inference, which is addressed thoroughly in chapter 1 (also see Taper and Lele 2004). Inference (inductive reasoning) is deriving logical conclusions from premises known or assumed to be true. Often, these premises are presuppositions, which, as employed here, simply are conditions taken for granted (but see Gauch 2003). Thus, I argue in this chapter that data constraints can lead to flawed premises, which can lead in turn to flawed conclusions. Finally, I present three situations, with extended examples, in which flawed conclusions can easily result from unrecognized constraints. These situations are rooted in the inherent complexity of interesting ecological research questions. I suggest that flawed conclusions engendered by data constraints may be more prevalent in ecological research than usually is granted, and provide some rough suggestions about how they can be avoided.

2.2 Ecological data constraints

2.2.1 *Values and biases*

All scientists are forced to make value judgments. "Science may be morally neutral, but so is a traffic light; car drivers and scientists are not" (Jackson 1986, pp. 1–2). Yet, it is tempting for scientists to think that their work is value-free. If scientists think that their work is value-free, then it follows that they think science is completely factual, that their work is without presuppositions, that value judgments do not structure their methods, or—if they do recognize the value judgments in their work—that their scientific conclusions are not highly sensitive to them. Such thinking underlies, for instance, arguments by conservation scientists that they are unable to support conservation actions, for fear of damaging the credibility of their findings (see Van Dyke 2008).

If scientists indeed do tend to downplay the importance of value judgments in their work, why should this be so? The reasons have been the subject of continuing discussion and debate. A scientist may tacitly subscribe to the fact–value dichotomy, which is described and discussed by Hanson (1958), Polanyi (1959, 1964), Toulmin (1961), and Kuhn (1970). The fact–value dichotomy is the belief that facts and values are completely separable, and that facts exist that are not value-laden. Applied to ecology, this claim means that accounts of community structure, for example, ought to consist of factual and neutral observations. Yet, simply altering the context of an analysis of the forces shaping community structure—from a closed system to an open system—can affect the conclusions that are reached. Belief in the fact–value dichotomy is incompatible with formulation of any scientific theory or analysis, because these formulations require one automatically to make value judgments. One must evaluate which theoretical or analytical options to present; and, as Tool (1979) has noted, these normative judgments are outside the scope of purely empirical inquiry. The traditional positivist and hypothetico-deductivist goal of eliminating values in science can be viewed as a noble ideal, because value-free observations, if they existed, would guarantee the complete objectivity of one's scientific research. Values compromise objectivity, however, only if they alone determined the facts, but they do not. Both our values and the action of the external world on our senses are responsible for our perceptions, observations, and facts. Just because facts are value-laden, this does not mean that science is completely subjective or relative (see Shrader-Frechette and McCoy 1993).

A useful classification of values has been presented by Longino (1990). On her view, values can be divided into three basic types: bias values, contextual values, and methodological values. Bias values occur in science whenever researchers deliberately misinterpret or omit data, so as to serve their own purposes. These types of values obviously can be avoided, but several famous scientists, such as Gregor Mendel (Moore 1993) and Robert Millikan (Jackson 1986), may have fallen victim to their allure; although conclusions about their behaviors remain controversial. Hull (1988) presents a compelling account of bias values that apparently influenced refereeing of papers by cladists, as opposed to non-cladists, for the journal *Systematic Zoology*. Siegelman (1991) presents an account of biased refereeing for the journal *Radiology*. Contextual values are more difficult to avoid in research than are bias values. Scientists subscribe to particular contextual values whenever they include personal, social, cultural, ethical, or philosophical emphases in their judgments. For instance, interruption of funding is considered an important issue facing the scientific community, suggesting that the need for funding may provide contextual values that drive some of what is done in the name of science (Erman and Pister 1989). A related contextual value is the need to fit research programs into the limited time periods afforded by funded research programs, advanced degree programs, and tenure accrual (Hilborn and Stearns 1982). Methodological values are impossible to avoid in science. For instance, whenever one uses a particular research design, with its attendant assumptions, hypotheses, and analyses, one must make methodological value judgments about the adequacy of the research design (Jackson 1986). Even gathering data requires use of methodological value judgments, because one must make evaluative assumptions about which data to collect and which to ignore, how to interpret the data, how to simplify one's task so as to make it tractable, and how to avoid erroneous interpretations (see Fagerström 1987; Miller 1993; Ottaviani et al. 2004).

Value judgments may be expressed as biased behaviors. Biased behaviors can be motivated by efforts to satisfy personal or group needs, such as self-esteem, power, prestige, achievement, and to belong (Heider 1958); but, they can also result from ingrained thinking habits, at least some of which appear to be "hard-wired" in humans (Baron 2007). I shall return to this latter kind of *cognitive bias* shortly. When researchers have choices about how to gather and analyze data, they may consciously or unconsciously make those choices so as to push the results in a favorable direction. Biases of this sort, such as ceasing data accumulation when just enough data have been gathered to reach a conclusion ("repeated peeks bias") or when the calculated p-value drops just below α ("optional stopping"), can influence the reported outcomes of research projects. For instance, they can lead to an overabundance of reported p-values slightly below 0.05 ("p-hacking," see Masicampo and Lalande 2012). The extent of these problems, if any, has not been explored in ecological research. Below, I list a selection of possible biases that have been pointed out specifically in the ecological literature. The list is not necessarily a complete one, the biases I do list are not at all unique to ecology, and at least some of the biases are avoidable. Examples were chosen to illustrate the fact that these biases are not particularly new to the discipline.

2.2.2 Biased behaviors in ecological research

- *Accepting a potential explanation based on incomplete and/or biased data.* Perhaps the most common methodological failing of scientists is to make inferences beyond what their data warrant. For example, interpretations about species distributions and community composition can be in error simply because incomplete and/or biased data were accepted at face value from the literature and unpublished sources (Kodric-Brown and Brown 1993). Although all scientists use induction to make generalizations on the basis of particular observations, the degree of reasonable belief in the correctness of those generalizations should be proportional to the support they have attained, preferably from testing.
- *Failing to see logical relationships among studies.* Just as scientists can err in making inductive leaps, they can also err in not examining the deductive relationships among hypotheses and conclusions in different studies. Even though two scientists are working on related problems, the connection between the studies may go unrecognized simply because no one is familiar enough with both of them. For example, detecting many temporal and spatial patterns from individual studies is difficult, requiring instead synthesis of large amounts of data, gathered from a variety of sources (Hom and Cochrane 1991).
- *Failing to consider as many realistic alternative explanations as possible.* Although informally reasoning that a particular observation can be explained by a particular hypothesis is essential to science, such reasoning can be problematic when a scientist jumps to the conclusion that it is the only hypothesis, or the most important hypothesis, or is sufficient to explain the observation. Descriptive studies so common in ecology, for example, may cause researchers not to consider alternative explanations for any patterns they may uncover, because such studies tend to be corroborative (Drew 1994).
- *Failing to report negative results ("trimming" and "cooking").* Trimming occurs when a scientist smooths irregularities to make data look accurate and precise, and cooking occurs when a scientist retains only those results that fit theory and discards other results. Failing to publish non-significant results, for example, could lead to serious misinterpretations of natural patterns and their causes (Csada et al. 1996).
- *Accepting a potential explanation without demanding "risky" evidence in its favor.* Corroboration of a hypothesis is relatively easy, and data frequently can support many different or even inconsistent hypotheses about the same observation (Popper 1959, 1963). It is too easy to accept a hypothesis as true because it has not been proved false or even just because no experiment has been done to determine whether it is false. In ecology, as in other sciences, the greatest confidence should be placed in hypotheses that have passed a large number of tough tests (Strong 1983).
- *Accepting the use of unclear terms and concepts.* This bias occurs whenever scientists use concepts, principles, or theories that are so vague, unclear, or inconsistent that it is impossible to tell even whether two scientists are studying the same phenomenon in the same way. Although scientists need not always have exactly the same conception of a particular term, they should clarify what they mean by the particular term when they use it. For example, the meaning of the term "stability" in ecology has been clouded historically by a lack of clarity (McCoy and Shrader-Frechette 1992).

- *Accepting the inappropriate use of stipulative or circular concepts and definitions.* This bias occurs whenever scientists treat what is vague as if it were more precise, and then try to fit it into an exact logical category. It is not as easily understood as the previous bias, but Cohen (1971, p. 675) referred to it when he wrote: "Physics-envy is the curse of biology. When someone else has done the dirty, tedious work of showing that a mathematically formulated physical principle leads to predictions correct to a specified number of decimal places in the boring world of Euclidean 3-space with Cartesian coordinates, theoreticians and textbook writers can axiomatize, generalize, and dazzle your eyes with the most coordinate-free, cosmically invariant representations you please." Although an ecological theory may be elegantly constructed, it may also lack operational definition of its basic elements; and, therefore, have little or no predictive ability (Peters 1977).
- *Accepting the potentially faulty work of others.* This bias is both epistemically and ethically problematic. It does not contribute to the give-and-take necessary for growth in knowledge; it neglects the duties that persons with expertise often have to society (Shrader-Frechette 1994; McCoy 1996); and it prevents professions from improving, from correcting their errors, and from retaining their credibility over the long term (Shrader-Frechette 1994). Wiens (1981) suggested that major flaws in published papers rarely are criticized, in part as a consequence of the structure of journals, which do not readily offer the opportunity for critical evaluation of previously published work, but also as a consequence of the politeness with which we approach science, regarding criticism as some sort of unseemly personal attack.

2.3 Potential effects of ecological data constraints

2.3.1 *Methodological underdetermination and cognitive biases*

Methodological value judgments, which make up the bulk of ecological data constraints, are impossible to avoid, and yet can be flawed. Perhaps the worst consequence of flawed methodological value judgments is failure to test a hypothesis that the researcher thought (assumed, hoped) was being tested. Failure to test the relevant hypothesis because of an inadequate research design is more formally called *methodological underdetermination*, which Mayo (1996; also see Loehle 1987, 2011; Forber 2009) defines as the situation in which evidence taken as a good test of (or good support for) a hypothesis could be taken as an equally good test of (or equally good support for) one or more rival hypotheses. Popper (1963) has noted that methodological underdetermination, such as that resulting from "hidden treatments" (Huston 1997; also see Marra and Holmes 1997), badly undercuts confirmation. One reason that the problem may be relatively common in ecology is that researchers are more-or-less free to choose methods of data gathering and analysis, and they sometimes choose methods that have a significant chance of yielding confirmation of a particular explanation but that are not good at eliminating competing— often inexplicitly stated—explanations. Chances are that these poor choices are not the result of any planned action, but rather are the product of one or more cognitive biases.

Cognitive biases, which are thinking habits that can lead to systematic errors in reasoning, may be particularly relevant to ecological research because many of them are related to judgments about probability and statistics. Examples include the tendency to attribute patterns and underlying causes to random events ("clustering illusion"), the tendency to think that past random events can influence current random events ("gambler's fallacy," "predictable-world bias"), and the tendency to depend on readily available information ("availability heuristic") or the most recently acquired information ("recency effect," "peak-end rule"), in making decisions. A particularly damaging cognitive bias—one that promotes methodological underdetermination directly—is the tendency to confirm rather than deny a current hypothesis (*confirmation bias*). Humans tend to seek solutions to problems that are consistent with hypotheses rather than attempt to refute those hypotheses. Thus, we are likely to ignore or neglect data that do not conform to our presuppositions.

2.3.2 *Cognitive biases in ecological research?*

An in-depth review of some of the controversies that have arisen in ecological research— many of which have yet to go away—has been presented by Keller and Golley (2000). At least some of these controversies would seem to have their origins in cognitive biases. I offer three examples of possible connections between controversies and cognitive biases in ecological research; these examples should be compared with the list of recognized biases in ecological research presented previously. Studying the role of cognitive biases in the development of ecology would seem to a potentially fruitful undertaking.

- *Living fossils*—As described by Gotelli and Graves (1996), the null-models revolution hit ecology in the 1970s. Among the early targets of this thought revolution were competitive interactions structuring communities (Connor and Simberloff 1979), species–area relationships (Connor and McCoy 1979), and latitudinal gradients (McCoy and Connor 1980). As is true of most revolutions, the causes mostly are now lost in history, and the null-models approach, despite its faults, has become reasonably common. One of the lesser-known early targets of the null-models approach was the idea that some organisms have special traits that allow them to persist seemingly unchanged for unusually long periods of time ("living fossils"); consider the Coelacanth, for example. Are these organisms really "special?" One way to address this question is to ask what might happen if lineages were created at random, using simple bifurcation rules: could "living fossils" appear by chance? According to Gould et al. (1977), the answer appears to be "yes." The conclusion to be drawn from these results is that particular traits that allow an organism to cope with changing environmental conditions are not necessary for it to attain a lengthy geologic history. In other words, no reason exists to attribute patterns and underlying causes to what may be a random event, unless the event can first be demonstrated not to be random ("clustering illusion").
- *Preference, particularly dietary*—No one would argue with the inherent value of replication (Underwood 1999; Johnson 2002). Proper intra-experimental replication provides the power necessary to reject a false null hypothesis without engendering too great a risk of rejecting a true null hypothesis. Proper inter-experimental replication provides evidence for or against the generality of some putative phenomenon. Exactly how the replication is undertaken can influence its realized value, however. Aberrant forms of replication, such as pseudoreplication (Hurlbert 1984) and quasireplication (Palmer 2000), undermine the importance of replication in general. One of the most pervasive examples of aberrant replication in ecological research is the sacrifical pseudoreplication often common to preference (choice, selection) experiments. The easiest way to appreciate this type of pseudoreplication is to try to answer the question: if a fox eats 49 mice and 1 hare, what percentage of its diet reflects a preference for mice? The answer is 50%, not 98%, because the objects of predator preference are the

prey categories—which, in the present case are two—not the individual prey items. Thus, the answer would be the same no matter how many mice or hares the fox was able to consume. Nevertheless, many ecologists continue to pseudoreplicate in preference experiments, perhaps because they have not considered the issue or because they worry that they will lose useful information. The fact that the fox ate 49× more mice than hares *must* be important in measuring preference. A tendency may exist for these ecologists to cling to their beliefs despite evidence to the contrary ("belief perseverance").

- *Error, probability, and cost*—Statistical analysis is prone to a variety of errors. Most ecologists are familiar with Type I (= concluding that the null hypothesis is incorrect when it is, in fact, correct) and Type II (= concluding that the null hypothesis is correct when it is, in fact, incorrect) errors. In ecological research, the probability of making a Type I error (α) typically is set in advance and at an extremely low level (i.e., 0.05), but the probability of making a Type II error (β), which increases as α is reduced, often is ignored (Shrader-Frechette and McCoy 1992; Fidler et al. 2006). Most ecologists know that β can be reduced by increasing sample size or by increasing treatment effect size, if possible. Type II errors can arise for other reasons, however, such as when continuous data are categorized (Streiner 2002). If the cost of making a Type I error far exceeds the cost of making a Type II error, then ignoring β may be justified. In applied ecological research, such as described by Shrader-Frechette and McCoy (1992) and Di Stefano (2001), or when analyzing encountered data in any sort of ecological research, such as described by Tukey (1977) and Ellison (1993), the cost of making a Type II error could approach, or even exceed, the cost of making a Type I error, however. Nevertheless, many ecologists— especially, it seems, when they take on the roles of reviewers and editors—continue to embrace the Type I error preference paradigm. A tendency may exist for these ecologists to look at things according to the conventions of the profession, ignoring a broader point of view ("déformation professionnelle").

2.4 Ecological complexity, data constraints, flawed conclusions

Thirty years ago, Hilborn and Stearns (1982) raised concerns about the role of reductionism in ecological research. They worried that a presupposition about causation, that every "properly analyzed" effect has a single cause, had biased ecological research. This presupposition, they claimed, had fostered a mindset among ecologists that simple hypotheses and experiments were to be preferred to complex ones. A principle danger of this mindset, they contended, is that "once a factor has been eliminated as a necessary cause [by a single-factor experiment], it is also eliminated from subsequent experiments designed with other single causes in mind, and is neither controlled nor monitored" (p. 154). They clearly show how the reductionist mindset can produce flawed conclusions.

Advances in ecological analysis over the last three decades have addressed some of Hilborn and Stearns' concerns. A particularly accessible description and illustration of some of these advances may be found in Hilborn and Mangel (1997). Yet, I contend that the basic problem they addressed—the reductionist mindset—still imposes a constraint on ecological research, because many ecologists seem to labor under the impression that the causes of ecological patterns are non-hierarchical and operate on local spatial scales and/or on short temporal scales. A constraint such as this could influence data gathering especially severely. I illustrate my contention with three situations in which ecological complexity, through data constraints, may lead to flawed conclusions. The extended examples accompanying these illustrations come from my own work, and they were chosen simply because I am familiar with their details and they prompted me to think about these situations in the first place; I make no claim that they are the best examples available.

2.4.1 *Patterns and processes at different scales*

One of the most vexing ecological questions is on what scale does a particular pattern effectively represent its underlying process? A research effort may yield data that elucidate a clear pattern but, because the process of interest is operating on a different scale, fail to reveal anything about the process. Furthermore, another ostensibly identical research effort may actually find a different pattern even though the process operated in the same way as in the first effort, simply because the scale of examination was different (e.g., McAuliffe 1984). Some particularly interesting and important examples from the ecological literature involve the effects of habitat patchiness—this connection is not particularly surprising, as the smaller the scale at which patterns are studied, the more likely the underlying processes operate at a different, mostly larger, scale. Doak et al. (1992), for instance, presented the results from a simple spatial model simulating the dispersal of animals in a landscape of stochastically clustered habitat patches. They found that the spatial scale at which clustering occurred was the most important feature in determining disperser performance in their model. Interestingly, they also concluded from a literature review that few studies have addressed the way in which spatial scale mediates the effects of patchiness on population dynamics. Since then, the importance of employing the proper scale when studying habitat patchiness, as well as habitat fragmentation, has been reinforced for a variety of specialized contexts, such as landscape ecology (e.g., Tischendorf 2001) and restoration ecology (e.g., George and Zack 2001), as well as for ecological systems in general (e.g., McMahon and Diez 2007).

One of the examples presented by Doak et al. to illustrate the importance of scale involved the leaf-top herbivore guild of *Heliconia* plants (McCoy 1984, 1985; figure 2.1). The first study was of host-plant colonization and the second of the interaction among the four members of the guild. Both studies took place within two cacao groves in Costa Rica, which were intended to be replicate locations. One site was surrounded by forest and was relatively humid, while the other site was on a bluff overlooking a river and was dried by winds blowing up from the river. Thus, despite both sites having virtually identical vegetation structures, they were quite different in ambient conditions. New habitat patches were created by pinning cacao leaves to *Heliconia* leaves, mimicking the natural debris that members of the leaf-top herbivore guild inhabit. Individuals of each guild member were placed singly under some cacao leaves and others were left unoccupied. In the second study, densities of the most common guild member were also manipulated, to examine possible density-dependent colonization and/or departure rates.

Colonization rates varied between the two sites. Initial colonization of new patches occurred more rapidly, and further colonization more frequently, at the forest site. The difference in colonization between sites may be attributable to the substantially larger pool of potential colonists at the forest site: about 20% of the naturally occurring debris patches were occupied at the forest site, but only 5% at the river site. These results and conclusions suggest that patterns of patch colonization depend in large part on processes working at the scale of sites, rather than processes working at, say, the scale of

Fig. 2.1 Cacao leaf patches on a *Heliconia* leaf (left) and larva of a hispine chrysomelid beetle (right). The larva is one of four members of the leaf-top herbivore guild; other members include the larva and adult of another chrysomelid beetle, and the larva of an unidentified lepidopteran. Leaf patches are colonized by all members of the guild, and individuals may remain in the same patches for long periods of time.

individual plants. Had the research presupposed that the "proper" scale was that of individual plants, the connection between pattern and process would not have been noted. Spatial scale also influenced interpretation of the pattern of interactions among colonizing individuals. *Heliconia* plants are characteristic of light gaps and other open areas. Colonization occurred only of a subset of patches, those on plants growing at the edges of the cacao groves. Potential colonists may not have viewed the interiors of the groves as places to look for host plants. Consequently, an apparent tendency existed for colonists to prefer already occupied patches if all plants were included in the analysis, but the tendency disappeared if only colonized plants were included.

2.4.2 *Discrete and continuous patterns and processes*

Over short time periods or across small geographical scales, only a portion of any underlying gradient is likely to be revealed. The hidden remainder of the gradient has been termed "veiled" by McCoy (2002; also see Allen and Starr 1982). If the pattern under investigation spans the entire gradient, but data gathering is limited to a single time–location, or even a few times–locations, then neither the true pattern nor the process responsible for the pattern is likely to be elucidated (this problem is akin to the range restriction problem recognized in other disciplines). Consider the response of a predator–prey system to habitat patchiness. Classical predator–prey theory suggests that stability (defined as simple persistence) can be conferred to the relationship between a predator and its prey by providing the prey with a refuge from the predator. Some early laboratory experiments supported the stabilizing influence of such a refuge (e.g., Huffaker 1958; Huffaker et al. 1963). A logical inference to be drawn from these findings is that habitat patchiness can provide the prey with a refuge, and, thereby, promote predator–prey stability (e.g., Vandermeer 1981; Begon and Mortimer 1986). Field experiments, such as those of Rey and McCoy (1979) and Kruess and Tscharntke (1994), have indicated that habitat patchiness indeed can provide the prey with a refuge, but Kareiva (1987; also see Cooper et al. 2012) contended that it need not necessarily lead to predator–prey stability. He suggested that any generalization about the role of patchiness in stabilizing the predator–prey relationship be tempered with details of the organisms' dispersal behaviors and demographies, a suggestion which has been reinforced many times (e.g., Hastings 1977; Kareiva and Odell 1987; Bernstein et al. 1999; Nachappa et al. 2011). Although the autecologies of organisms certainly must play an important role in determining the stability of predator–prey relationships, an extremely simple heuristic model can be constructed to show how veiled gradients might also play a role.

Assume that the probability of the prey finding and exploiting a habitat patch [$p(V)$] divided by the probability of the predator finding and exploiting the prey on a habitat patch [$p(P)$] equals N (for "normal"); that the balance of probabilities (N) promotes stability; and that, in keeping with classical predator–prey theory, decreasing the patchiness of the habitat [$p(V) / p(P) < N$] gives an advantage to the predator by making it easier to locate prey, and increasing the patchiness [$p(V) / p(P) > N$] gives an advantage to the prey by making it easier to find a refuge. Now, construct a gradient of habitat patchiness and place the predator–prey system on it at random. If the system lands in a position in which the balance of probabilities is > N, then the system should stabilize; but, if the balance is < N, then the system should destabilize (*ceteris paribus*). Thus, experiments that set initial levels of patchiness very low, such as those of Huffaker, are likely to see increasing stability as the system approaches N; while experiments that divide intact areas of habitat into smaller patches, such as those of Kareiva, are likely to see decreased stability as the system moves away from N. Perhaps I have described a reasonable possibility, but, in real situations, the position of the system along the gradient typically is unknown. In fact, in most cases, the researcher is unaware that an underlying gradient exists at all. The gradient, therefore, is veiled.

Real ecological examples of veiled gradients are provided by McCoy (1990, 2002; figure 2.2). The first study was of the distribution of insect species at 12 sites across Virginia and adjacent areas of Maryland and North Carolina (eastern US), ranging from 150 m to 1,650 m elevation. Maximum species richness tended to occur at mid-elevations. Maximum species richness also tended to occur at mid-elevations for 20 studies taken from the literature, and the elevation at which it did occur was related to latitude. Some of the studies that did not follow the general relationship with latitude may not have identified the true elevation of maximum richness, because they sampled a restricted range of elevations; that is, the gradients may have been veiled. The second study was of color morphs of the common meadow spittlebug (*Philaenus spumarius* (L.)) at the same 12 sites. Some color morphs of the meadow spittlebug are "melanic" and others "non-melanic" (see Stewart and Lees 1996). The relative frequency of color morphs often varies with elevation; in particular, the relative frequency of melanic morphs tends to increase with elevation. The frequency of melanic morphs cannot always be shown to increase with elevation, however. McCoy proposed that both temporally and spatially veiled gradients could be at least partially responsible for the different findings. The temporal veiling follows from the observations that melanic morphs typically are largely or wholly restricted to females, post-mating survival of females is greater than that of males, and life cycles are longer at lower elevations (allowing time for increased skewing of sex ratios). Thus, at any particular time during the reproductive season, the relative number of females would increase with decreasing elevation; and, the perceived pattern of phenotypic variation would be influenced by precisely when data were gathered. The spatial veiling comes from how the data collected from disparate locations are grouped. When sites from three distinct geographic regions were grouped by region, the relative number of females decreased with elevation in all three cases, as predicted. When all of the sites from the three regions were grouped together, however, the relationship disappeared. The easy explanation for this

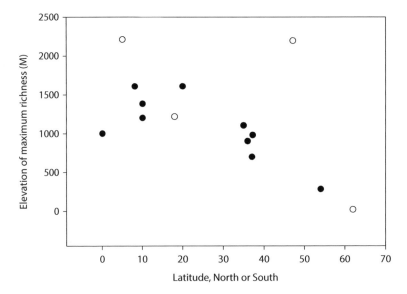

Fig. 2.2 Elevation of maximum species richness from studies of insect communities at various latitudes (data from McCoy 1990). Filled circles are studies in which maximum richness occurred at an elevation that was intermediate between the highest and lowest elevations sampled; open circles are studies in which maximum richness occurred at either the highest or lowest elevations sampled. One or more of the seemingly aberrant studies may not have sampled at elevations low enough to reveal the true elevation of maximum richness.

seeming paradox is that reproductive seasons were of different lengths in the different regions. Temporal and spatial veiling thus could inhibit recognition of possible relationships of melanism with elevation (e.g., an increase with elevation, "alpine melanism").

2.4.3 *Patterns and processes at different hierarchical levels*

Population-level attributes are sometimes used incorrectly to assess individual attributes, and vice versa. When this situation occurs, it is called the *ecological inference problem* (King 1997). The most familiar example of the problem probably is Simpson's paradox (Scheiner et al. 2000). The ecological inference problem apparently is not commonly recognized in ecology (but see Levins and Lewontin 1985), although it is well recognized in a variety of other disciplines. Public health researchers, such as Koopman (1977), Rose (1992), and Greenland (2004), for instance, have long found the problem to be a serious impediment to their research. Consider a couple of public health examples (from Schwartz and Carpenter 1999). Most studies of homelessness examine the characteristics that differentiate homeless from non-homeless persons (e.g., physical and mental health problems). While these characteristics may determine who becomes homeless, they do not determine the rate of homelessness. Data on inter-individual differences are unlikely to inform us as to how to reduce the rate of homelessness, even though they often are touted as being able to do so. Likewise, examining the characteristics that differentiate obese from non-obese persons (e.g., genetic factors that account for a significant proportion of the variance in body mass) is unlikely to stem the rate of increase in obesity. This example of the ecological inference problem, resulting from incorrect representation of the situation of interest, has been dubbed "Type III statistical error" (Kimball 1957; Schwartz and Carpenter 1999). Although I use *Type III error* in the way that I have just described, other meanings may be found in the literature. One meaning, for example, is incorrectly inferring the direction of a correctly rejected null hypothesis (Leventhal and Huynh 1996), and another is incorrectly inferring "importance" of a correctly rejected null hypothesis, because of extraordinarily large sample sizes (Crawford et al. 1998).

An ecological example of a Type III statistical error is provided by McCoy (2008, figure 2.3). This study was of the susceptibility of different-sized individuals of the gopher tortoise (*Gopherus polyphemus* Daudin) to the effects of upper respiratory track disease (URTD). URTD is an emergent disease of tortoises caused by species of *Mycoplasma*, and it is easily transmitted by direct contact between individuals. Thus, size-specific differences in behavior could promote greater or lesser exposure of non-infected individuals to infected individuals. To examine the possible size-related differences in the effects of URTD, two previous studies had compared the mean size of individuals that had died from URTD to the mean size of live individuals in the same population. One study found no significant difference for either males or females, but the other study concluded that the mean size of dead females was smaller than the mean size of live females. The question

Fig. 2.3 A newly hatched gopher tortoise, of about 50 mm carapace length. Growth to sexual maturity could take ten or more years, and adults may reach a carapace length of more than 385 mm. Upper respiratory tract disease may have been associated with the gopher tortoise for an extremely long time, but has emerged as a result of increased stress on the remaining individuals. The disease seems to contribute more to morbidity than to mortality within established populations. The disease-causing organism is spread by direct contact; so, individuals that have more direct contact with other individuals, such as males just reaching sexual maturity, may be more prone to contracting the disease.

of interest in both studies was whether the risk of death per individual was independent of body size, but the researchers had no direct measure of individual risk; so, they substituted the size distribution of dead individuals as surrogate data. The test then applied to the surrogate data was a comparison of the mean sizes of dead individuals and live individuals. The implicit assumption of this substitution and comparison is that risk of death per individual and the size distribution of dead individuals have the same cause, URTD. McCoy showed, however, that the difference between mean sizes of dead and live individuals depends in large part on the size distribution of live individuals in a population, which was different in the two studies. This error is a Type III error.

The Type III error problem is insidious not only because one correctly answers the wrong question, but also because the problem tends to bear substantial associated costs (Shrader-Frechette and McCoy 1992). For the gopher tortoise / URTD example, for instance, Type III error could prevent increased understanding of the transmission of URTD. No matter how many additional studies employing the same design ("replicates") were undertaken, they would not strengthen the answer to the question of interest. Worse, the problem could be directly detrimental to the well-being of the gopher tortoise. One might easily infer from the different conclusions of the two studies that no general relationship between risk of death from URTD and body size exists, when, in fact, URTD may selectively target relatively small adults. The ultimate result of this false inference could be failure to prevent the deaths of some of the most demographically valuable individuals.

2.5 Conclusions and suggestions

The process of gathering ecological data is influenced by many constraints. Some of these constraints are more easily avoided than others; some—methodological value judgments—are inescapable. Methodological value judgments can impose substantial restrictions on the gathering of ecological data; so much so that the resulting data may not be what are needed for strong tests of hypotheses ("methodological underdetermination"). The nature of ecological research itself may make methodological underdetermination of particular concern. The inherent complexity of ecological research, because of issues of scale, for example, forces large numbers of methodological value judgments, which increase the probability of flawed premises and, in turn, flawed conclusions.

So, what can you, the ecological researcher, do about the great variety of potential data constraints? How can you deal with these constraints so as to reduce the likelihood of flawed conclusions? Simply recognizing and criticizing one's presuppositions, values, and biases will go a long way toward improving data gathering in ecology (e.g., McCoy and Shrader-Frechette 1992, 1995; Haila and Kouki 1994; Power et al. 1996; Hurlbert 1997). Several organized strategies have been proposed for conducting this kind of self-reflection, such as strong inference (Platt 1964) and devil's advocacy (Schwenk 1990), although none is suitable in all circumstances. One might rely on safeguards, such as replication by other parties, peer reviewing, expert panels, and even meta-analysis, to deal with the problems engendered by data constraints. Apart from the obvious shortcomings with at least some of these institutional approaches, they are post hoc safeguards, becoming useful at all only after problems have arisen.

Far more preferable would be a method for preventing problems beforehand. Little can be suggested in this regard, however, save due diligence. What entails due diligence in this case? (1) After noting the presuppositions, values, and biases that you bring to the data gathering process (see de Winter and Happee 2013), reflect on how you could err because of these constraints. Ask yourself how the resulting data might be different under different sets of reasonable constraints. (2) Ensure that you have conducted a thorough literature synthesis (de Winter and Happee 2013). (3) Be open to different methods of analyzing data. Keep abreast of advances in statistical thinking which could influence the way in which you gather data (e.g., Gelman and Tuerlinckx 2000; Wagenmakers 2007). (4) Be aware of the many potential errors of interpretation waiting to trap the unwary. For instance, in addition to Type I, II, and III errors, statisticians sometimes informally identify a "Type II 1/2 error," which is confirming the null hypothesis, statistically, but then behaving as if it were absolutely correct. This error may be more clearly called "affirming the consequent," and it can be avoided by clearly distinguishing "no statistically discernible effect" from "no effect." Statisticians also identify Type IV error. Type IV error is incorrectly interpreting a correctly rejected null hypothesis. Marascuilo and Levin (1970) illustrate this error as a physician's correctly diagnosing an ailment (e.g., viral infection), but then prescribing the wrong treatment (e.g., antibiotics). Type IV error is an important consideration in ANOVA, and arises because higher-order interactions may fail to be properly rejected as a result of low power (Law et al. 2000). (5) Overall, pay as much attention to how you conduct your research as you do to the sort of research you do.

Acknowledgments

I thank the many colleagues with whom I have had conversations about these topics covered in this chapter over the years. I am especially indebted to Kristen Shrader-Frechette for sharing her insights into the roles of values and biases in science in general and ecology in particular. Mick McCarthy, Henry Mushinsky, Michael Reed, and Jessica Gurevitch's graduate students at Stony Brook University commented on draft versions of the chapter. I thank the editors for allowing me to put together this unusual (for a book about statistics) chapter, and for their constructive comments concerning presentation.

CHAPTER 3

Likelihood and model selection

Shane A. Richards

3.1 Introduction to likelihood and model selection

Ecologists study the way biotic and abiotic interactions affect the distribution and abundance of organisms. Interpreting ecological studies, though, is generally not straightforward because most biological patterns of interest are affected by complex webs of interactions (Hilborn and Mangel 1997). Often, numerous simplifying biological hypotheses can be proposed that describe key processes or relations among variables, and the objective of an ecological study is to identify which hypothesis (or set of hypotheses) is best supported. To achieve this it is usually necessary to formalize the hypotheses using mathematical models and then fit them to the data at hand (Richards 2005; Bolker 2008). Model selection is the process of determining which model(s) best describe the data, and, in turn this provides a way for determining which biological hypotheses have support.

There are two key problems associated with model selection. First, how can biological hypotheses be translated into mathematical models that can then be fit to the data observed? Second, what metric should one use to assess the performance of a model once it has been fit to the data? In this chapter I provide an introduction to model selection by introducing the notion of a likelihood function and I show how it can address both of these key problems.

The likelihood function provides a natural way of linking a hypothesis with a study; however, there is no consensus as to how the likelihood function can then be best used to assess model performance (Bolker 2008). A reasonable property of a good model is that it explains the data well and at the same time it leaves out any unnecessary details. Such models are termed parsimonious with respect to the data. Parsimonious models are useful as they are less prone to over-fitting, which can lead to poor model predictions under novel conditions (figure 1.7). Parsimonious models also tend to have reduced uncertainty in their parameter estimates.

Three approaches for estimating model parsimony are commonly adopted in the ecological literature, and they differ in the way they interpret models and data. In brief, the approaches incorporate either null hypothesis testing (NHT), Bayesian statistics (BS), or information theory (IT). In this chapter I focus on the NHT and the IT approaches. First, I show how NHT can be used to select a parsimonious model, but I also highlight its potential limitations. I then provide a detailed explanation of what is meant by 'best model' under the IT framework and I present an IT approach that can be used to select parsimonious models. Throughout this chapter I illustrate key concepts using

Ecological Statistics: Contemporary Theory and Application. First Edition. Edited by Gordon A. Fox, Simoneta Negrete-Yankelevich, and Vinicio J. Sosa. © Oxford University Press 2015. Published in 2015 by Oxford University Press.

simple ecological examples. I conclude the chapter by providing suggestions for good practice when performing model selection.

3.2 Likelihood functions

Suppose a biological variable (Y, say) is of interest and a number of hypotheses have been proposed that link it with other variables ($X_1, X_2, \ldots$). For example, the hypothesis might state, or imply, a relation between the expected value of Y (say, the number of flowers per plant) for a given set of observed X_i (such as plant size and flowering date). Suppose the hypothesis also describes the uncertainty in Y for a given set of X_i. In this case, it is often possible to write down the probability (or probability density; see chapter 1 and book appendix) of observing a set of data if the hypothesis were true, which I denote Pr(data|hypothesis). Pr is a probability if the response variable is categorical (e.g., count data) and a probability density if the response variable is continuous.

To illustrate this idea of a probability of the data, consider linear regression with a single covariate, X. The hypothesis is that the expected value of the response variable, Y, is linearly related to X, and variation in the response data about this expectation has a Normal distribution with a fixed standard deviation σ. The relation between the expected value of the response variable and the covariate can be written as

$$E[X] = \mu(x) = \beta_0 + \beta_1 x, \tag{3.1}$$

where β_0 and β_1 are constants, and x is the value of the covariate. Suppose the data consists of n (x, y) pairs and all the pairs are independent. Let x_i and y_i denote the data for the ith pair, and let $\mathbf{x}$ and $\mathbf{y}$ denote the set of x- and y-values. In this case the probability density of all the data, given the hypothesis, is

$$Pr(data|\boldsymbol{\theta}) = \prod_{i=1}^{n} f_n(y_i|\mu(x_i), \sigma), \tag{3.2}$$

where data = $\{\mathbf{x}, \mathbf{y}\}$ and $\boldsymbol{\theta} = \{\beta_0, \beta_1, \sigma\}$ is the set of model parameters that define the hypothesis. In this chapter the terms model and hypothesis are considered synonymous. The function $f_n(y|\mu, \sigma)$ is the probability density of observing y when the expected value is μ and the residuals—the difference between the observation and expectation—have a Normal distribution with standard deviation σ. This function can be written

$$f_n(y|\mu, \sigma) = \frac{1}{\sqrt{2\pi\sigma^2}} e^{-\frac{1}{2}\left(\frac{y-\mu}{\sigma}\right)^2}. \tag{3.3}$$

The hypothesis, in this case described by equations (3.1–3.3), incorporates assumptions about both the expectation of the response variable and the nature of the variation in the response variable about the expectation. These equations also incorporate the assumption of independence among data points, because the probability of the data is simply the product of the probabilities associated with observing each datum. As we will see, this assumption can be modified for more complicated data structures like repeated sampling of subjects (also see chapters 6, 10, and 11). Importantly, the probability function formally links the hypothesis with the study data. Having calculated the probability density of the data, given the model, we can then ask: what set of model parameters make it most likely that we'd observe the data?

The likelihood of the model is proportional to the probability (or probability density) of the data. In this chapter I set the constant of proportionality to unity, so that

$L(\theta|\text{data}) = \text{Pr}(\text{data}|\theta)$. In this case, the term probability is used when the parameters, θ, are considered fixed, whereas the term likelihood is used when the data are considered fixed. L is referred to as the likelihood function; a thorough exploration of the properties of L can be found in Pawitan (2001). Some sets of model parameter values will result in a greater likelihood, indicating that the hypothesis described by those values is more consistent with the given data. For example, if $L(\theta_1|\text{data})/L(\theta_2|\text{data}) = 2$ then the hypothesis described by the set of parameters, θ_1, is twice as likely as the hypothesis described by θ_2, given the data. The parameter values that maximize the likelihood function are referred to as the maximum-likelihood parameter estimates (MLPEs), denoted θ^*.

Usually, the likelihood becomes vanishingly small as more data are fit to the model, which makes it difficult to evaluate θ^* using $L(\theta)$. To alleviate this numerical issue it is more convenient to work with the natural logarithm (ln) of the likelihood, which I refer to as the log-likelihood and denote LL. The issue of a likelihood rounding to zero is avoided because the natural logarithm of an increasingly small number is an increasingly negative number. Importantly, ln is a monotonic, increasing function, so the set of parameter values that maximizes the likelihood will also maximize the log-likelihood. For the linear regression example presented above the log-likelihood is:

$$LL\,(\theta|\text{data}) = \sum_{i=1}^{n} \ln f_n\,(y_i|\mu(x_i), \sigma)$$

$$= -\frac{n}{2}\ln(2\pi) - n\ln\sigma - \frac{1}{2\sigma^2} \sum_{i=1}^{n}(y_i - \mu(x_i))^2. \tag{3.4}$$

Equation (3.4) shows that for simple linear regression the maximum likelihood estimate minimizes the sum of squares.

The example presented above could easily be modified to describe an alternative hypothesis where the expected relation is non-linear (e.g., $\mu(x) = \beta_0/(1 + \beta_1 x)$) or where the variation in the differences between the expectation and the data follows some other distribution. The book appendix provides a list of other possible probability density functions that may be substituted for f_n in Equation (3.2). The list includes the log-normal and gamma distribution, which are appropriate when the residuals are expected to have a positively skewed distribution. This list also includes useful probability functions that might be appropriate when the response variable is described by count data. Modifying the hypothesis to incorporate non-independencies in the data, however, is often less straightforward and will be discussed further below.

As the likelihood function describes the level of support associated with a set of parameter values, it can be used to calculate confidence intervals for parameters. To see how, consider the following simple study. Suppose the objective is to quantify seed set for a species of interest. The study involves observing N randomly selected flowers in a population and noting the number that produce seed by the end of the season. The hypothesis is that all flowers in the population randomly set seed with probability q. Under this hypothesis, the probability of observing n of the N flowers setting seed is given by the binomial distribution. If n flowers were observed to set seed, then the likelihood that flowers in the population set seed with probability q is:

$$L(q|n, N) = P_{bn}(n|N, q) = \binom{N}{n}q^n(1 - q)^{N-n}. \tag{3.5}$$

For this simple example it can be shown that the likelihood is maximized when $q = q^* = n/N$. Figure 3.1A shows the likelihood function (also referred to as the likelihood profile) for q if $n = 4$ flowers were observed to set seed among $N = 12$ sampled flowers. In

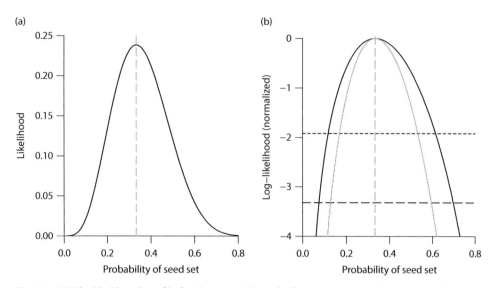

(a)

(b)

Fig. 3.1 (a) The likelihood profile for the probability of a flower setting seed, q, when 4 of 12 flowers are observed to set seed. (b) The corresponding log-likelihood profile (black solid line), and the log-likelihood profile when the sample size is doubled and 8 of 24 flowers are observed to set seed (gray solid line). In both cases the log-likelihood profile has been normalized by subtracting the maximum log-likelihood. Horizontal dashed lines indicate thresholds of 1.92 and 3.32, which can be used to estimate the 95% and 99% CI, respectively.

this example the most likely probability of seed set is $q^* = 1/3$, and values of q less than 0.1 or greater than 0.6 are over five times less likely. It makes sense that a confidence interval for the parameter q will be the range of q-values for which the likelihood is greater than some threshold.

Venzon and Moolgavkar (1988) show how the likelihood function can be used to estimate confidence intervals (CIs) for parameters in general, and this is referred to as the likelihood profile approach. An approximate $(1-\alpha)$% confidence interval for a parameter θ_0 (say) is the range of θ_0 where the difference between the maximum log-likelihood, $LL(\theta^*)$, and the log-likelihood, having been maximized over the remaining model parameters, is less than half the $(1-\alpha)$-th quantile of the chi-squared distribution with 1 degree of freedom. For approximate 95% CI and 99% CIs, the threshold differences are 1.92 and 3.32, respectively. As the model used to describe the seed set data above has only a single parameter, CIs for q can be calculated by identifying the range of q where the log-likelihood is within a threshold of the maximum. The approximate 95% and 99% CI for the probability of seed set are [0.11, 0.62] and [0.07, 0.70] (figure 3.1B). Suppose the number of flowers sampled doubled to $N = 24$ and a third were again observed to set seed. In this case the additional data further support the original findings and result in a tightening of the 95% and 99% CI for q so that they are now [0.16, 0.54] and [0.12, 0.60] (figure 3.1B).

3.2.1 *Incorporating mechanism into models*

Often, the objective of an analysis is simply to look for evidence of relations between biological variables (see Introduction). In this case, simple relations between variables (e.g., linear) can usually be incorporated into commonly adopted statistical analyses (see chapters 6 and 10). However, hypotheses based on biological mechanisms often lead to the expectation of non-linear relations among variables. These mechanistic hypotheses usually involve parameters describing biological rates and probabilities (Mangel 2006; Richards 2008). The advantage of developing more mechanistic models is that model parameters are usually easier to interpret; they also can provide a more straightforward way to use data to assess the relative merit of distinct biological hypotheses (Hobbs and Hilborn 2006). The drawback is that it is usually harder to derive a likelihood function when it incorporates a more mechanistic description of a system, and these likelihoods may not correspond with the functional forms provided by many statistical routines.

A more mechanistic hypothesis for the seed set example might be to propose that for a self-compatible plant, seed set comes from two sources: self-pollination and outcross-pollination via animal vectors. The degree to which both of the pollination mechanisms lead to seed set may be of interest. A biological hypothesis is that flowers set seed with probability s in the absence of animal pollination, and each visit to a flower by a pollinator results in it setting seed with probability w. Under this hypothesis, the probability that a flower visited x times by a pollinator will set seed is

$$q(x|s, w) = 1 - (1 - s)(1 - w)^x. \tag{3.6}$$

Support for this hypothesis could be examined by randomly selecting plants and controlling the number of visits to flowers. Suppose a visitation treatment is set up whereby flowers on a plant are allowed to be visited either $x = 0$, 1, 2, 3, or 4 times. Also, suppose that T randomly chosen plants are allocated to each treatment level, and N randomly chosen flowers on each plant are monitored for seed set. Thus, the study design involves $5T$ plants and seed set data come from $5TN$ flowers. An example data set for this study design is presented in figure 3.2 when $T = 5$ and $N = 6$. The data suggest a weak increase in the probability of a flower setting seed as it is visited more times.

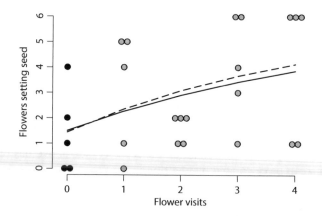

Fig. 3.2 The number of flowers (of six sampled per plant) that set seed in relation to the number of times the flowers on the plant are visited. Five plants are allocated to each visit category. Black and gray symbols indicate unvisited and visited plants, respectively. The solid line is the maximum-likelihood fit when plants are assumed identical, whereas the dashed line assumes that there is variation in the probability of seed set among plants and it is beta-distributed. These means are calculated using: $y = Nq(x|s^*, w^*)$ (see equation 3.6).

If plants are assumed to be identical in terms of their propensity to set seed, then the probability of observing the seed data can be written down in terms of binomial probabilities. Let $y_{x,i}$ be the number of the N sampled flowers that go on to set seed on the ith plant sampled among the T in the group having their flowers visited x times. The log-likelihood of the model described by $\theta = \{s, w\}$, given the data, is:

$$LL(s, w) = \sum_{x=0}^{4} \sum_{i=1}^{T} \ln P_{bn}\left(y_{x,i} | N, q(x)\right) \tag{3.7}$$

(for clarity I have removed reference to the data in the LL term and the parameters used to calculate q). Numerical routines, such as `optim` in R and `Solver` in Excel, can be used to show that, for the data in figure 3.2, equation (3.7) is maximized when $s = s^* = 0.25$ and $w = w^* = 0.17$. Only a quarter of unvisited flowers are expected to set seed and just over a half of the flowers visited four times are expected to set seed (figure 3.2).

From here on I adopt a notation to distinguish among alternative models (hypotheses) I develop to describe the seed set data. Let M(v) denote the model described above that is defined by $\theta = \{s, w\}$; v indicates that seed set is dependent on visit number. A null model for this study, denoted M(null), is that visit number has no effect on seed set. The models M(null) and M(v) are related, as M(null) is identical to M(v) when w is fixed at zero. Models are termed nested if the simpler model is the result of fixing one or more parameters of the more complex model. In this case M(null) is nested within M(v). This nesting relation among models is important when performing model selection (see later). The likelihood profile approach when applied to M(v) estimates the 95% CI for s and w to be [0.12, 0.40] and [0.08, 0.26], respectively. As the 95% CI for w does not include 0, it suggests that animal pollinators contribute to seed set (i.e., model M(v) is a better descriptor of the data than M(null)).

3.2.2 Random effects

Up until this point examples have been presented where the data are assumed to be independent. However, many observational and experimental studies involve data collection where subsets of the data are expected to be correlated to some degree, often as a result of the study design. For example, suppose study subjects are randomly chosen, subjects are then randomly allocated to treatments, and data are repeatedly taken on the chosen subjects (e.g., before–after experiments). Often, the subjects will vary in ways that have some effect on the variable of interest. In this case, even if all else is equal (e.g., the treatments have no effect), the data may appear to differ when grouped according to the subjects. This additional variation due to unknown variation among the subjects is referred to as a *random effect* (chapter 13). It is important to recognize when this additional variation is present, as ignoring it may lead to false inference regarding treatment effects.

In fact, the seed set data presented in figure 3.2 are suggestive of a random effect due to non-independence. Under the assumption that all plants are identical, the variation in the number of flowers setting seed for each visitation level should be consistent with the binomial distribution. The standard deviation of the binomial distribution, when parameterized according to equation (3.5), is $\sqrt{Nq(1-q)}$, which is less than half that observed (figure 3.3). If the variance of a data set is greater than the variance predicted by a probability distribution, then the data are referred to as *overdispersed* with respect to the distribution (chapter 12). In this case the seed set data are overdispersed with respect to the binomial distribution, which suggests that plants differ in their propensity to set seed.

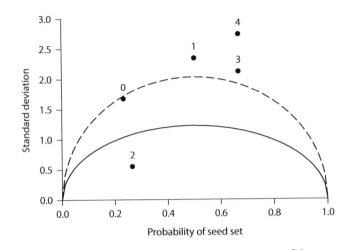

Fig. 3.3 Observed and predicted standard deviations of the number of flowers setting seed. Circles show the mean proportion of flowers setting seed and the standard deviation for each treatment level, *x*. The solid curve is the relation predicted by the binomial distribution ($N = 6$), which assumes all plants are identical. The dashed line is the relation for the beta-binomial distribution ($N = 6, \phi = 0.54$), which assumes that variation in the probability of flowers setting seed varies among plants according to a beta-distribution.

The reasons why the response variable varies across subjects are usually not known (and sometimes are not of interest); instead, understanding the general effects of some treatment(s) is the focus of the analysis. Mathematically, this extra variation can be thought of as being caused by nuisance variables, and it can be accounted for by incorporating a compound probability distribution into the likelihood function (Bolker 2008; Richards 2008; also see chapter 12). Compound probability distributions assume that a random variable is drawn from a parameterized distribution in which one or more of its parameters (usually the mean) are drawn from other probability distributions. By explicitly describing the source of the additional uncertainty (e.g., random variation about the mean response among subjects), compound probability distributions can often account for overdispersed data (chapter 12).

It is likely that plants will differ in their propensity to set seed due to factors that have not been measured; for example, seed set may be limited by the availability of some unmeasured resource that differs among plants. Suppose that the probability that flowers on a plant set seed when visited *x* times is now drawn from a beta distribution with mean $q(x)$ and variance $q(x)[1 - q(x)]\phi/(1 + \phi)$. Let $f_b(p|\mu, \phi)$ denote the probability density of the beta distribution having mean μ and variance $\mu(1 - \mu)\phi/(1 + \phi)$ (see book appendix). This new hypothesis has an additional parameter, ϕ, which quantifies the random variation among plants, and its likelihood is given by:

$$L(s, w, \phi) = \prod_{x=0}^{4} \prod_{i=1}^{T} \int_{p=0}^{1} f_b\left(p|q(x), \phi\right) P_{bn}\left(y_{x,i}|N, p\right) dp. \tag{3.8}$$

In this case the integral can be solved analytically to give the beta-binomial distribution, which is denoted P_{bb} (see book appendix). The log-likelihood of the model simplifies to

Table 3.1 Summary of the maximum-likelihood fits to the data presented in figure 3.2. Bracketed terms indicate 95% CIs for model parameters, k is the number of estimated parameters, and LL^* is the maximum log-likelihood. CIs have been estimated using the likelihood profile approach

Model	s^*	w^*	ϕ^*	k	LL^*
M(v)	0.25 [0.12, 0.40]	0.17 [0.08, 0.26]	NA	2	−59.1
M(v + r)	0.24 [0.07, 0.48]	0.20 [0.04, 0.35]	0.54 [0.22, 1.17]	3	−45.4

$$LL(s, w, \phi) = \sum_{x=0}^{4} \sum_{i=1}^{T} \ln P_{\text{bb}}\left(y_{x,i} | N, q(x), \phi\right). \tag{3.9}$$

As this model incorporates random variation among subjects it is denoted M(v + r) to distinguish it from the previous model, M(v).

The variance of the beta-binomial distribution when it is described by $P_{\text{bb}}(y|N, q, \phi)$ is $vNq(1 - q)$, where $v = 1 + (N - 1)\phi/(1 + \phi)$ is the variance inflation factor (Richards 2008). The maximum likelihood parameter estimate for ϕ when M(v + r) is fit to the seed set data is $\phi^* = 0.54$ (table 3.1), which gives $v = 2.75$ (i.e., the variance in the seed set data is over twice that predicted by the binomial distribution). The standard deviation of the observed data is much better predicted by M(v + r) when compared with M(v) (figure 3.3). Interestingly, the estimates of s and w are comparable with those estimated by M(v) (table 3.1), suggesting that although M(v) was unable to describe the variation in the data it could estimate the probabilities of seed set associated with selfing and outcrossing. Both models produce similar expectations with regard to the number of flowers setting seed across treatment levels (figure 3.2). This type of consistency, however, should not be expected in general. Models that underestimate the variance in the data will often result in biased parameter estimates, as they more heavily penalize large residuals and thus try to accommodate them more in their fit. Although M(v + r) provides a much better overall fit than M(v) (the log-likelihood is increased by nearly 15), the inclusion of random variation among plants has resulted in wider 95% CIs for s and w (table 3.1).

Ecological studies often involve the collection of count data where, unlike the example above, there is no theoretical fixed upper bound to the data. For example, insect pitfall traps could result in the number of captures ranging between 0 and some large number. For unbounded count samples, the response variable can be described in terms of rates and how ecological variables affect the rates. Such hypotheses usually predict that the data will be consistent with the Poisson probability distribution (Mangel 2006). A property of the Poisson distribution is that the mean and variance are the same; however in ecological data the variance is typically greater than the mean. A common way to account for this form of overdispersion is to assume that subjects (e.g., pitfall traps) vary in their mean rate according to a gamma distribution, which results in the data having a negative binomial distribution (Richards 2008; book appendix; chapter 12). Lindén and Mäntyniemi (2011) provide an excellent summary of how to incorporate the negative binomial distribution into ecological analyses.

3.3 Multiple hypotheses

The development of the seed set example has resulted in multiple hypotheses being proposed, formally described using mathematics, and linked to the study design. In practice, many hypotheses will be proposed to explain a data set, as it is common that information on a number of potentially influential variables will also be collected. Thus, it is usual for a number of models to be fit to the data. How do we deem which model is best?

To illustrate a more realistic scenario suppose that the pollination treatment was performed using two pollinator species. An example data set is presented in figure 3.4A in which plants are exposed to one of two pollinator species (A or B). Now, it could be proposed that the probability a flower visit results in seed set depends on the pollinator species; when visited by pollinator species l this probability is w_l. Thus, in addition to the four models we have already considered {M(null), M(v), M(r), and M(v + r)}, there are two more: M(s) assumes pollinator species effects alone, and M(s + r) assumes effects of both pollinator species as well as individual plant variation.

The six models and their nesting relations are presented in figure 3.5. Note that the beta-binomial distribution reduces to the binomial distribution in the limit as ϕ approaches zero, so models described with an r-term are more complicated than those without. The most complicated model, M(s + r), assumes species-specific probabilities of seed set and assumes that the variation in seed set among plants follows a beta distribution. The log-likelihood associated with this model is given by

$$LL(s, w_A, w_B, \phi) = \sum_{i=1}^{T} \ln P_{\text{bb}}(y_{0,i}|N, s, \phi) + \sum_{l=A,B} \sum_{x=1}^{4} \sum_{i=1}^{T} \ln P_{\text{bb}}\left(y_{x,i,l}|N, q\left(x|s, w_l\right), \phi\right), \quad (3.10)$$

where the additional l subscript associated with y indicates which pollinator species visited the plant. The first term refers to the T unvisited plants.

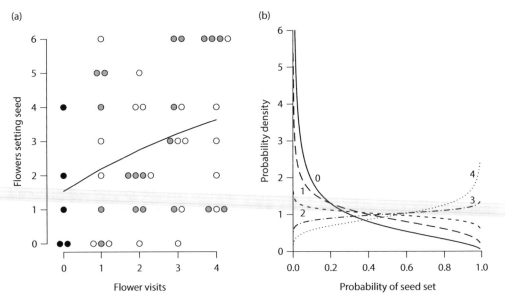

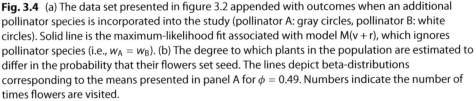

Fig. 3.4 (a) The data set presented in figure 3.2 appended with outcomes when an additional pollinator species is incorporated into the study (pollinator A: gray circles, pollinator B: white circles). Solid line is the maximum-likelihood fit associated with model M(v + r), which ignores pollinator species (i.e., $w_A = w_B$). (b) The degree to which plants in the population are estimated to differ in the probability that their flowers set seed. The lines depict beta-distributions corresponding to the means presented in panel A for $\phi = 0.49$. Numbers indicate the number of times flowers are visited.

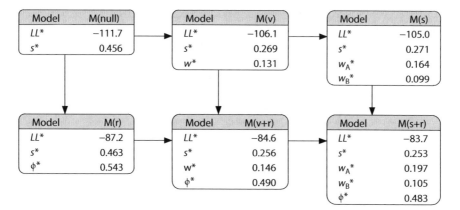

Fig. 3.5 Summary of the six model fits to the data presented in 3.4A. Arrows indicate the nesting structure by showing the models that have added complexity. The maximum log-likelihood, LL^*, and the maximum likelihood parameter estimates are presented for each model.

Maximum log-likelihoods and maximum-likelihood parameter estimates for each of the six models are presented in figure 3.5. The most complicated model, M(s + r), has the highest log-likelihood, LL^*. A model should never have a lower maximum log-likelihood than any of the simpler models that are nested within it because it must fit the data at least as well as cases when some of the model's parameters are fixed (you can check that this is indeed the case in figure 3.5). Adding among-plant variation greatly increases the fit (LL^*) and removing visit effects ($w = 0$) increases the estimate of selfing, s. Although model M(s+r) has the highest LL^*, the data presented in figure 3.4A shows no clear differences in seed set numbers between the two pollinator species. In the following sections I suggest how these maximum likelihoods, in conjunction with the nesting structure, can be used to select the better models.

3.3.1 *Approaches to model selection*

Models that have been fit to data can be useful in at least two ways. First, they can help quantify how consistent a data set is with the finite number of assumptions incorporated into the model. For example, the reduction in the maximized model likelihood when an ecological factor or process is removed reveals its potential importance for understanding the study system. Secondly, fitted models also allow one to predict new data, possibly under novel conditions. It may be tempting to predict new data using the most complex model considered, as all the factors considered in a study are usually thought to play some role. However, uncertainty in the estimates of model parameters usually increases as the ratio of model parameters to independent data declines; thus, it may be best to predict new data with a simpler model (Burnham and Anderson 2002). In both situations a good model should be parsimonious; it should make predictions that are consistent with the data but it should not be overly complex.

As the utility of a model depends on context, it is not possible to give a general definition of best model. In fact, there are a number of ways to interpret a model, which further complicates the notion of best. For example, frequentists assume that the set of data collected during a study is one realization of a potentially infinite number of repetitions of the study, each capable of producing a statistically independent set of data. In this case, truth can be thought of as a probability distribution that describes the probability of observing each potential set of data recorded during a study if the study were repeated keeping the conditions as consistent as practically possible. On the other hand, Bayesians might reject the notion that studies could have multiple realizations. Instead, they will propose a set of models that make claims (either specific or probabilistic) about the study and its data, include prior evidence for each model proposed, and then use the likelihood of each model, given the data, to update the weight of evidence for each model and the uncertainty in each of their parameters (Kass and Raftery 1995; also see chapter 1). For example, consider a case where the problem is to determine if a study site is occupied by a particular species. For this type of problem there are two mutually exclusive hypotheses of which one must be true—the species is present or it is absent. In this case the Bayesian approach is most appropriate for evaluating the relative evidence for both hypotheses (McCarthy 2007). The conceptual distinctions adopted by frequentists and Bayesians lead to different approaches to model selection and their interpretation (Kass and Raftery 1995; chapter 2).

Below I outline how null hypothesis tests can be used to select among a set of models and I mention some limitations of the approach. I then focus on the information-theoretic approach to model selection, which overcomes some of the limitations of null hypothesis testing. Both of these approaches adopt the frequentist viewpoint. I present an interpretation of best model under the information-theoretic paradigm and I suggest why it may be a useful definition of best. I then present some simple rules that can be used to seek this best model. As mentioned above, Bayesian methods offer an alternative approach to model selection; however, they will not be considered in detail here. McCarthy (2007) provides an introduction to model selection for ecologists using Bayesian statistics, and Hobbs and Hilborn (2006) summarize the main conceptual distinctions between the frequentist and Bayesian approaches to model selection. These three general approaches will often result in similar inference, however their estimation of parameter and prediction uncertainty can differ markedly (chapter 2; Kass and Raftery 1995; Hobbs and Hilborn 2006).

3.3.2 Null hypothesis testing

Suppose a set of competing models is proposed to explain a data set and all the models can be organized according to a nested hierarchy. Knowledge of the nesting relations among the competing models provides a framework for using null hypothesis tests to select a best model. To see how, consider the most complex model in the hierarchy. One could ask if the model's consistency with the data is severely reduced if any of its assumptions are removed (e.g., inclusion of predictor or nuisance variables). The models that correspond to removing an assumption are the nested models. For each pairing of nested models a likelihood ratio test can be performed. The null hypothesis for this test is that the data were generated by the simpler model and the test statistic is $G = 2[LL_1^* - LL_0^*]$, where LL_1^* and LL_0^* are the maximum log-likelihoods of the complex and simpler model, respectively (Sokal and Rohlf 1995). Note that the test statistic, G, can also be calculated as the *deviance* of the simpler model minus the deviance of the more complex model. The deviance of a fitted model is often included in the output produced by statistical programs. The deviance of model M is defined as $-2(LL^*(M) - LL_S^*)$, where LL_S^* is the maximum log-likelihood of the saturated model (see section 3.3.5). Under the null hypothesis the test statistic G is chi-squared distributed with degrees of freedom equal to the number of additional parameters estimated by the more complex model. If any of these tests are considered not

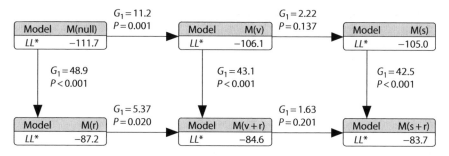

Fig. 3.6 Likelihood ratio tests that are incorporated in backwards and forwards stepwise model selection. Arrows indicate the two models considered and each arrow is associated with the test statistic and p-value.

significant then the most complex model is not the most parsimonious descriptor of the study and it can be replaced by the model associated with the least significant p-value. This process could be repeated for the replacement model until a model is found where all of the possible likelihood ratio tests are considered statistically significant. This final model is considered the best model and the process is referred to as backwards stepwise model selection.

To see how this process works, consider the six models proposed to explain the seed set data and their nesting structure depicted in figure 3.6. Starting with the most complex model, M(s + r), there are two nested models: M(v + r) removes the assumption that pollinator species is important, and M(s) removes the assumption that unexplained random variation among plants needs to be accounted for. The two likelihood ratio tests associated with M(s + r) are presented in figure 3.6; only the test that assumes M(v + r) generated the data results in a p-value that is not significant when a type I error of 5% is deemed unacceptable. Thus, model M(v + r) may be considered a more parsimonious descriptor of the system rather than M(v+s). Again, there are two possible likelihood ratio tests for model M(v+r): M(r) removes the assumption that pollinators influence seed set, and M(v) removes the assumption of between plant variation. Both of these tests are statistically significant, so the algorithm stops and we are left with model M(v + r) as the best model.

The same type of approach could be used to select a model by comparing a model with the models that add complexity rather than removing it. In this case the algorithm starts with the simplest model proposed. If any of the likelihood ratio tests are significant, then there is reason to reject the simpler model. In this case, the new simpler model is the one associated with the most significant p-value and the process is repeated. The best model is the one for which no likelihood ratio tests are considered significant. This approach is referred to as forward stepwise model selection.

For the seed set example the starting model is M(null), which assumes that all flowers set seed with equal probability, irrespective of the number of times they are visited by pollinators. There are two models that add complexity to this model: M(v) incorporates the effect of pollinator visits, and M(r) incorporates random between-plant variation in the probability of seed set (figure 3.6). Both likelihood ratio tests are statistically significant, the one associated with the random effects being more so. When M(r) is considered to be the simpler model only model M(v + r) adds complexity and the test is significant, which provides support for M(v + r). Now, setting M(v + r) as the simpler model, the only comparison is with the most complex model M(s + v) and this test is not significant. Thus, the algorithm stops at model M(v + r) and it is deemed the best model.

For the above example both the forward and backward stepping approaches lead to the same final best model, M(v + r). Unfortunately, this is not always the case, and is one reason why stepping between models is sometimes considered a poor approach for model selection (Whittingham et al. 2006), although many ecologists routinely use stepwise selection. Another concern with the approach is that some stopping rule regarding statistical significance needs to be stated and it is unclear which type I error cut-off is best. In practice the 5% threshold is often used but as multiple statistical tests are being applied the final conclusion does not correspond to a type I error of only 5%.

Despite the clear problems associated with stepwise approaches to model selection, if used carefully, it will often lead to a model providing useful inference about the system studied, provided the selected model makes predictions that are consistent with the data (Zuur et al. 2007). It is important to recognize that, because not all models are necessarily investigated, the approach may not select the best model. I mention this stepwise approach because the idea of repeatedly comparing pairs of models is commonly adopted in statistical analyses. For example, data involving more than one factor are often modeled with the analysis of variance (ANOVA) approach, resulting in F-tables with multiple p-values. Each p-value is associated with a pairwise comparison between a simple and a more complex model, and correct interpretation of these tests relies on an understanding of which models are being compared (Grafen and Hails 2002). The order with which these pairwise tests are evaluated is also important for correct interpretation. When investigating support for an interaction it is usual to examine first the F–test statistic produced by comparing the model that includes the interaction with the model excluding the interaction. If this test is deemed not statistically significant, then the interaction term is removed from subsequent models and additional F tests are performed investigating the statistical significance of removing main effects. This common approach is akin to backwards model selection.

3.3.3 *An information-theoretic approach*

An alternative approach to model selection that avoids the need for models to be nested is to quantify model parsimony using Akaike's Information Criterion (AIC). This approach is widely adopted by ecologists; however, exactly what AIC is estimating is often not appreciated. According to information theory, truth can be thought of as a distribution describing the probabilities of outcomes of the study if the study were repeated. Note that here a frequentist assumption is being made about the data (see chapter 1). For example, suppose that the set of data collected during a study, referred to as the study outcome, consists of n counts: $\mathbf{y} = \{y_1, \cdots y_n\}$. Consider again the pollination example when only a single pollinator species was investigated, but now suppose that the plant species is self-incompatible ($s = 0$). Also, suppose that $N = 3$ flowers were monitored on each of four plants and a single plant is allocated to each of four treatment levels: $x = 1, 2, 3,$ or 4 visits to flowers. For this study there are $4^4 = 256$ possible outcomes, which can be ordered: $\{0,0,0,0\}, \{0,0,0,1\}, \cdots \{3,3,3,3\}$, where the ith count indicates the number of flowers setting seed for the plant having its flowers visited i times. Truth, in this case, is the probability distribution describing the probability of observing each of the 256 study outcomes, and is denoted $\mathbf{p} = \{p_1, p_2, \cdots p_{256}\}$.

In order to illustrate some key concepts, suppose that the unknown true probability of seed set when a flower is visited i times is $q(i)$, such that $q(1) = 0.2$, $q(2) = 0.34$, $q(3) = 0.45$, and $q(4) = 0.51$. These probabilities reflect a decrease in the probability of seed set for each additional flower visit, which might occur if pollen tubes clog the stigmas. If flowers set

(a)

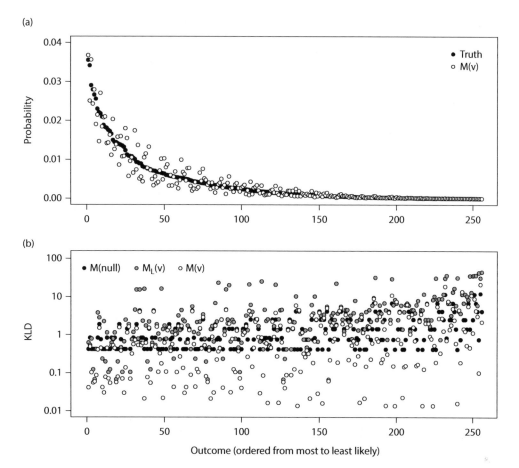

Fig. 3.7 (a) True probabilities of observing each of the 256 possible outcomes of a hypothetical pollination study (black circles). These data represent the number of flowers sampled that develop into fruit when flowers are visited 1, 2, 3, or 4 times by pollinators. $N = 3$ flowers were sampled for each visitation treatment. Note that the outcomes have been ordered from most to least likely. Open circles are predicted probabilities when the most mechanistic model, M(v), is fit to the most likely study outcome: {0, 1, 1, 2}. (b) Kullback–Leibler distances calculated for three models: M(null), M(v), and M_L(v) (see text for model details) when they are fit to each of the 256 possible study outcomes.

seed independently, then the true probability of observing study outcome $\{y_1, y_2, y_3, y_4\}$ can be calculated using

$$\Pr\left(\{y_1, y_2, y_3, y_4\}\right) = \prod_{x=1}^{4} P_{\text{bn}}\left(y_x | 3, q(x)\right). \tag{3.11}$$

In this case, the most likely outcome is {0, 1, 1, 2} and occurs with probability 0.036 (figure 3.7A).

As the plant is known to be self-incompatible a model of seed set is that flowers visited i times set seed with probability $q(i) = 1 - (1 - w)^i$. It can be shown that if outcome {0, 1, 1, 2} is observed, then the MLE for w is $w^* = 0.161$. Substituting this estimate into $q(x)$ gives: $q(1) = 0.16$, $q(2) = 0.30$, $q(3) = 0.41$, and $q(4) = 0.50$. The probability of observing each possible study outcome according to these probabilities is presented in figure 3.7A (open circles). Note that despite the model leaving out the reduction in seed set with each additional flower visit, both the true and the model predicted probability distributions are similar. One way to quantify the difference between the two probability distributions presented in figure 3.7A is to use the Kullback–Leibler divergence (KLD). If $\pi = \{\pi_1, \pi_2, \cdots \pi_{256}\}$ denotes the probability distribution according to the proposed model, and $\mathbf{p} = \{p_1, p_2, \cdots p_{256}\}$ gives the probability distribution of the true underlying mechanism, then the KLD is

$$I(\mathbf{p}, \pi) = \sum_i p_i \ln\left(\frac{p_i}{\pi_i}\right). \tag{3.12}$$

For the two distributions presented in figure 3.7A the KLD is $I(\mathbf{p}, \pi) = 0.041$. In general, I is non-negative, and I only equals zero when the two distributions $\mathbf{p}$ and π are identical. Equation (3.12) is the expectation of $\ln(p_i/\pi_i)$ with respect to the truth, and I is referred to as the information lost when π is used to approximate $\mathbf{p}$. If, instead, the observations are a continuous variable, then the summation term in equation (3.12) is replaced by integration (Burnham and Anderson 2002).

A model might be considered desirable if it has a low KLD whenever its parameters are estimated from likely study outcomes. Thus, we can quantify how well different models approximate the truth with the expected value of the KLD when we consider all possible outcomes:

$$E_{\mathbf{p}}[I(\mathbf{p}, \pi)] = \sum_i p_i I(\mathbf{p}, \pi(\theta_i^*)), \tag{3.13}$$

where θ_i^* is the set of model parameters estimated using maximum likelihood when study outcome i is observed. This approach to model selection does not assume that one of the proposed models is true (i.e., generated the data), which distinguishes it from most Bayesian approaches to model selection (Kass and Raftery 1995).

Now, suppose that two additional models are used to describe the seed set data. The simplest model, denoted M(null), assumes that the probability of seed set is independent of visit number and implies that $q(x) = \omega$ (say). Alternatively, if little thought is given to the biological mechanism leading to seed set, then $q(x)$ could simply be assumed to be described by the logistic equation: logit $q = \ln(q/(1-q)) = \beta_0 + \beta_1 x$. Let M(v) denote the first model we considered and $M_L(v)$ denote the logistic model—v is used to indicate that the model incorporates visit number. Figure 3.7B shows the KLD for each of the three models when fit to each of the 256 study outcomes. Model M(v) has the lowest EKLD and model $M_L(v)$ has the greatest EKLD (table 3.2). Thus, the best model from an information-theoretic perspective is the one that assumes each flower visit results in a fixed probability of fruit development.

Table 3.2 Summary of AIC and AICc estimates of the relative EKLD for the hypothetical pollination study presented in figure 3.7. $E_{\mathbf{p}}[.]$ denotes the expectation with respect to the probability distribution $\mathbf{p}$, and SD is the standard deviation of AIC (and AICc) among hypothetical replicated studies. k is the number of estimated model parameters

Model	k	$2(E_{\mathbf{p}}[I(\mathbf{p}, \pi)] - c)$	$E_{\mathbf{p}}$[AIC]	$E_{\mathbf{p}}$[AICc]	SD
M(null)	1	11.2	11.1	13.1	2.1
M(v)	1	10.5	10.3	12.3	2.1
$M_L(v)$	2	13.4	11.1	23.1	2.2

The pollination example illustrates how information theory quantifies model parsimony. Variation in the KLD for the two simplest single parameter models, M(null) and M(v), when calculated for the most likely outcomes, is lower than the variation in the KLD associated with the more complex two-parameter logistic model, $M_L(v)$ (figure 3.7B). Thus, although model $M_L(v)$ may often have a high likelihood when fit to a study outcome, it often over-fits the data as it poorly identifies other outcomes that are truly likely. In general, the EKLD will be higher for models that incorporate unnecessary complexity.

Since the truth, **p**, is generally unknown in practice, how can we estimate the EKLD? Akaike (1973) derived an approximate relation between the EKLD of a model and its maximum log-likelihood. For a proposed model, M, its AIC value is calculated using:

$$AIC(M) = -2LL(\theta^*) + 2k, \tag{3.14}$$

where θ^* is the set of k model parameters estimated from the data using maximum-likelihood. Akaike (1973) showed that

$$AIC(M) \approx 2(E_{\mathbf{p}}[I(\mathbf{p}, \pi)] - c), \tag{3.15}$$

where $c = \sum p_i \ln p_i$ is a model-independent constant that depends only on the truth defined by **p**. Since c is the same for all models, we do not need to know it to compare models: the difference between the AICs for two models, $AIC(M_1) - AIC(M_2)$ is the same whether we include c in the calculation or not. AIC differences estimate how much closer a model is to the truth relative to the other models but it does not quantify how close the models are to the truth. The expected AIC value for each of the three pollination models (i.e., the mean AIC if the study were repeated indefinitely) is presented in table 3.2. We can see that in this case AIC provides a good estimate of the relation given in equation (3.15). Note that the least accurate estimate is associated with $M_L(v)$, which had the highest EKLD. This result is not surprising as the derivation of AIC assumes that the proposed model is a good approximation of the truth (see Pawitan 2001 for more details).

3.3.4 *Using AIC to select models*

The best model from an information-theoretic perspective is the one having the lowest AIC. Importantly, as AIC is only related to an estimate of the EKLD (see equation 3.15), a model not having the lowest AIC may in fact have the lowest EKLD. For example, the pollination example shows that the variation in AIC across repeated simulations of the study is high relative to its expectation (table 3.3). Thus, some additional rules are needed to increase the chance that the true lowest EKLD model is selected.

Model selection using AIC proceeds by identifying the model with the lowest AIC, and then retaining all models that have an AIC within some threshold of this lowest value. The difference between the AIC value of a model and the lowest calculated is denoted ΔAIC. Richards (2005, 2008) used simulations to show that, for a range of ecological studies, a ΔAIC threshold of around six is required to be 95% sure of retaining the model with the lowest EKLD. However, some further refinement to selection is necessary in order to avoid retaining overly complex models. To see why, consider the case when a model that incorporates an additional, and unnecessary, factor is fit to the data. As the factor has no biological relevance, this model will often have a maximum-likelihood that is equal to, or only marginally higher than, the model that ignored it. Clearly, this overly complex model should not be retained and used for inference; however, it will be retained if the simple threshold rule using a cut-off of six is adopted because the additional parameter will only increase AIC by at most two (i.e., it will have a ΔAIC ≤ 2). Richards (2008) suggested that to avoid this problem of selecting overly complex models, models should be removed from the selected set if a simpler nested version of the model has also been se-lected and has a lower AIC. This nesting rule has been shown to remove the vast majority of unnecessarily complex models from the initially selected set and is unlikely to remove the model having the lowest EKLD (Richards 2008).

The model selection rules advocated above will usually result in more than one model being selected. If the purpose of the analysis is prediction, then the best AIC model can be used for prediction. On the other hand, if biological inference is the focus of the study, then multiple selected models may conflict in their specifics, making inference unclear. In this case, a very conservative approach is to base inference on the simplest model selected. An alternative approach is to consider strong support for a factor (or mechanism) if it is present in all models selected. Factors included in the best AIC model but not included in all models selected can be considered to have some support, and factors only included in models other than the best AIC model can only be considered to have very weak support (Richards et al. 2011).

3.3.5 *Extending the AIC approach*

The book by Burnham and Anderson (2002) has played a major role in popularizing the use of AIC as a model selection metric when analyzing ecological data. These authors have made a number of further suggestions for how an AIC analysis should be performed, which have been widely adopted in the ecological literature. Specifically, Burnham and Anderson (2002) advocate the use of a small sample correction to AIC, referred to as AICc. Secondly, they advocate multi-model inference, whereby models are combined us-ing model weights. They also suggest the use of quasi-AIC (which is defined below) when proposed models cannot account for the degree of variation observed in the data. The reliability of these suggestions is mixed, as I show below.

Often, only small sample sizes are possible when performing an ecological study. Hur-vich and Tsai (1989) showed that when sample size is small and the proposed model is linear with homogeneous, Gaussian-distributed residuals, then AIC may have negative bias, which can be reduced by including an additional term. The small-sample corrected version of AIC, denoted AICc, is given by

$$\text{AICc(M)} = \text{AIC(M)} + \frac{2k(k+1)}{n-k-1}, \tag{3.16}$$

where n is the sample size used to estimate the model parameters. Although Burnham et al. (1994) found the AICc estimate to be robust when applied to likelihood functions involving the product of multinomial distributions, Richards (2005) showed that when data were few, AICc was no more reliable at estimating the relative EKLD when the study involved count data and the fitted model was non-linear. The pollination study presented here also demonstrates the unreliability of AICc when the assumptions of linearity and Gaussian-distributed variation are not met. In this case, $n = 4$ as the models are fit to four counts. AICc positively biases the estimate and the bias is substantial for the logistic model, $M_L(v)$ (table 3.3). I do not recommend that AICc be adopted if any of the models are non-linear, or if the residuals are clearly non-Gaussian.

Model uncertainty occurs when model selection results in more than one model being selected. Akaike (1983) proposed that the likelihood of a model, M, having the lowest EKLD, is proportional to $\exp(-\Delta\text{AIC(M)}/2)$, where $\Delta\text{AIC(M)}$ is the difference between the AIC of the model and the best AIC model. This leads to the notion of model weights. If a set of I models, denoted $M_1 \ldots M_I$, are fit to data, then the weight of model M_i is

$$w(\mathrm{M}_i) = \frac{\exp\left(-\frac{1}{2}\Delta(\mathrm{M}_i)\right)}{\sum_{j=1}^{I} \exp\left(-\frac{1}{2}\Delta(\mathrm{M}_j)\right)}.$$

(3.17)

The best way to interpret these weights is unclear (Richards 2005; Bolker 2008). Hobbs and Hilborn (2006) claim that $w(\mathrm{M}_i)$ approximates the proportion of times M_i would be chosen as the best model. The weights for models M(null), M(v), and $\mathrm{M}_L(v)$, if the most likely outcome occurred, are 0.31, 0.41, and 0.27, respectively. If the study were repeated, then 0.31, 0.56, and 0.14 are the corresponding proportions of times that these models would be the best AIC model. More importantly, 94% of the time model M(v) would have had the lowest KLD (figure 3.7B). Thus, the weights are relatively unreliable estimates of the probability that the model will be judged to be the best AIC model and they also poorly reflect the probability that the model has the lowest KLD under repeated sampling. Whether these poor relations are due to a small sample size is unclear; however, similar results have been shown for other scenarios when data are less sparse (Richards 2005; Richards et al. 2011). Note also that equation (3.17) ignores the effect of model nesting on inference, which has been shown above to reduce the reliability of AIC when used to select the best EKLD model. In addition, these weights cannot be interpreted as the probability of the model having the lowest EKLD because it is unclear what constitutes a repeated analysis that could result in another model having the lowest EKLD. For the pollination example, M(v) is the model with the lowest EKLD, given the proposed hypotheses and the study design. As with AICc, caution is suggested when deciding on whether to adopt model weights.

If predictions are of interest and more than one model is selected, then an appealing idea is to reduce prediction error and bias by averaging across the selected models, which is referred to as multi-model inference. Buckland et al. (1997) suggest model averaging by using model weights to calculate unconditional regression coefficients when the models are linear, and Burnham and Anderson (2002) suggest that model weights be used to combine model predictions in the case of non-linear models. However, (Richards (2005) and Richards et al. (2011) have tested these suggestions using simulation studies and found that bias and various measures of prediction error (i.e., KLD, mean squared errors in effect sizes) were not reliably reduced when adopting model averaging. It is not hard to see why model averaging can be unreliable. If, by chance, data are collected such that the coefficient associated with an important factor is underestimated, then the model that ignores the factor will also likely have significant weight. In this case, model averaging will further weaken the importance of the factor. Alternatively, if by chance the data overestimates the role of a factor, then models that include the factor will have high weight and model averaging will have little effect on reducing the bias. For the pollination example, model averaging could be applied to $q(x)$. However, in this case the model-averaged $q(x)$ only results in a lower KLD 56% of the time when compared with using the $q(x)$ calculated from the best AIC model (i.e., model averaging was worse nearly half the time). Unfortunately, it appears that little is likely to be gained from the simple model averaging approaches described above (also see Richards et al. 2011).

Lebreton et al. (1992) and Burnham and Anderson (2002) point out that AIC can result in the selection of overly complex models if the proposed models all fail to explain the degree of variation in the data. This can happen when data distributions with restricted variance, such as the binomial distribution and the Poisson distribution, are included in the models. In this case the data are said to be overdispersed with respect to the restricted distribution (chapter 12). In section 3.2.2, I suggested that a solution to overdispersion is to develop a likelihood function for the study that incorporates compound distributions.

An alternative is to keep the variance-limited distributions in the models and modify their AIC value by using quasi-likelihoods (chapter 6) rather than likelihoods. The resulting quasi-AIC is given by

$$\text{QAIC(M)} = -\frac{2}{\tilde{v}} LL(\theta^*) + 2k, \tag{3.18}$$

where $\tilde{v}$ is the estimated variance inflation factor (VIF). The VIF is estimated from the data by comparing the likelihoods of the most complex model proposed (C) with the saturated model (S), which is the most complex model that can be fit to the data. The estimate is given by

$$\tilde{v} = \frac{2}{\text{df}} \left[LL(\theta_S^*) - LL(\theta_C^*) \right], \tag{3.19}$$

where df is the additional number of parameters needed to be estimated by the saturated model, compared with the most complex model proposed (Burnham and Anderson 2002). The AIC analysis can proceed as described above but with AIC replaced by QAIC. The advantage of this approach is that it may be easier to estimate the VIF rather than incorporate the extra variation in the likelihood function. Richards (2008) provides more details on the use of QAIC and uses simulations to show that a QAIC analysis usually results in similar inference when compared with an AIC analysis that explicitly describes the sources of overdispersion. Thus, QAIC is recommended when overdispersion is apparent for the best-fit models and overdispersion cannot be accounted for using compound probability distributions.

   The final step of any analysis involving AIC should be to check that the best AIC model produces outcomes consistent with those observed in the data. This check is necessary as the accuracy of an AIC estimate is usually reduced for those models that more poorly reflect the truth (table 3.2). Although the IT approach to model selection acknowledges that all models are wrong (i.e., none could exactly match **p** for any set of parameter values), little confidence can be had for an AIC analysis when all the proposed models poorly reflect the data (Burnham and Anderson 2002). Unfortunately, many published AIC analyses do not demonstrate that their best AIC model is consistent with the data, which may be due to the difficulty in demonstrating consistency. One approach is to present a goodness-of-fit test. However, if much data have been collected, then the null hypothesis that the model generated the data will often be rejected, despite the model adequately describing the biology so that its AIC estimate is accurate. In this case, simulating data according to the best AIC model and simply comparing the distribution of data with the observed distribution can usually give a sense of whether or not a model is adequate (Hobbs and Hilborn 2006).

3.3.6 A worked example

Consider again the seed set study involving two pollinator species (figure 3.4A). The AIC values associated with each of the six models are presented in table 3.3 and figure 3.8. In this example model M(v + r) is the best AIC model. Models M(r) and M(s + r) have a ∆AIC less than 6, but model M(s + r) can be discounted because model M(v + r) is simpler and has a lower AIC. Thus, two models are parsimonious candidates: M(v + r) and M(r). As the random effect is included in both of these models, the analysis provides strong support for some unknown factor influencing seed set among plants. More interestingly, the analysis provides some support for the hypothesis that the number of pollinator visits to a flower influences seed set; however, there is no support for seed set being influenced

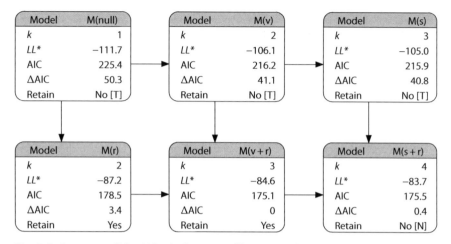

Fig. 3.8 Summary of the AIC calculations and how AIC values and model nesting are used to select models considered to be parsimonious with the data. Bracketed letters indicate the reason for model rejection: [T] the AIC value of the model was 6 or more than the lowest AIC calculated, [N] the model is associated with a simpler nested model having a lower AIC value.

Table 3.3 Maximum likelihood values for six models fit to the data from the two-pollinator experiment presented in figure 3.4A. Also presented are the corresponding AIC and ΔAIC values. Bold ΔAIC values indicate the selected models. k is the number of estimated model parameters

Model	s^*	$w_A{}^*$	$w_B{}^*$	ϕ^*	k	LL^*	AIC	ΔAIC
		Model parameters						
M(null)	0.456	0	0	–	1	−111.7	225.4	50.3
M(v)	0.269	0.131	0.131	–	2	−106.1	216.2	41.1
M(s)	0.271	0.164	0.099	–	3	−105.0	215.9	40.8
M(r)	0.463	0	0	0.543	2	−87.2	178.5	**3.4**
M(v + r)	0.256	0.146	0.146	0.490	3	−84.6	175.1	**0.0**
M(s + r)	0.253	0.197	0.105	0.483	4	−83.7	175.5	0.4

by the species of the pollinator. Biologically, it is highly likely that pollinator species do differ in their propensity to transport pollen, but given the data it is not possible to reliably estimate that difference because it appears to be small.

Of course, the claim that models M(v + r) and M(r) are parsimonious with the data requires demonstration that they predict patterns of seed set that are consistent with the data. Figure 3.4A shows that this is the case for model M(v + r). Previously, Model M(v + r) was also shown to predict variation in seed set that is consistent with the data (figure 3.3). Thus, this model could be used to make quantitative predictions, although caution is needed when using the model to extrapolate beyond four flower visits. The corresponding beta distributions associated with each number of flower visits are extremely wide (figure 3.4B), which is consistent with the conclusion that some unknown factors are causing large variation in seed set among plants.

In this example, the biological interpretation when using the AIC analysis and the null-hypothesis approach are very similar, as they both choose model M(v + r) as the best. However, the AIC approach is more conservative: it does not discount the possibility that the processes generating the data can be best approximated by M(r), which ignores visit-dependence. An AIC analysis, when using the rules of thumb described above and having checked for consistency, provides a conservative but justifiable means for quantifying the relative support for all the models that are initially proposed to be reasonable explanations of the system studied.

This example illustrates the potential consequence of ignoring overdispersion when performing model selection. If only the binomial models were fit to the data (i.e., models M(null), M(v), and M(s)), then model M(s) would be the best AIC model, and only model M(v) would also have been selected (table 3.3). In this case, the interpretation is that pollinator species differ in their effects on seed set and there is strong evidence that seed set is related to visitation number, as model M(null) is not selected. Looking at the data (figure 3.4A) it seems hard to defend a claim that flower visits by pollinator species A is 65% more likely to result in seed set ($w_A^* = 0.164, w_B^* = 0.099$; see Table 3.3).

An alternative approach to model selection is to ignore models M(r), M(v + r), M(s + r), and select among models M(null), M(v), and M(s) based on their QAIC values. The details of a quasi-AIC analysis are presented in appendix 3A. The estimated variance inflation factor when comparing the fits of M(s) and the saturated model is $\tilde{v} = 3.46$. Model M(v) has the lowest QAIC and M(null) is also included in the selected set, but M(s) is not selected (table 3A.1 in appendix 3A). Thus, there is some support for visit effects on seed set, but species-specific differences, if present, appear to be small; these conclusions are the same as those resulting from the AIC analysis that compared all six models.

3.4 Discussion

Here, I have presented two approaches to model selection that employ likelihood functions. I have suggested adopting the AIC approach over the stepwise regression approach, and I have provided a well-defined interpretation of best model when employing IT to select among models. However, a potential drawback with the AIC approach is that it may be computationally expensive, as all the models considered need to be fit to the data. The potential advantage of the stepwise approach is that not all of the models have to be fitted and the model selection process can often be automated using statistical software. A large number of models will need to be investigated when the analysis involves identifying the role of many covariates, especially if interaction effects are deemed to be potentially important. Many statistical packages present the AIC value associated with a model, and these values, in principle, could be used to perform stepwise model selection. For example, when performing backwards stepwise regression, if the removal of a variable does not increases the AIC value, then the simpler model could be chosen over the more complex model. In this case, the algorithm would stop when all the simple nested models have a higher AIC. Despite some cautious recommendations for using AIC in a stepwise selection context (e.g., Kabacoff 2011) I am unaware of the theoretical justifications for using AIC in this manner. On the other hand, here I have provided theoretical and empirical support for using AIC to select a parsimonious set of models based on ΔAIC values and model nesting.

Sometimes finding the set of parameter values that maximize the likelihood function is difficult. Numerical routines may produce poor parameter estimates if the likelihood function contains either multiple peaks or long flat ridges in parameter space, which are more likely to be present if the model contains many covariates, especially if they are correlated. If predictor variables are found to be highly correlated, it is best if biological considerations make it possible to discount all but one as being causative. Often, however, it will not be possible to discount all but one correlated covariate, in which case one could restrict the fitted models to those that do not contain highly correlated covariates. In this case, if AIC were used to select among the fitted models then many models may be in the selected set and they may differ markedly in the covariates they contain, making inference difficult (and rightly so). Unexpectedly wide confidence intervals for model parameters calculated using the likelihood profile approach might indicate the presence of long flat ridges in the likelihood function that are caused by correlated predictors.

Maximization routines often have the option to include initial parameter estimates, and confidence that MLPEs have been found can be boosted by checking that they are insensitive to the initial estimates. Finding good initial estimates can be made easier by carefully constructing the likelihood functions. For example, in a regression analysis with multiple covariates, it may be best to define the mean of the response variable in terms of the means and standard deviations of the covariates. For a linear model with I covariates, the mean when the covariates take the values x_i, can be written:

$$\mu\left(x_1, \cdots, x_I\right) = \beta_0 + \sum_{i=1}^{I} \beta_i \left(\frac{x_i - \bar{x}_i}{s_i}\right),$$ (3.20)

where the $\bar{x}_i$ and s_i are the mean and the standard deviation of covariate i. The parameter β_0 is the mean response for a typical set of covariates and the β_i are dimensionless and therefore directly comparable when investigating relative effect sizes. In this example, the MLPEs for each of the β_i will usually be similar across models that include mixtures of the remaining covariates. Thus, by first fitting the simpler models having only a single covariate, their MLPEs can be used as initial parameter estimates for the more complex models that include multiple covariates.

In some cases it may not be possible to write down the likelihood function. Even if the function can be formulated, its evaluation may be too computationally expensive to allow the use of maximization routines. This may occur when the model incorporates multiple random effects that require numerical integration. Bayesian approaches to model selection may be a better approach in such cases (McCarthy 2007; chapter 2), although convergence issues can still occur with the Monte Carlo methods they typically employ.

Not surprisingly, there is no consensus regarding the best way to perform model selection. I have provided recommendations when using AIC to select among models, which may be considered somewhat conservative. For example, I do not find strong support for using AICc in general, nor do I find compelling reasons for applying model weights and model averaging. Given the above-mentioned list of potential difficulties when seeking the set of MLPEs I also suggest caution if employing automated procedures for model selection; always have expectations regarding plausible parameter estimates, and always check plausibility of the maximum likelihood values produced by numerical routines by noting the nesting relations among the models. One should always try to produce a figure that presents the data in a manner that supports the conclusion drawn from an AIC analysis. I have found the AIC approach to be flexible enough to apply to a wide range of ecological studies, and, when the selection rules advocated above are adopted, I also find

CHAPTER 4

Missing data: mechanisms, methods, and messages

Shinichi Nakagawa

4.1 Introduction to dealing with missing data

In an ideal world, your data set would always be perfect without any missing data. But perfect data sets are rare in ecology and evolution, or in any other field. Missing data haunts every type of ecological or evolutionary data: observational, experimental, comparative, or meta-analytic. But this issue is rarely addressed in research articles. Why? Researchers often play down the presence of missing data in their studies, because it may be perceived as a weakness of their work (van Buuren 2012); this tendency has been confirmed in medical trials (Wood et al. 2004), educational research (Peugh and Enders 2004), and psychology (Bodner 2006). I speculate that many ecologists also play down the issue of missing data.

The most common way of handling missing data is called *list-wise deletion*: researchers delete cases (or rows/lists) containing missing values and run a model, e.g., a GLM (chapter 13) using the data set without missing values (known as *complete case analysis*). While common, few researchers explicitly state that they are using this approach. Another common practice that is usually not explicit involves statistics performed on pairs of points, like correlation analysis. For example, in analyzing the correlations among *x*, *y*, and *z*, we may be missing some data for each variable. Missing a value for *x* in some cases still allows one to use *y* and *z* from those cases. This is called *pair-wise deletion*, and it can often be noticed by seeing that there are different sample sizes for different correlations. List-wise and pair-wise deletion are often the default procedures used by statistical software.

What is wrong with deletion? The problems are twofold: (1) loss of information (i.e., reduction in *statistical power*) and (2) potential *bias* in parameter estimates under most circumstances (*bias* here means systematic deviation from population or true parameter values; Nakagawa and Hauber 2011).

To see the impact on statistical power, imagine a data set with 12 variables. Say only 5% of each variable is missing, without any consistent patterns. Using complete case analysis, we would lose approximately 43% of all cases. The resulting reduction in statistical power is fairly substantial. To ameliorate the reduction in power, some researchers use stepwise regression approaches called *available case analysis*, where cases are deleted if they are missing values needed to estimate a model, but the same cases are included for simpler models not requiring those values. For example, a full model would contain 12 variables with

Ecological Statistics: Contemporary Theory and Application. First Edition. Edited by Gordon A. Fox, Simoneta Negrete-Yankelevich, and Vinicio J. Sosa. © Oxford University Press 2015. Published in 2015 by Oxford University Press.

~43% of cases missing, while a reduced model might have 3 variables with ~10% missing. Are parameter estimates or indices like R^2 from these different models comparable? Certainly one cannot use information criteria such as AIC (Akaike Information Criteria; see chapter 3) for model selection procedures because these procedures require a complete case analysis. Such model selection can only be done using *available variable analysis* that only considers variables with complete data. However, this approach can exclude key information (e.g., Nakagawa and Freckleton 2011).

With regard to the bias problem in deleting missing data, cases are often missing for underlying biological reasons, so that parameter estimates from both complete case and available case analyses are often biased. For example, older or "shy" animals are difficult to catch in the field (Biro and Dingemanse 2009), so that their information may be systematically missing, leading to biased parameter estimates.

Some researchers "fill-in" or impute missing values to circumvent these problems. You may be familiar with filling missing values with the sample mean value (*mean imputation*). Indeed, in comparative phylogenetic analysis it has been common to replace missing values with taxon means (see Freckleton et al. 2003). Alternatively, missing data imputation can be slightly more sophisticated, using regression predictions to fill in missing cases (*regression imputation*). However, these methods, known as *single imputation* techniques, result in uncertainty estimates that do not account for the uncertainty that the missing values would have contributed (e.g., too small a standard error, or too narrow a confidence interval; McKnight et al. 2007; Graham 2009, 2012; Enders 2010). Thus, the rate of Type I error (chapter 2) increases; I call this phenomenon *biased uncertainty estimates*. These simple fixes using single imputation will yield biased parameter estimates.

The good news is that we now have solutions that combat missing data problems. They come in two forms: *multiple imputation* (MI), and *data augmentation* (DA; in the statistical literature, the term data augmentation is used in different ways, but I follow the usage of McKnight et al. 2007). The bad news is that very few researchers in ecology and evolution use such statistical tools (Nakagawa and Freckleton 2008). MI and DA have been available to us since the late 1980s, with some key publications in 1987 (Allison 1987; Tanner and Wong 1987; Little and Rubin 1987; Rubin 1987). In the beginning, few of us could use such techniques as they were not implemented in statistical packages or programs until the late 1990s. There are now *R* libraries (e.g., norm and pan; Schafer 1997, 2001) that make MI and DA relatively easy to use for many analyses (for reviews of statistical software for treating missing data, see Horton and Kleinman 2007; Yucel 2011). Why the lag in using such important statistical tools? Many of us may have never heard about *missing data theory* until now because it is not a part of our general training as ecologists and evolutionary biologists. However, the main reason may be psychological. It certainly feels a bit uneasy for me to "make up" data to fill in gaps! We are not alone: medical and social scientists have also been exposed to methods for handling missing data, but they have also been slow to adopt them (Raghunathan 2004; Graham 2009; Sterne et al. 2009; Enders 2010). Researchers often may see procedures such as data imputation and augmentation as cheating, or even as something akin to voodoo. It turns out that our current quick fixes are a lot more like voodoo! As Todd Little (cited in Enders 2010) puts it: "For most of our scientific history, we have approached missing data much like a doctor from the ancient world might use bloodletting to cure disease or amputation to stem infection (e.g., removing the infected parts of one's data by using list-wise or pair-wise deletion)." It is high time for us to finally start using missing data procedures in our analyses. This is especially so given the recent growth in the number of *R* libraries that can handle missing data appropriately using MI and DA (Nakagawa and Freckleton 2011; van Buuren 2012).

In this chapter, I explain and demonstrate the powerful missing data procedures now available to researchers in ecology and evolution (e.g., Charlier et al. 2009; González-Suárez et al. 2012). I first describe the basics and terminology of missing data theory, particularly the three different classes of missing data (*missing data mechanisms*). I then explain how different missing data mechanisms can be detected and, at least for some of the classes, how to prevent it in the first place. The main section will cover three types of methods for analyzing missing data (deletion, augmentation, and imputation), with emphasis on MI, practical issues associated with missing data procedures, guidelines for the presentation of results, and the connection between missing data issues and other chapters in this book.

4.2 Mechanisms of missing data

4.2.1 *Missing data theory, mechanisms, and patterns*

Rubin (1976) and his colleagues (e.g., Little and Rubin 1987, 2002; Little 1992, 1995) established the foundations of missing data theory. Central to missing data theory is his classification of missing data problems into three categories: (1) missing completely at random (*MCAR*), (2) missing at random (*MAR*), and (3) missing not at random (*MNAR*). These three classes of missing data are referred to as *missing data mechanisms* (for a slightly different classification, see Gelman and Hill 2007). Despite the name, these are not causal explanations for missing data. Missing data mechanisms represent the statistical relationship between observations (variables) and the probability of missing data. The other term easy to confuse with missing data mechanisms is *missing data patterns*; these are the descriptions of which values are missing in a data set (see section 4.3.1).

4.2.2 *Informal definitions of missing data mechanisms*

Here, I use part of a data set from the house sparrow (*Passer domesticus*) population on Lundy Island, UK (Nakagawa et al. 2007a; Schroeder et al. 2012; I will use a full version of this data set in section 4.4.5). Male sparrows possess what is termed a "badge of status," which has been shown to reflect male fighting ability (Nakagawa et al. 2007b). The badge of status is a black throat patch, which substantially varies in size, with larger badges resenting superior fighters. Table 4.1 contains the information on badge size (Badge) and male age (Age) from 10 males, and also on the three missing data mechanisms in the context of this data set. The MCAR mechanism occurs when the probability of missing data in one variable is not related to any other variable in the data set. The variable, Age$_{[MCAR]}$ in table 4.1 is missing completely at random (MCAR) because the probability of missing data on Age is not related to the other observed variable, Badge.

The MAR mechanism is at work when the probability of missing data in a variable is related to some other variable(s) in the data set. If you are wondering "So how is this missing at random?" you are not alone: the term confuses many people. It is helpful to see MAR as "conditionally missing at random"; that is, missing at random after controlling for all other related variables (Graham 2009). In our sparrow example, Age$_{[MAR]}$ is missing at random (MAR) because the missing values are associated with the smallest three values of Badge. Once you control for Badge, data on Age are missing completely at random, MCAR. This scenario may happen, for example, if immigrants to this sparrow population (whose ages are unknown to the researcher) somehow have a smaller badge size.

Table 4.1 Badge size (mm) and Age (yr) information for 10 house sparrow males. Age consists of 4 different types of data sets according to the mechanism of missing values (–): Complete data, MCAR data, MAR data, and MNAR data

Bird	Badge	Age			
(Case)	Complete	Complete	MCAR	MAR	MNAR
1	31.5	1	1	–	1
2	33.5	2	–	–	2
3	34.4	3	3	–	3
4	35.1	1	–	1	1
5	35.4	2	2	2	2
6	36.7	4	4	4	–
7	37.8	2	2	2	2
8	38.8	4	4	4	–
9	40.3	3	3	3	3
10	41.5	4	–	4	–

The MNAR mechanism happens when the probability of missing data in a variable is associated with this variable itself, even after controlling for other observed (related) variables. $Age_{[MNAR]}$ is missing not at random because the three missing values are 4-year old birds, and it is known that older males tend to have larger badge sizes. Such a scenario is plausible if a study on this sparrow population started 3 years ago, and we do not know the exact age of older birds.

4.2.3 Formal definitions of missing data mechanisms

Now to provide more formal/mathematical definitions of the missing data mechanisms, I introduce relevant notation and terminology from missing data theory.

- **Y** is a matrix of the entire data set (including response and predictor variables) that can be decomposed into $\mathbf{Y_{obs}}$ and $\mathbf{Y_{mis}}$ (the observed and missing parts of the data);
- **R** is a *missingness* matrix—these are indicators of whether the corresponding locations in **Y** are observed (0) or missing (1); and
- **q** is a vector of parameters describing the relationship between missingness, **R** and the data set, **Y** (table 4.2; see Little and Rubin 2002; McKnight et al. 2007; Molenberghs and Kenward 2007; Enders 2010; Graham 2012). Importantly, **q** is known as the mechanism of missing data and provides the basis for distinguishing between MCAR, MAR, and MNAR. An intuitive interpretation of **q** is that the content of **q** indicates one of the three missing data mechanisms.

It is the easiest to begin with the description of MNAR data because it includes all these mathematical terms. The three mechanisms, in relation to the concepts of missingness and ignorability (discussed later in this section), are summarized in figure 4.1.

Following Enders (2010), the probability distribution for MNAR can be written as:

$$p(\mathbf{R}|\mathbf{Y_{obs}}, \mathbf{Y_{mis}}, \mathbf{q}). \tag{4.1}$$

This says that the probability of whether a position in **R** is 0 or 1 depends on both $\mathbf{Y_{obs}}$ and $\mathbf{Y_{mis}}$, and this relationship is governed by **q**. In table 4.2, $\mathbf{v_2}$ is MNAR, if missing

Table 4.2 An illustrative example of a data set $\mathbf{Y_{obs}}$ with three variables ($\mathbf{v_1}$–$\mathbf{v_3}$; Mis = missing observations and Obs = observed values) and its missingness, $\mathbf{R}$ (the recording of $\mathbf{v_1}$–$\mathbf{v_3}$ into binary variables, $\mathbf{m_1}$–$\mathbf{m_3}$); modified from Nakagawa and Freckleton (2011). Note that $\mathbf{v_3}$ is not measured; it is included here for illustrative purposes but would not usually be a part of $\mathbf{Y}$ and $\mathbf{R}$

Case	Data [$\mathbf{Y} = (\mathbf{Y_{obs}}, \mathbf{Y_{mis}})$]			Missingness [$\mathbf{R}$]		
	$\mathbf{v_1}$	$\mathbf{v_2}$	$\mathbf{v_3}$	$\mathbf{m_1}$	$\mathbf{m_2}$	$\mathbf{m_3}$
1	Obs	Mis	Mis	0	1	1
2	Obs	Obs	Mis	0	0	1
3	Obs	Obs	Mis	0	0	1
4	Obs	Mis	Mis	0	1	1
5	Obs	Obs	Mis	0	0	1
6	Obs	Obs	Mis	0	0	1
7	Obs	Obs	Mis	0	0	1
8	Obs	Mis	Mis	0	1	1
9	Obs	Mis	Mis	0	1	1
10	Obs	Obs	Mis	0	0	1

values depend on $\mathbf{v_2}$ itself. Such missing values can (but need not) be related to $\mathbf{v_1}$, a completely observed variable. In this particular case, the probability of MNAR missingness depends completely on $\mathbf{Y_{mis}}$, i.e., $p(\mathbf{R} \mid \mathbf{Y_{mis}}, \mathbf{q})$, which is a special case of missing values that are related to both $\mathbf{v_1}$ and $\mathbf{v_2}$ (i.e., equation 4.1). Another more complicated, form of MNAR is when $\mathbf{v_2}$ depends on a completely unobserved variable, for example $\mathbf{v_3}$ in table 4.2. For a concrete example in table 4.1, the MAR missing values in Age would become MNAR, if we had no measurement of badge size (Badge). In practice one can only suspect or assume MNAR, because it depends on the unobserved values in $\mathbf{Y_{mis}}$ (but see section 4.4.5).

The probability distribution for MAR can be written as:

$$p(\mathbf{R} \mid \mathbf{Y_{obs}}, \mathbf{q}). \tag{4.2}$$

This means that missingness depends only on $\mathbf{Y_{obs}}$, and this relationship is governed by $\mathbf{q}$. In table 4.2, $\mathbf{v_2}$ is MAR if its missing values depend on the observed variable $\mathbf{v_1}$.

Finally, the probability distribution for MCAR is expressed as:

$$p(\mathbf{R} \mid \mathbf{q}). \tag{4.3}$$

This says that the probability of missingness does not depend on the data (neither $\mathbf{Y_{obs}}$ nor $\mathbf{Y_{mis}}$), but that whether positions in $\mathbf{R}$ take 0 or 1 is still governed by $\mathbf{q}$. So $\mathbf{v_2}$ is MCAR if its missing values do not depend on an observed variable ($\mathbf{v_1}$) or the values of v_2 itself.

Ignorability is another important concept; note that the same word is used with other meanings in other statistical contexts (for examples, see Gelman and Hill 2007). MNAR missingness is "non-ignorable" whereas MAR and MCAR are "ignorable" (Little and Rubin 2002). Ignorability refers to whether we can ignore the way in which data are missing when we impute or augment missing data; it does not imply that one can remove missing data! In the MAR and MCAR mechanisms, imputation and augmentation do not require that we make specific assumptions about how data are missing. On the other hand, non-ignorable MNAR missingness requires such assumptions to build a model to fill in missing values (section 4.4.8).

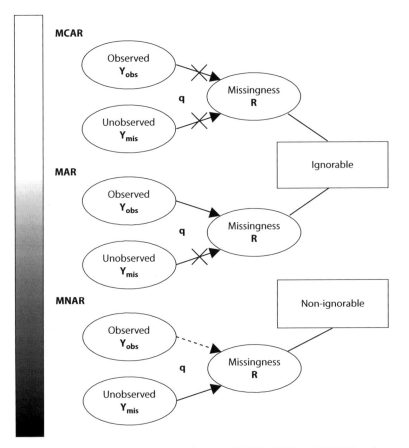

Fig. 4.1 The three missing data mechanisms (MCAR, MAR, and MNAR) and ignorability (whether we need to model the mechanism of missing data) in relation to observed data (Y_{obs}), missing data (Y_{mis}), the missingness matrix (**R**), and their relationships (**q**; parameters that explain missingness, i.e., mechanism). The solid arrows, dotted arrows, and arrows with crosses represent "connection," "possible connection," and "no connection," respectively. The lines connecting ignorability and missingness group the three mechanisms into the two ignorability categories. Also no pure forms of MCAR, MAR, and MNAR exist, and all missingness can be considered as a form of MAR missingness; this is represented by the shaded continuum bar on the left. Modified from Nakagawa and Freckleton (2011).

4.2.4 *Consequences of missing data mechanisms: an example*

Figure 4.2 shows the three different mechanisms of missing data in a bivariate example in two situations where % missing values are different (40% and 80% from the sample size of 200). The missing values are all in Variable 2 (plotted as a response variable; analogous to v_2 in table 4.2) but not in Variable 1 (analogous to v_1 in table 4.2). The population true mean (μ) and standard deviation (σ) for Variable 2 are 0 and 1.41 (variance, $\sigma^2 = 2$), respectively, while the true intercept (α), slope (β) and residual variance (σ_e^2) for the linear relationship between Variable 1 and Variable 2, are 0, 1, and 1, respectively. Parameter estimates from analysis from "observed" data of three missing data mechanisms (i.e., complete case analysis) are summarized in table 4.3.

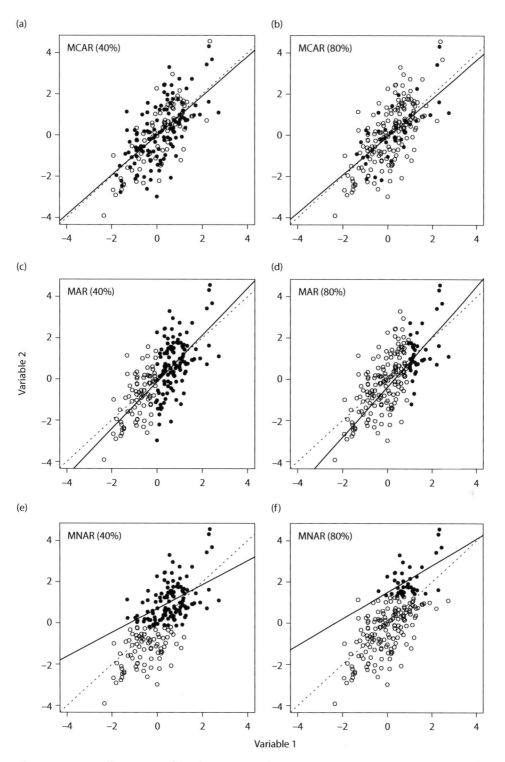

Fig. 4.2 Bivariate illustrations of the three missing data mechanisms and consequences for a) MCAR with 40% missing values (40%), b) MCAR with 80% missing values (80%), c) MAR (40%), d) MAR (80%), e) MNAR (40%), and f) MNAR (80%). Solid circles are observed data and empty circles are missing data; dotted lines represent "true" slopes while solid lines were estimated from observed data.

Table 4.3 The estimates of descriptive statistics for Variable 2 (see the main text) and the estimates from regression analysis of Variable 2 against Variable 1 (complete case analysis), using the complete data set and the three types of data sets with missing values (MCAR, MAR, and MNAR) in two scenarios where 40% or 80% of Variable 2 are missing (the total sample size, $n = 200$; no missing values in Variable 1). The true value for each parameter is $\mu = 0$, $\sigma = 1.414$, $\alpha = 0$, $\beta = 1$, and $\sigma_e = 1$; the mean (μ) and standard deviation (σ) are for Variable 2, α and β are the intercept and slope respectively, and σ_e is the residual standard deviation. For corresponding plots, see figure 4.2

Missing data mechanisms (% missing data)	$\hat{\mu}$	$\hat{\sigma}$	$\hat{\alpha}$	s.e.	$\hat{\beta}$	s.e.	$\hat{\sigma}_e^2$
No missing data	0.091	1.464	−0.052	0.075	1.069	0.079	1.055
MCAR (40%)	0.129	1.415	−0.019	0.101	0.961	0.106	1.092
MCAR (80%)	0.189	1.351	−0.063	0.155	0.930	0.145	0.950
MAR (40%)	0.723	1.308	−0.139	0.170	1.136	0.177	1.131
MAR (80%)	1.355	1.185	−0.374	0.510	1.219	0.341	1.038
MNAR (40%)	1.040	0.942	0.700	0.095	0.580	0.098	0.831
MNAR (80%)	2.093	0.811	1.499	0.186	0.645	0.163	0.691

As we would expect, parameter estimates from the regression, using the complete data set are close to population true values (table 4.3). As theory suggests, no obvious bias in the parameter estimates from the MCAR data sets can be detected, although standard errors for regression estimates increased (i.e., there is less statistical power). In general, many parameter estimates from the MAR data sets seem to be biased to some certain extent. Noticeably, many parameter estimates from the MNAR data sets seem to be severely biased. In the data sets of all the three mechanisms, deviations from true estimates usually increase when the percentage of missing values is raised, i.e., from 40% to 80% (all relevant R code is provided in appendix 4A).

In real data sets, the consequences of missing data will be further complicated by the existence of more than two variables and the presence of missing values in more than one variable. Furthermore, it is usually impossible to unambiguously classify cases into the three mechanisms (Graham 2009, 2012). For example, it is hard to imagine missing data that are entirely unrelated to other variables in the data set, i.e., purely MCAR. Missing data in real data sets are somewhere on a continuum from MCAR through MAR and to MNAR, as depicted in figure 4.1. In a sense, it may be easiest to think of all missing data as belonging to MAR to some degree because MAR resides in the middle of this continuum. Further details can be found in Nakagawa and Freckleton (2008, 2011).

4.3 Diagnostics and prevention

4.3.1 *Diagnosing missing data mechanisms*

In this and the next section (section 4.4), I will use snippets of R code along with example data sets. The full R code, related data sets, and more detailed explanations of these are all found in appendix 4B.

It is straightforward to visualize missing data patterns with the aid of R functions. As an example, I again use a part of the Lundy male sparrow data (table 4.1). The `missingmap` function in the `Amelia` library (Honaker et al. 2011) produces figure 4.3, which is a visual representation of missing data patterns or in fact, a matrix, **R** (missingness). Plotting

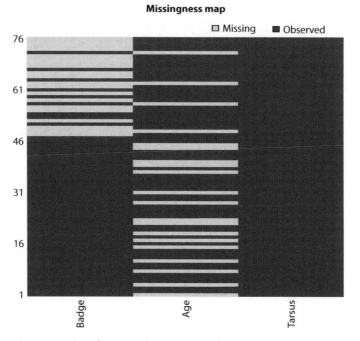

Fig. 4.3 A plot of missing data patterns of the three variables (Badge, Age, and Tarsus), produced by the `missmap` function in the `Amelia` library (Honaker et al. 2011). See text for more details.

missing data patterns can sometimes reveal unexpected patterns such as a cluster of missing values, which were not noticeable during the data collection stages. Then we can ask why such patterns exist. However, missing data patterns alone do not tell us about which missing data mechanism(s) underlie our data.

By deleting cases where missing values exist (complete case analysis), we implicitly assume MCAR. There are a number of ways to diagnose whether or not missing data can be classified as MCAR (reviewed in McKnight et al. 2007). However, as we have learned, MCAR is an unrealistic assumption because such precise missingness is implausible (Little and Rubin 2002; Graham 2009, 2012; see figure 4.1) and also because biological and/or practical reasons generally underlie missingness (Nakagawa and Freckleton 2008). MAR—for which the pattern of missingness is ignorable—is a more realistic assumption. In fact, the MAR assumption is central to many missing data procedures (section 4.4). My main recommendation is to deal with missing values under the assumption of MAR even when all missing data are diagnosed as MCAR (see Schafer and Graham 2002; Graham 2009, 2012; Enders 2010).

When is it really useful to identify missing data mechanisms? You may want to see MCAR diagnostics if you have to resort to missing data deletion. The simplest method is to conduct a series of t tests on values between observed and missing groups in each variable (0 being the one group and 1 the other in missingness **R**; see table 4.2), which assess mean difference in the other variables in the data set. If all t-tests are non-significant, then you can say missing values in that data set are MCAR; if not, they are MAR or MNAR. However, as the size of the matrix grows, performing and assessing multiple t-tests gets tedious very quickly and also may result in Type I errors. Little (1988) proposed a multivariate version of this procedure, which produces one statistic (a χ^2 value) for the entire data set (for details, see Little 1988; McKnight et al. 2007; Enders 2010). This extension of the *t*-test approach can be carried out by the `LittleMCAR` function in the `BaylorEdPsych` R library (Beaujean 2012).

For the example data set (`PdoDataPart`, see appendix 4B), the test produces $\chi^2_5 = 36.65$ and $p < 0.0001$. We can conclude that this data set contains non-MCAR missingness. This test has the advantage of being simple, but has two major shortcomings: (1) the data set may often have weak statistical power, especially when the observed and missing groups are unbalanced and (2) a non-significant result can be obtained even if missingness is MAR or MNAR. This occurs when, for example, missing values in a variable are related to the high and low values of another variable.

There are neither statistical tests nor visual techniques to distinguish between MAR and MNAR (McKnight et al. 2007; van Buuren 2012). This is not surprising given that the probability distributions for MAR ($p(\mathbf{R} \mid \mathbf{Y_{obs}}, \mathbf{q})$) and MNAR ($p(\mathbf{R} \mid \mathbf{Y_{obs}}, \mathbf{Y_{mi}}, \mathbf{q})$) differ only in that MNAR depends on $\mathbf{Y_{mi}}$ (unobserved values), and we have no way of knowing what unobserved values were. Rather, we need to ascertain whether or not missing values are considered MNAR from our understanding of the biological systems under investigation. For example, in the MNAR example in table 4.1, age information was missing from the oldest birds, because of the limited duration of the study.

Graphical methods for diagnosing missingness are generally much more useful. Visualizations of the relationship between the original data set and missingness (e.g., $\mathbf{m_1}$ in table 4.2) is easily done in *R*, using built-in functions and the `pairs.panels` function from the `psych` library (Revelle 2012).

```
> Missingness <- ifelse (is.na (PdoPartData) == TRUE, 0, 1)
# create the missingness matrix
> MissData <- data.frame (PdoPartData, Missingness)
# combine the original dataset with the missingness matrix
> library (psych) # loading the psych library
> pairs.panels (MissData, ellipses = FALSE, method = "spearman")
```

The resulting figure (figure 4.4) contains visual information on all the original variables and missingness variables, as well as information about all the correlations among these variables. I encourage the reader to study this figure to identify non-MCAR missingness.

4.3.2 *How to prevent MNAR missingness*

As the father of modern statistics, Ronald A. Fisher is reported to have said, "the best solution to handling missing data is to have none," but this is probably not the easiest solution (McKnight et al. 2007). Missing data prevention requires careful planning and execution of studies and experiments, as well as a good understanding of the biological systems at hand, and even then, missing data are often unavoidable (Nakagawa and Freckleton 2008). However, there is a trick that you can use to make missing values much easier to handle. The trick is to begin your study with a data collection plan, wherein you will turn MNAR missingness into MAR missingness. In other words, this means altering non-ignorable missing values to make them ignorable; missing values can then be handled with ordinary missing data procedures such as multiple MI (or without making special assumptions due to MNAR; see Schafer and Graham 2002; Graham 2009, 2012).

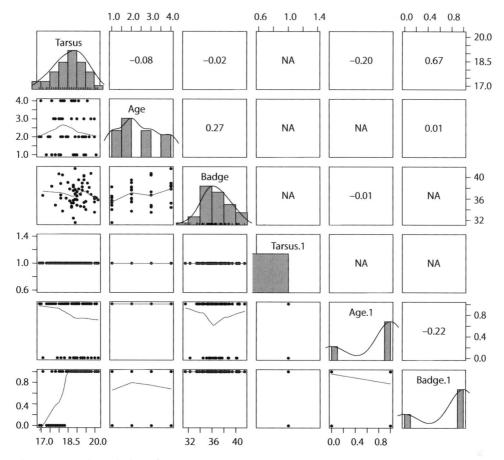

Fig. 4.4 Paired panel plots of the data matrix **Y** and missingness matrix **R** for the house sparrow data set, created by the `pairs.panels` function in the `psych` library (Beaujean 2012). Tarsus, Age, and Badge are numerical values in **Y**, while Tarsus.1, Age.1, and Badge.1 indicate missingness for these values, respectively. The upper triangle panels show Spearman correlations (NA means "not available"), while the lower triangle panels show scatterplots with lowess (locally weighted scatterplot smoothing) lines. The diagonals show histograms. There is some evidence for MAR because the correlation between Tarsus and Badge.1 is high ($r_S = 0.67$). Similarly, the moderate correlation between Tarsus and Age.1 ($r_S = -0.20$) suggests that we may have missing data in Age when birds have smaller tarsus size.

When you have a good understanding of your biological system, you usually know which variables will be likely to have missing values. If you collect data on known correlates of these missing-prone variables your missing values will be more likely to be MAR than MNAR. These correlates are called *auxiliary variables* in the missing data literature. An extension of this idea is the *planned missing data design*, in which you make the use of the MAR assumption to deliberately incorporate MAR missingness in your data collection. This may seem very strange at first, but think of a situation where Measurement A is very expensive to collect and is a variable of interest, while Measurement B is very cheap to measure but is not of interest (e.g., A may be a biochemical marker of oxidative stress while B is the color of a trait, which is correlated with this marker). If A and B are correlated, you can collect B for all subjects, while you can only collect A for a random subset (i.e., creating missing values on purpose). Given missing values in A are MAR, missing data procedures can actually restore the statistical power of your statistical models as if you had collected A for all subjects! This design is called two-method measurement design (Graham et al. 2006; Enders 2010). Investigations into planned missing data design are relatively new and an active area of research (Baraldi and Enders 2010; Graham 2009, 2012; Rhemtulla and Little 2012), but I expect that developments will enormously benefit research planning in ecological and evolutionary studies in the near future.

4.4 Methods for missing data

4.4.1 *Data deletion, imputation, and augmentation*

Three broad categories of methods for handling missing data are: deletion, imputation, and augmentation (McKnight et al. 2007; see also Nakagawa and Freckleton 2008). Data imputation has two subcategories: single imputation and multiple imputation (MI). Schematics in figure 4.5 provide conceptual representations of the four ways of handling missing data (i.e., data deletion, single imputation, MI and DA).

Here I focus on MI under the MAR assumption, because I believe that MI methods are currently the most practical and useful for ecologists and evolutionary biologists. Further, many recent software developments have focused on MI methods (van Buuren 2012), so *R* has a number of libraries available. Despite this focus, I will also provide brief pointers for non-ignorable (MNAR) missing data and sensitivity analysis (in section 4.4.8).

4.4.2 *Data deletion*

Data deletion methods such as list-wise and pair-wise deletion (section 4.1) are efficient ways of dealing with missing data as long as missing data are MCAR (figure 4.5A). Then, relevant analysis (e.g., complete or available case analysis) will produce unbiased parameter estimates with tolerable reductions in statistical power (cf. figure 4.2). If, say, only 1% of cases have missing values, then deletion would certainly offer the quickest way to deal with missing data. However as the fraction of missing cases grows, problems will quickly arise. I would follow Graham's (2009) recommendation that, if 5% or more of cases are missing, one should use multiple imputation or data augmentation.

4.4.3 *Single imputation*

Single imputation (figure 4.5B) has often been used because this procedure will result in a complete data set. There are many commonly used methods for single imputation, such as mean imputation and regression imputation (section 4.1). Other single imputation methods include hot- and cold-deck, and last and next observation carried forward, to name a few (reviewed in McKnight et al. 2007; Enders 2010). These methods often result in severe bias in parameter estimates, especially when missing data are not MCAR, so I will not discuss them further. However, stochastic regression imputation is worth mentioning, as it forms the basis of some missing data procedures introduced below. Like regression imputation, this method uses regression predictions to fill in missing values in a variable by using observed variables, but it incorporates noise in each predicted value by adding error based on a residual term. Under the MAR assumption, parameter estimates from single imputation by stochastic regression are unbiased (for more details, see Gelman and Hill 2007; Enders 2010). Unfortunately, they suffer from biased uncertainty estimates—for example, *s.e.* values are too small or unrealistically precise.

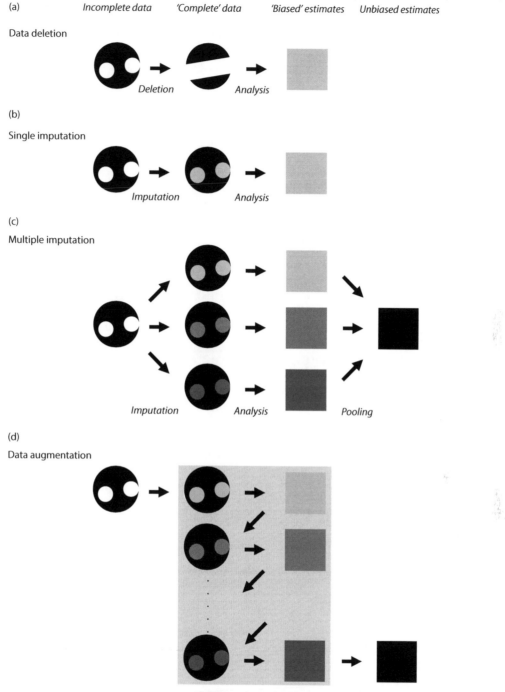

Fig. 4.5 Diagrams illustrating the process of a) data deletion, b) single imputation, c) multiple imputation, and d) data augmentation. "Biased" estimates mean biased parameter estimates, biased uncertainty estimates, or both. A circle represents a data set, and holes in the circle represent missing values. Such holes can be deleted (a) or filled in (b–d). A square represents a set of estimated parameters; the degree of bias in estimation is represented by a gray scale, with darker shades being less biased. See text for details (this figure was modified from Nakagawa and Freckleton 2008).

4.4.4 *Multiple imputation techniques*

Multiple imputation (MI) creates more than one filled-in completed data set. By doing so, MI, proposed by Rubin (1987), has solved the problem of biased uncertainty, which troubles all the available single imputation methods. MI has become the most practical and the best-recommended method in most cases (Rubin 1996; Schafer 1999; Allison 2002; Schafer and Graham 2002; McKnight et al. 2007; Graham 2009; Enders 2010; van Buuren 2012). Among imputation techniques that can generate unbiased parameter estimates under the MAR assumption, most relevant and useful are two methods, expectation maximization (EM) algorithms and Markov chain Monte Carlo (MCMC) procedures. These methods form the basis of multiple imputation.

EM (expectation maximization) algorithms are a group of procedures for obtaining maximum likelihood (ML; chapter 3) estimates of statistical parameters when there exist missing data and unobserved (unobservable underlying or latent; section 4.4.7) variables (for accessible descriptions, see McKnight 2007; Molenberghs and Kenward 2007; Graham 2009; Enders 2010: for more formal treatments, see Dempster et al. 1977; Schafer 1997; Little and Rubin 2002). The EM algorithm that estimates the descriptors of a multivariate matrix, a vector of means ($\mathbf{m}$), and a variance-covariance matrix ($\mathbf{V}$) consists of a two-step iterative procedure (E-step and the M-step). First, the E-step will use a very similar method to stochastic regression imputation to estimate $\mathbf{m}$ and $\mathbf{V}$ ($\hat{\mathbf{m}}$ and $\hat{\mathbf{V}}$) from observed values and then "expect" (or fill in) missing values. Next in the M-step, these complete data are used to estimate $\mathbf{m}$ and $\mathbf{V}$ and fill in missing values again. The two steps are repeated until $\hat{\mathbf{m}}$ and $\hat{\mathbf{V}}$ converge to ML estimates. However, the EM algorithm does not provide uncertainty estimates (*s.e.*) for $\hat{\mathbf{m}}$ and $\hat{\mathbf{V}}$. To obtain *s.e.*, bootstrapping (i.e., sampling observed data with replacement) can be combined with the EM algorithm to obtain frequency distributions for $\hat{\mathbf{m}}$ and $\hat{\mathbf{V}}$. This combined procedure is termed the EMB algorithm (Honaker and King 2010; Honaker et al. 2011; see figure 4.6A). I note that the `Amelia` library mentioned above employs the EMB algorithm to conduct MI.

One restriction to the EM and EMB algorithms is the assumption of multivariate normality, or $\mathbf{Y} \sim \text{MVN}(\mathbf{m}, \mathbf{V})$, where all variables come from one distribution. That is why this type of approach is called *joint modeling*. MCMC procedures circumvent this restriction by using a *fully conditional specification* where each variable with missing values can be treated or imputed separately when it is conditioned on other values in the data set (i.e., using Gibbs sampling; van Buuren et al. 2006; van Buuren and Groothuis–Oudshoorn 2011; van Buuren 2012). In this process each variable can have a different distribution and different linear modeling. For example, the algorithm can apply a binomial and a Poisson generalized linear model (chapter 6) for a binary and count variable respectively. This type of procedure is also called *sequential regression imputation* (Enders 2010).

MCMC procedures (and also Gibbs sampling) are often called Bayesian methods (chapter 1) because their goal is to create the posterior distributions of parameters, but methods using MCMC have much wider applications than Bayesian statistics). The MCMC procedure, is akin to the EM algorithm (Schafer 1997) in that it uses a two-step iterative algorithm to find $\hat{\mathbf{m}}$ and $\hat{\mathbf{V}}$. The imputation step (I-step) uses stochastic regression with observed data. Next, the posterior step (P-step) uses this filled-in data set to construct the *posterior distributions* of $\hat{\mathbf{m}}$ and $\hat{\mathbf{V}}$. Then, it uses a Monte Carlo method to sample a new set of $\hat{\mathbf{m}}$ and $\hat{\mathbf{V}}$ from these distributions. These new parameter estimates are used for the subsequent I-step. Iterations of the two steps create the Markov chain, which eventually converges into fully fledged posterior distributions of $\hat{\mathbf{m}}$ and $\hat{\mathbf{V}}$ (figure 4.6B). These

(a)

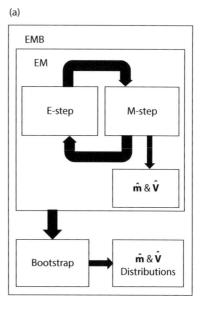

(b)

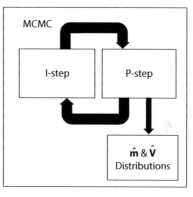

Fig. 4.6 Schematics illustrating the process of a) the EM (expectation maximization) and EMB (expectation maximization with bootstrapping) algorithm with the E-step (expectation) and M-step (maximization), and b) the MCMC procedure with the I-step (imputation) and the P-step (posterior). $\hat{m}$ is a vector of the means and $\hat{V}$ is a variance–covariance matrix. Thicker arrows represent iterative processes. See text for details.

distributions are, in turn, used for multiple imputation (for more details, see Schafer 1997; Molenberghs and Kenward 2007; Enders 2010). The two R libraries, `mice` (van Buuren and Groothuis–Oudshoorn 2011) and `mi` (Su et al. 2011), are notable here because they both implement MCMC procedures using a fully conditional specification, known as *multivariate imputation by chained equations* (MICE). In the statistical literature (e.g., Schafer 1997), this MCMC procedure is often referred to as data augmentation (see below).

4.4.5 *Multiple imputation steps*

There are three main steps in MI: imputation, analysis, and pooling (figure 4.5C). In the imputation step, you create m copies of completed data set by using data imputation methods such as the EM/EMB algorithms or the MCMC procedure. In the analysis step, you run separate statistical analyses on each of m data sets. Finally, in the pooling step, you aggregate m sets of results to produce unbiased parameter and uncertainty estimates. This aggregation process is done by the following equations (which are automatically calculated in R):

$$\bar{b} = \frac{1}{m} \sum_{i=1}^{m} b_i, \tag{4.4}$$

$$v_W = \frac{1}{m} \sum_{i=1}^{m} s.e._i^2, \tag{4.5}$$

$$v_B = \frac{1}{m-1} \sum_{i=1}^{m} (b_i - \bar{b})^2, \tag{4.6}$$

$$v_T = v_W + v_B + \frac{v_B}{m}, \tag{4.7}$$

where $\bar{b}$ is the mean of b_i (e.g., regression coefficients), which is a parameter estimated from the ith data set (i = 1, 2, . . ., m), v_W is the within-imputation variance calculated from the standard error associated with b_i, v_B is the between-imputation variance estimates, and v_T is the total variance ($\sqrt{v_T}$ is the overall standard error for $\bar{b}$). This set of equations for combining estimates from m sets of results is often referred to as Rubin's rules, as it was developed by Rubin (1987).

Statistical significance and confidence intervals (CIs) of pooled parameters are obtained as:

$$df = (m-1) \left(1 + \frac{mv_W}{(m+1)v_B} \right)^2, \tag{4.8}$$

$$t_{df} = \frac{\bar{b}}{\sqrt{v_T}}, \tag{4.9}$$

$$100(1-\alpha)\% \text{ CI} = \bar{b} \pm t_{df,(1-\alpha/2)} \sqrt{v_T}, \tag{4.10}$$

where df is the number of degrees of freedom used for t-tests or to obtain t values and CI calculations, and α is the significance level (e.g., 95% CI, $\alpha = 0.05$).

To illustrate the three steps in multiple imputation, I again use the house sparrow data set but this time with the seven variables (EPP, Age, Badge, Fledgling, Heterozygosity, Tarsus, Wing, and Weight). The question this time is which male non-morphological characteristics (i.e., Age, Fledgling, and Heterozygosity) best predict extra-pair paternity (EPP). EPP is a common phenomenon in the animal kingdom, especial among bird species, where males often have offspring outside their social bonds (Griffith et al. 2002). Nakagawa and Freckleton (2011) used the Amelia library (i.e., the EMB algorithm) for MI with this data set. Here I use the mice library (MCMC algorithm) to carry out the three steps of MI.

```
> library (mice) # loading the mice library
# the imputation step with 5 copies
> imputation <- mice (PdoData, m = 5, seed = 7777)
> analysis <- with (imputation, glm (EPP ~ Age + Fledgling
   + Heterozygosity, family = quasipoisson)) # the analysis step
   with a GLM (see chapters 6 & 12)
> pooling <- pool (analysis) # the pooling step
> summary (pooling)
```

With this three-step MI process, we obtain unbiased parameter and uncertainty estimates (table 4.4; for individual outputs, see the online appendix 4C).

Table 4.4 Results of analyses for the house sparrow data, using complete case analysis, `mice`, and `mi` (the latter two are multiple imputation via MCMC procedures). Estimates from `mice` and `mi` are the pooled model-averaged parameter estimates from the five imputed data sets ($m = 5$), pooled regression coefficients ($\bar{b}$), overall standard error, s.e. ($\sqrt{v_T}$), 95% confidence intervals (CI), and the fraction of missing information (γ). For details, see appendix 4C

Procedure	Predictor	Estimate	s.e.	Lower CI	Upper CI	γ
Complete case analysis	Intercept	−1.733	2.391	−7.009	2.466	–
	Age	0.479	0.273	−0.062	1.020	–
	Fledgling	0.090	0.132	−0.155	0.368	–
	Heterozygosity	0.167	2.232	−3.929	4.899	–
`mice`	Intercept	−3.389	2.406	−8.343	1.565	0.335
	Age	0.750	0.214	0.315	1.184	0.258
	Fledgling	−0.040	0.094	−0.227	0.148	0.099
	Heterozygosity	1.605	2.258	−3.102	6.312	0.392
`mi`	Intercept	−3.624	2.416	–	–	–
	Age	0.782	0.236	–	–	–
	Fledgling	−0.046	0.099	–	–	–
	Heterozygosity	1.844	2.221	–	–	–

In addition, you will get a value for each regression coefficient, labeled as "fmi," which stands for the *fraction (or rate) of missing information*, γ. This index γ varies between 0 and 1, and is a very important feature of MI, because it reflects the influence of missing data on uncertainty estimates for parameters. The fraction of missing information is defined by:

$$\gamma = \frac{v_B + v_B / m + 2/(df + 3)}{v_T}, \tag{4.11}$$

where all components are defined as in equations 4.5, 4.7, and 4.8. As you can see, the fraction of missing information, γ, reflects not only the fraction of missing values, but also the importance of missing values in relation to the complete information (McKnight et al. 2007; Enders 2010). There are two more indices in the missing data literature: ρ (the relative increase in variance due to missing data) and λ (the fraction of missing information assuming m is very large). They can be expressed as:

$$\rho = \frac{v_B + v_B / m}{v_W}, \tag{4.12}$$

$$\lambda = \frac{v_B + v_B / m}{v_T}. \tag{4.13}$$

Also, γ is often written using ρ, as:

$$\gamma = \frac{\rho + 2 / (df + 3)}{1 + \rho}. \tag{4.14}$$

The importance of γ can be more easily appreciated by examining λ (equations 4.7, 4.13) because λ is the ratio of variance due to missing data (between-imputation variance, v_B), in relation to the total variance (v_T).

This index γ has two practically useful properties. First, when missing data are non-ignorable (MNAR), γ will be large (McKnight et al. 2007), although there is no definite test to distinguish between MAR and MNAR (section 4.3.1). Li et al. (1991) proposed that γ up to 0.2 can been seen as "modest," 0.3 as "moderately large" and 0.5 as "high." Although these benchmarks should not be used as absolute (analogous to Cohen's benchmarks, 1988), it is true that when $\gamma > 0.5$, the way missing data are handled will impact the final parameter estimates and statistical inferences (van Buuren 2012).

Second, γ can be used to quantify the efficiency of MI. The relative efficiency (ε) quantifies the errors due to MI, relative to its theoretical minimum (which occurs when $m = \infty$)

$$\varepsilon = \left(1 + \frac{\gamma}{m}\right)^{-1}. \tag{4.15}$$

For example, at $m = 3$ and $\gamma = 0.5$, the efficiency is 85.71 % while at $m = 10$ and $\gamma = 0.5$, the efficiency is 95.24%. So even in the latter case there is still much room for improvement in efficiency. Although Rubin (1987) suggested that m between 3 and 10 would be sufficient. Given that m can be easily increased with the use of R, we should aim for over 99% ($m = 50$ with $\gamma = 0.5$ produces $\varepsilon = 0.9901$). However, for practicality, we can use $m = 5$ during the analysis step, and only use high m for the "final" three steps of MI (van Buuren 2012). Other recommended rules of thumb or guidelines on the number of m can be found elsewhere (e.g., Graham et al. 2007; von Hippel 2009).

It is important to check the results from the MI models of your choice. One way of doing this (sensitivity analysis) is to run MI using a different library. The three-step MI process can be done using the `mi` library (Su et al. 2011), which uses a different version of MCMC procedure from the `mice` library.

```
> library (mi)
# get information on each variable
> info <- mi.info (PdoData)
# EPP and Fledgling are count data
> info <- update (info, "type", list (EPP = "count", Fledgling
    = "count"))
# the imputation step with 5 copies
> imputation <- mi (PdoData, info = info, n.imp = 5, seed = 777)
# the analysis step (with GLM) and the pooling step
> AandP <- glm.mi (EPP ~ Age + Fledgling + Heterozygosity,
    family = quasipoisson, mi.object = imputation)
> display (AandP)
```

The results are very similar for analyses using `mice` and `mi` (table 4.4).

The R code for both libraries gives the impression that MI procedures may be very simple and straightforward. In one sense, this is true, but there are many practical pitfalls, which need consideration before and during MI (e.g., convergence of the imputation steps and which variables should be included for MI). I will cover such practical considerations in section 4.5.1.

4.4.6 *Multiple imputation with multilevel data*

Multilevel structures in ecological and evolutionary data are common because biological processes by nature occur in hierarchies; therefore an ability to handle missing data for multilevel data sets will prove extremely useful. So-called multilevel or hierarchical data are modeled by linear and generalized linear mixed-effects models (LMM and GLMM respectively; chapter 13; Bolker et al. 2009; O'Hara 2009). However, proper missing data procedures for multilevel data are still in their infancy (van Buuren 2011, 2012). Available

R functions are currently very limited in both number and capacity. I will introduce some extensions of the above MI methods but great care needs to be taken when applying them.

Data are frequently arranged in clusters or groups (e.g., sibships, stands of trees, and the like), each of which has its own mean (and therefore intercept and sometimes slope). Handling of missing data in such cases is not straightforward because the imputation needs to account for this clustering (Graham 2009, 2012; van Buuren 2011, 2012). In other words, you have multiple levels of vectors of means and variance–covariance matrices (**m** and **V**; section 4.4.4).

Longitudinal data are a case in point; imagine growth data of house sparrow chicks. Half the broods are fed extra food every second day (this was our treatment and what we were interested in); tarsus measurements (a good size indicator) of chicks were taken at 6 different time points (2, 4,. . .12 days after hatching). Here, each chick is a cluster and also, each brood acts as a higher-level cluster (usually 3–5 chicks). Typical to such data, some tarsus measurements are missing because some chicks died/disappeared due to adverse weather, predation etc. This data set of the seven variables (ChickID, Treatment, Age, Tarsus JulianDate, BroodID, and Year) includes 273 chicks from 76 broods, with 403 measurements missing out of 1638 (see Cleasby et al. 2011 and appendix 4C). Let us see how a normal MI procedure performs using the `mi` library. The coding will be exactly the same as the previous example, but I will introduce a LMM in the analysis step using the function `lmer.mi`.

```
# get information on each variable
> info <- mi.info (PdoGrowthData)
> imputation <- mi (PdoGrowthData, info = info, n.imp = 5, seed = 777)
# the imputation step with 5 copies (the default)
> AandP <- lmer.mi(Tarsus ~ Treatment + I (Age - 12) + (I (Age - 12)|
    ChickID) + (1 | BroodID), mi.object = imputation)
# the analysis step (with LMM; see Chapter 13) and the pooling step;
# note that I(Age-12) makes treatment effect be assessed at 12 days
# after hatching
> display (AandP)
```

This process gives us some (sensible) results (table 4.5; detailed results are in appendix 4C) and similar approaches have been often used. However, the validity of performance without explicitly specifying clustering and its consequences are not well studied (van Buuren 2011). In the `mice` library, we can actually specify grouping by incorporating the `pan` library, which uses a special MI procedure designed for two-level clustered data (Schafer 2001; Schafer and Yucel 2002). A current limitation is that only one grouping variable is allowed.

```
> preparation <- mice (PdoGrowthData1, maxit = 0)
# running an empty imputation for the two objects below as preparation.
#   Also variables in PdoGrowthData were turned numerical
# the predictor matrix
> predictor <- preparation $ predictorMatrix
# the vector of imputation methods
> imputation <- preparation $ method
# specify ChickID as a grouping factor
> predictor ["Tarsus", "ChickID"] <- -2
> imputation ["Tarsus"] <- "2l.pan"
```

```
# using the 2-level mixed modeling method from the pan library
> imputation <- mice (PdoGrowthData1, m = 5, seed = 7777)
# the imputation step with 5 copies
> analysis <- with (imputation, lmer (Tarsus ~ Treatment + I
(Age - 12) + (I (Age - 12) | ChickID) + (1 | BroodID))
# the analysis step with a LMM (see Chapter 13)
> pooling <- pool (analysis) # the pooling step
> summary (pooling)
```

The preparation is a little involved, but the three-step MI process is the same as above. The results from `mice` specifying grouping in this data set resemble those from `mi` (table 4.5). This is encouraging, but recall that we were unable to include brood identities (i.e., correlated structure) as a grouping factor, so one should draw conclusions cautiously.

There is another important problem in multilevel data: there are multiple levels of predictors, so missing data processes can operate at different levels. Consider two-level data; if the response is weight at time t_i, predictors can be height at time t_i (level 1) and sex (level 2). If weights are taken at 6 different occasions (t_1–t_6), missing data on sex for one individual can appear as missing values in 6 cells. If we subject this data set to normal MI procedures, these 6 cells may be assigned different sexes! Where multiple types of predictors are present, Gelman and Hill (2007) suggest data imputation should be carried out separately for each level (e.g., time and sex). The `mice` library has this capability, but it is currently limited to only two levels (i.e., only one clustering variable is allowed).

Table 4.5 Results of analyses for house sparrow data, treated as a multilevel data set. Estimates are from four procedures: complete case analysis, MI using both the mice and mi libraries, and DA using MCMCglmm. Estimates from mice and mi are the pooled model-averaged parameter estimates from the five imputed data sets ($m = 5$), pooled regression coefficients ($\bar{b}$), overall standard error, s.e. ($\sqrt{v_T}$), 95% confidence intervals (CI), and the fraction of missing information (γ). For MCMCglmm, the estimates are posterior means, s.e. are standard deviation of the posterior distributions of the estimates, and CI represents credible intervals. Only the results from the fixed factors are presented. For details, see appendix 4C

Procedure	Predictor	Estimate	s.e.	Lower CI	Upper CI	γ
Complete case analysis	Intercept	18.110	0.152	17.807	18.413	–
	Treatment	0.167	0.160	−0.15	0.486	–
	Age	1.143	0.011	1.122	1.165	–
mice	Intercept	17.880	0.362	17.168	18.592	0.020
	Treatment	0.316	0.229	−0.133	0.766	0.028
	Age	1.157	0.009	1.139	1.174	0.097
mi	Intercept	18.147	0.208	–	–	–
	Treatment	0.259	0.241	–	–	–
	Age	1.147	0.010	–	–	–
MCMCglmm	Intercept	18.015	0.402	17.171	18.720	–
	Treatment	0.355	0.247	−0.106	0.837	–
	Age	1.169	0.011	1.149	1.192	–

Other issues associated with imputation in multilevel data are described in van Bu-uren (2011; see also Raudenbush and Bryk 2002; Daniels and Hogan 2008; Enders 2010; Graham 2012).

4.4.7 *Data augmentation*

The processes and results of data augmentation (DA; Graham 2009, 2012; Enders 2010) are similar to those of MI. The main difference is that in MI, the user will see the replaced missing values, while DA internalizes the three-step procedures, including Rubin's rules with $m = \infty$, and feedback between the imputation and analysis steps (figure 4.5D; *sensu* McKnight et al. 2007). DA is superior to MI because a DA procedure is akin to the number of data imputations (or augmentations) being infinite, and also because there is a feedback process between missing data and parameter estimation (Nakagawa and Freckleton 2008). However, MI has an advantage: DA can only use variables that are in the model, while MI can include auxiliary variables, which may often be required to convert MNAR missingness into MAR (section 4.3.2). Therefore, in most cases, MI procedures are recommended over DA (Graham 2009, 2012).

In the case of multilevel data, DA procedures may sometimes be preferable. If the response variable is the only variable with missing data, as is the case with the sparrow growth data used in section 4.4.3, DA can treat such missing values appropriately by taking all the clustering groups (e.g., individuals, broods, and families) into account. In Bayesian statistical packages, such features are usually included as the default. Here, I use the MCMCglmm library (Hadfield 2010).

```
> library (MCMCglmm)
# run a Bayesian LMM (see Chapter 13)
> model <- MCMCglmm (Tarsus ~ Treatment + I (Age - 12), random = ~ us
    (I (Age - 12)) : ChickID + BroodID, data = PdoGrowthData,
    verbose = FALSE)
> summary (model)
```

In this case, the results are very similar to those from mi and mice (table 4.5). Note that the MCMCglmm function will not tolerate missing values in predictors. However, if multiple variables with missing data are all entered as responses (i.e., multi-response models; Hadfield 2010), DA will handle missing values for all these response variables. As an example in which we used this strategy in a bi-response/bivariate meta-analysis, see Cleasby and Nakagawa (2012). It is worth mentioning that multi-response (or multivariate) models are closely related to structural equation modeling (SEM), which is sometimes referred to as latent variable modeling, path analysis, or causal modeling (chapter 8). Missing data in such models are briefly discussed later (section 4.5.3).

4.4.8 *Non-ignorable missing data and sensitivity analysis*

As mentioned above, there are no tests to detect MNAR (non-ignorable) missingness, so we need to rely on our understanding and knowledge of the biological systems at hand. We can, however, suspect that MNAR missingness is possible, especially when the fraction of missing information (γ) is high ($\gamma > 0.5$). Two main methods exist for non-ignorable (MNAR) missingness: *selection models* and *pattern-mixture models*. The details of the MNAR methods are beyond the scope of this chapter, so I refer readers to accessible accounts elsewhere (Allison 2002; Molenberghs and Kenward 2007; Enders 2010). However, I will mention some main aspects of these models. Both models require constructing specific assumptions with regard to MNAR missingness. If these assumptions are incorrect, these non-ignorable models may perform worse than the models for ignorable missingness (i.e., MI and DA). To put it simply, a good MAR model may be better than a bad MNAR model (Schafer 2003; Demirtas and Schafer 2003).

The main problem of non-ignorable missing data is that there are an infinite number of ways in which such missingness can occur. Naturally, very few generally applicable software implementations are able to cope with infinitely different manifestations of non-ignorable missingness (Allison 2002). However, there is an *ad hoc* sensitivity analysis to explore the possible impacts of non-ignorable missingness on the pooled estimates from MI under MAR (Rubin 1987). For example, you might suspect the age variable in the sparrow data to be MNAR rather than MAR (section 4.3.1). It is possible younger birds (or older birds) are selectively missing. Such MNAR missingness can be explored by first adding (or subtracting) imputed values under MAR. We can then compare pooled estimates from this sensitivity analysis (a MNAR model) to the original estimates under MAR. Rubin (1987) suggested a 20% decrease or increase in imputed values would be a sufficient sensitivity test, but this is an arbitrary suggestion. Enders (2010) suggests ±0.5 standard deviation of the variable should be added. This sensitivity method can be easily implemented using the `mice` library. You will find an example analysis in appendix 4C.

4.5 Discussion

4.5.1 *Practical issues*

There are several practical considerations to consider prior to using MI or DA procedures, and I discuss five of them here. First, is there a minimum requirement for sample size? This question is hard to answer. Of course, larger samples are desirable, because missing values in a small data set further decrease the amount of information, which is already limited (Graham 2009). However, Graham and Schafer (1999) conducted a simulation study where they showed that a MI procedure, which assumes multivariate normality, performed very well with up to 18 predictors and 50% missing data; this means that the data set only had around 15 degrees of freedom. They also demonstrated that a joint modeling approach with the multivariate normal assumption did well with non-normal data (a version of the `norm` library was used in this study; Schafer 1997) although such an approach would be limited compared to sequential regression imputation (used in the `mi` and `mice` libraries).

This leads to my second point. For MI procedures assuming a multivariate normal distribution such as in the `norm` and `Amelia` libraries, non-normal data should be transformed first. Indeed, the `Amelia` library comes with various transformation options (Honaker et al. 2011). Back-transformation can be used to recover the original scale. A related issue is whether imputed data should be rounded when the original data are integers. Generally it is not a good idea to do so, unless an imputed variable is a response variable to which a Poisson regression (chapter 6) will be applied (Graham 2009, 2012; Enders 2010; van Buuren 2012). Furthermore, if you are using MI procedures with the multivariate normal assumption, categorical variables should probably be turned into binary variables using *dummy coding*. For example, if you have a categorical variable with four levels, this variable can be recoded into three binary (dummy) variables. More generally, p levels in a categorical variable can be turned into $(p-1)$ dummy variables. Note that coding dummy variables from a categorical variable can be easily done in *R* using `dummy.code` in the `psych` library (Revelle 2012; see an example in the online appendix 4C). If you are using sequential regression imputation such as in the `mi` and `mice` libraries, you need to make sure missing values in categorical data are imputed with techniques for categorical data (e.g., logistic and multinomial regression).

Third, for MI, it is important to check for convergence in the imputation step. Convergence here means that an imputation step reaches a set of stable values for a vector of means (**m**) and a variance–covariance matrix (**V**) (section 4.4.4). There are graphical functions to assess convergence in the two *R* libraries mentioned (`Amelia` and `mi`; see appendix 4C). If you have trouble with convergence in MI, transformation of skewed data may help, as skewed data could be slowing down imputation processes (Graham 2009, 2012).

Fourth, when your statistical models include interaction terms, such terms should be included in the imputation step in MI procedures (von Hippel 2009; Graham 2009, 2012; Enders 2010; van Buuren 2012). Interaction terms usually come in two forms: the product of two continuous variables, or the product of one continuous variable and one categorical (dummy) variable (e.g., males and females). When creating interaction terms, a continuous variable needs to be centered (i.e., subtracting the mean from each value). In fact, centering or scaling (i.e., *z*-transformation) of all continuous variables is very frequently a good idea in regression modeling because this process can make linear models more interpretable (e.g., the intercept will be located at the means of predictors; Schielzeth 2010). Inclusion of interactions in the imputation step is necessary, because if you do not consider a particular interaction in the imputation step, the effect of this interaction can be lost even when missing data are MCAR. This is because data imputation is carried out assuming such an interaction does not exist (Enders 2010; Enders and Gottschall 2011). The same applies to a quadratic term, as it can be seen as an interaction with itself. These derived terms (i.e., terms created by existing variables) should be handled by *passive imputation* rather than included as extra variables in the data matrix. Passive imputation maintains relationships between original and derived variables during the imputation process (von Hippel 2009; van Buuren and Groothuis–Oudshoorn 2011). Examples for these processes are found in appendix 4C and in van Buuren and Groothuis–Oudshoorn (2011).

Fifth, our "expert" knowledge is useful during MI. The ranges, or possible maxima and minima, for variables with missing data can be included as *ridge priors* in a MI procedure, such as that in the `Amelia` library (Honaker et al. 2011; see Nakagawa and Freckleton 2011). This process potentially reduces bias, especially when the fraction of missing information (γ) is at least moderately large. Unfortunately, the ridge prior functionality is not implemented in the `mice` and `mi` libraries (but see the argument `squeeze` in `mice`). I recommend more than one library be used to run MI for a data set as a form of sensitivity analysis (section 4.4.5). If the results from different libraries disagree, one likely explanation is that the imputation step did not converge.

4.5.2 *Reporting guidelines*

For publication, it is advisable to provide details and rationale of your missing data procedures, because such procedures will probably look foreign and even outlandish to potential editors and reviewers. Here, I will present the reporting guidelines for missing data analysis from van Buuren (2012). His list consists of 12 items that should be included, when reporting results obtained from MI procedures.

(1) *Amount of missing data*: Give the ranges of % missing values in all variables and the average % in your data set.

(2) *Reasons for missingness*: Give reasons why such missing values were present.

(3) *Consequences*: Report known differences between subjects with and without missing values.

(4) *Method*: Describe which method was used, and under what assumptions (e.g., a MCMC procedure for MI under MAR).

(5) *Software*: Name the software libraries (e.g., `Amelia`) along with descriptions of the important settings.

(6) *Number of imputed data sets*: This is m in the imputation step (see section 4.4.5).

(7) *Imputation model*: Report the variables included in the imputation step (i.e., the imputation model) and whether any transformations were applied.

(8) *Derived variables*: Mention what kind of derived variables (e.g., interaction and quadratic terms) were included in the imputation step.

(9) *Diagnostics*: Report on diagnostics for convergence of the methods used, methods (section 4.5.1), and for checking whether imputed data are plausible.

(10) *Pooling*: Explain how pooling of results was done (usually pooling of m estimates by Rubin's rules; section 4.4.5), if possible along with related indices including, most importantly, the fraction of missing information (γ) and the relative efficiency (ε).

(11) *Complete case analysis*: Report results from complete case analysis, and compare with those from proper missing data procedures (i.e., MI and DA).

(12) *Sensitivity analysis*: Conduct sensitivity analyses, and report the results. Sensitivity analysis can be in the forms of Rubin's ad hoc adjustment or the use of different software packages.

van Buuren (2012) considers items 1, 2, 3, 4, 6, and 11 are essential, but that the others can be reported in an appendix or online materials. Although his list was tailored for MI, I believe that following his guidelines will be useful even when using DA and other missing data procedures. Such details will be certainly helpful for editors and reviewers who are unfamiliar with missing data methods.

4.5.3 *Missing data in other contexts*

Here I provide you with connections between this chapter and other chapters in this book. As mentioned in section 4.1, missing data procedures may be essential for model selection (chapter 3; Nakagawa and Freckleton 2011), although there is surprisingly little research on this relationship (but see Claeskens and Hjort 2008). Some procedures for censored or truncated data (chapter 5) involve an imputation step. In addition to the imputation used in the NADA library discussed in that chapter, the kmi library (Allignol 2012) uses a Kaplan–Meier estimator to impute missing censoring times.

Different forms of linear models, such as GLMs (chapter 6), models with overdispersion (chapter 12), and mixed models (chapter 13), can be integrated within MI procedures at the analysis step. However, special care is required for multilevel data (Raudenbush and Bryk 2002; van Buuren 2011). All the regression models can be seen as special cases of structural equation modeling, SEM (causal modeling or mediation analysis; chapter 8). SEM has a long history of missing data methods (reviewed in Allison 2002; Enders 2010), and the majority of stand-alone SEM software libraries (e.g., Mplus and AMOS) come with missing data procedures (MI or DA). There are a number of other *R* libraries available for missing data in SEM, including bmem (Yuan and Zhang 2012) and rsem (Zhang and

Wang 2012). Meta-analysis (chapter 9) is a type of weighted regression model. Therefore, missing data procedures described in this chapter are applicable for, at least, predictors (called moderators in the meta-analysis literature; Pigott 2009, 2012). However, treating potential missing data in the response variable (i.e., effect size statistics) has attracted much research, and has its own unique techniques, some of which are akin to selection models for MNAR missingness (Sutton 2009).

Hadfield (2008) utilizes missing data theory in evolutionary quantitative genetic contexts. He showed that MNAR missingness could be converted to MAR missingness using pedigree information, which can be included as a correlation matrix in mixed models. Genetic relatedness can act as a kind of auxiliary variable; siblings must share similar morphological characters. In a similar manner, spatial correlation (chapter 10) and phylogenetic correlation (chapter 11) can inform missing values in associated models because these different types of correlations are, in fact, the same (or very similar) mathematically in terms of specifying relationships among data points in the response variable (Ives and Zhu 2006; Hadfield and Nakagawa 2010). Interestingly, phylogenetic comparative analysis by Fisher et al. (2003) was the very first case of using MI in evolutionary biology, but few followed their initiative. The shortcomings of ignoring missing data are now, however, starting to be recognized in comparative analysis (Garamszegi and Møller 2011; González-Suárez et al. 2012), with some implementations of missing data procedures appearing (e.g., `PhyloPars`; Bruggeman et al. 2009). We can expect a rapid future integration and development of missing data procedures in this and related areas of research.

4.5.4 *Final messages*

Missing data are pervasive, and pose problems for many statistical procedures. I hope I have convinced you that we all should be using methods that treat missing data properly (i.e., MI or DA), rather than deleting data or using single imputation. Importantly, it is not difficult to implement these missing data procedures (in particular, MI) with the aid of *R*. I also hope that you will now think about the missingness mechanisms when planning studies (i.e., collecting auxiliary variables). Especially, I think that ecologists and evolutionary biologists can probably benefit a lot from learning the planned missing design (Baraldi and Enders 2010; Graham et al. 2006; Graham 2009, 2012; Rhemtulla and Little 2012), although such a concept is nearly unheard of in our field.

I also presented you with some current difficulties associated with missing data. There are no easy solutions for missing values in multilevel data, especially when missing values occur in multiple levels and when clustering occurs at more than two levels. Nor is the implementation of MNAR models straightforward. But missing data theory is an active area of research, so who knows what the future will bring to us and to *R*? Enders (2010) comments that "Until more robust MNAR analysis models become available (and that may never happen), increasing the sophistication level of MAR analysis may be the best that we can do."

Acknowledgments

I thank Losia Lagisz for help with figure preparation. I also thank Shane Richards, Gordon Fox, Simoneta Negrete, Vini Sosa, and Alistair Senior for their very useful and constructive comments on earlier versions of this chapter. I gratefully acknowledge support from the Rutherford Discovery Fellowship (New Zealand).

CHAPTER 5

What you don't know can hurt you: censored and truncated data in ecological research

Gordon A. Fox

Censored and truncated data are quite common in ecological settings, including remote sensing, survival and other life history studies, mark–recapture, telemetry, fire history, dendrochronology, and animal behavior. Environmental measurements (like soil or water contents) are frequently censored or truncated as well. Unfortunately, data of this kind are poorly understood; many ecologists handle them in ways that lead to severely problematic parameter estimates and significance tests. Indeed, a major problem in community ecology—estimating community composition when the uncommon species are likely to be missed—has been discussed since Preston's (1948) paper on the topic, but ecological data have generally been analyzed without use of methods that are designed for this sort of problem. Fortunately, there are a number of methods for handling censored and most truncated data problems. Because I aim to convince ecologists of the importance of these problems, I rely here mainly on simulated data: we know the correct answers, and can compare how well different statistical approaches approximate those answers.

Censorship and truncation sound similar; they are related to but distinct from one another, and I will discuss them separately in this chapter. The distinction is this: data points are *censored* if we know only that the observation is larger than, or smaller than, a given value, or is between two given values. Mathematically, this is expressed as an inequality such as $y_i > c$. If we also lack information on some predictors x_i for cases where $y_i \leq c$, then the data set is *truncated*. In this chapter, I use y to describe observed values; y^* refers to the true (*latent*) variable, which is observed only for uncensored cases.

5.1 Censored data

5.1.1 *Basic concepts*

A microbial ecologist wants to estimate the density of a bacterial species in a river, and to compare the bacterial densities before and after the planned release of impounded water from an upstream dam. She takes three daily samples for two weeks prior to, and two weeks following, the release. Her goal is to account for both random variation between

Ecological Statistics: Contemporary Theory and Application. First Edition. Edited by Gordon A. Fox, Simoneta Negrete-Yankelevich, and Vinicio J. Sosa. © Oxford University Press 2015. Published in 2015 by Oxford University Press.

samples, and random fluctuations over time, in making the before–after comparison. Each sample involves collecting the same amount of water, taking it back to the lab, and plating several samples on an agar medium. The number of colonies formed on the plate is counted two days later.

On some plates, she finds no colonies. It might be tempting to treat these as samples that had zero bacterial density, but doing so can be a big mistake: the assay used has an intrinsic lower limit (often called the *limit of detection,* sometimes written as LOD) c, such that it can estimate densities of c or more, but densities less than c are measured as 0. Data points of this sort are called censored: we have partial information about them (in this case, y is between 0 and c, that is, $0 \leq y_i < c$), but not a precise measurement.

Data censorship is a very common (and under-recognized) problem in ecological data. Censorship is a characteristic of individual data points, and it is defined by generalizing this example: a censored value is one for which we know an inequality, but not the actual value. Specifically, the true value is less than or greater than some cutoff values, or is between two values. These cutoff values—*censorship levels*—may vary among data points within a study.

Censorship arises by a number of different mechanisms. The most obvious case occurs when there are some limits to detection for certain observations, as in the microbial density example. Often our instruments (including our eyes) make it impossible to measure things that are too small (or large), or too near (or far). A second source of censorship arises from research design. For example, a survival study may end (perhaps because of funding limitations) prior to the death of all individuals. Not surprisingly, censorship can easily occur in certain spatial studies as well. Consider a study aimed at estimating the distribution of primary dispersal distances for wind-dispersed seeds, using direct observation of seeds released from a point. The researcher can only distinguish travel distances up to some limiting distance c that his eyes allow. For any seed traveling farther, he can only know that its dispersal distance is greater than c. Over the course of the study, c may vary because light conditions vary. Any study involving longitudinal measurements—whether individuals are followed through time or over space—is especially likely to involve censorship. A third source of censorship is random; more than once I have accidentally disturbed a plant in a study before the event of interest (say, flowering) occurred.

5.1.2 *Some common methods you should not use*

Returning to the bacterial density example, some commonly used approaches (which are generally bad ideas) include (a) treating the data values measured as 0 at face value—that is, as 0; (b) treating them as values measured at c; and (c) estimating their value as $c/2$. These different approaches can result in very different estimates; worse yet, they all lead to *inconsistent* estimates (Breen 1996; Long 1997; Greene 2005). As discussed at length by Kendall (chapter 7) bias is a problem of small samples, while inconsistent estimators fail to converge to the true value even with infinitely large samples. As an illustration, consider the example in figure 5.1 and appendix 5A. Here I have simulated data from a bivariate Normal distribution, for which the true relationship is $\log(y^*) = x/2$. With a sufficiently large sample, we would hope to estimate the intercept as close to 0 and the slope as about 1/2. For the 500 random points used in figure 5.1, linear regression for the complete data set comes quite close, estimating the intercept as 0.002 and the slope as 0.514. Next, points with $y^* \leq 0.5$ are censored; these amount to 16% of the data set. Treating the censored data points as having the value $c/2$ does a fairly good job in this case, estimating the intercept as 0.038 and the slope as 0.489. The other methods perform much more

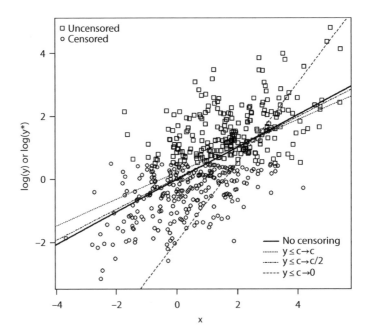

Fig. 5.1 Using 500 simulated points with $\ln(y^*) = x/2$, linear regression estimates the intercept and slope of the uncensored data at 0.002 and 0.514, respectively. Censoring all data with $y^* < c = 0.5$, assigning the values c, $c/2$, and 0 to censored points leads linear regression models to a variety of estimates.

poorly. As we will see in section 5.2, still another commonly used approach—deleting the censored data—can lead to even worse estimates.

Unfortunately, all of these ad hoc substitutions have serious problems. Figure 5.2 shows the data recoded for each of these methods; in each case there is pattern that was not present in the original data. Regression diagnostics suggest that using these ad hoc methods of substituting values for censored data can lead to more problems than just inaccuracy. Plots of the residuals against the fitted values (figure 5.3) reveal a substantial amount of pattern for the cases where censored values were set to c, $c/2$, and 0. This suggests that these methods may (at least in this case) violate the assumption of homoscedasticity: one should regard error estimates (and consequently, P-values) with some skepticism, and any predictions made from models using these approaches with even more skepticism. How well do these approaches conform to the assumption that residuals are normally distributed (chapter 6)? Quantile–quantile (Q–Q) plots (not shown, but the relevant R code is provided in appendix 5A) of the residuals suggest that this assumption is very strongly violated when censored values were set to 0, and violated (but much less severely) when censored values were set to c; the $c/2$ estimate meets this assumption reasonably well. In summary, data substitution imposes a pattern on the error distribution that was not present in the uncensored data; it is a bad idea (Helsel 2012).

Unfortunately, these are not general results: we cannot assume that assigning the value $c/2$ will work well in general—in fact, it is known not to do so (Helsel 2012). It certainly cannot work when the censorship is for a maximum value (that is, when we know only that $y_i > c_i$). Moreover, it may be impossible to use methods like this simple ad hoc substitution when the data have a more complicated structure. Special methods—based on

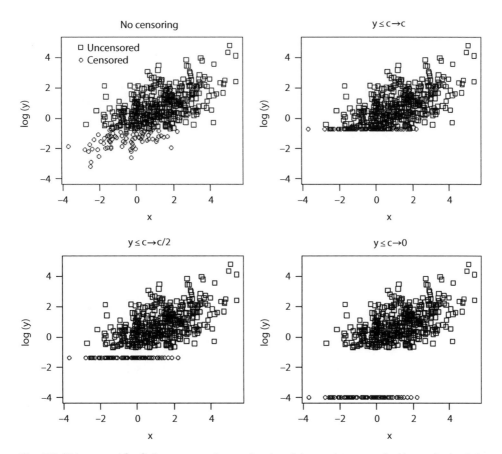

Fig. 5.2 Values used for fitting a regression to simulated data, using several ad hoc substitution methods of handling censorship. The lower-right panel actually involves substitution of a very small value (here, –4) for log(y), because log(0) = –∞. Each of the ad hoc substitutions introduces pattern that was not present in the original (no censoring) data set.

maximum likelihood estimation (chapters 1, 3)—designed for censored data are generally more appropriate. Before we consider those methods, though, it is useful to examine censoring more closely.

5.1.3 *Types of censored data*

Understanding problems with censored data requires a few new ideas. Most of them are fairly simple, but of course there is an attendant terminology. The terminology is not entirely standardized, as in many areas of statistics. The key is to master the ideas.

If a censored data value is one for which we know an inequality, there are three possibilities: we know that it is less than or equal to some value ($y_i \leq c_i$), greater than some value ($y_i > c_i$), or between two values ($b_i < y_i < c_i$). These are called, respectively, *left censorship*, *right censorship*, and *interval censorship*.

- Left-censored data in ecological studies occur most often when there are lower limits to detection (as in the bacterial example); these are often called *non-detects*. Left censorship

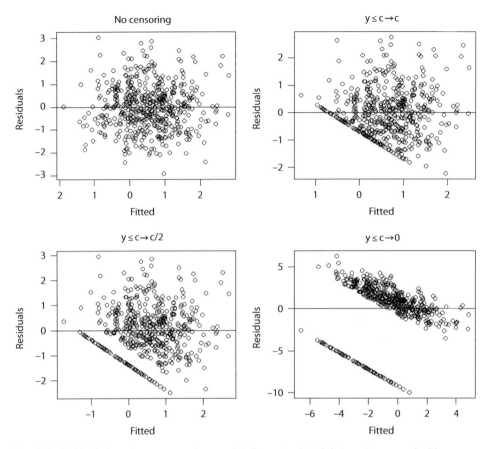

Fig. 5.3 Residuals from linear regression models fit to simulated data, using several ad hoc substitutions for censored values. There is substantial pattern in these residuals.

also occurs in some longitudinal studies. For example, in a study of timing of bird nesting, some nests may already be present when the researchers arrive at the field site.

- Right-censored data arise frequently in ecological studies of event times. For example, some animals may engage in a behavior under study only after the end of data acquisition. The example of a study of wind-dispersed seeds is also a case of right censorship.

- Interval censorship typically occurs in ecology as a consequence of sampling at long intervals. For example, a graduate student may visit her study site every second month to record the responses of plants to her nutrient-addition treatments. Some plants that were leafless when she visited in June are leafy when she records data in August. We do not know the true date of leaf flush, but we can put bounds on it.

Does this mean that all data are interval censored? After all, we never know exact values of data points. If we are working on an event–time problem, shouldn't we consider all cases in which we do not know the instant of the event as interval censored? A moment's thought should point to the answer: this is a problem of scale. It is important to treat a data point as interval censored if the range of possible values is large relative to the scientific question being studied. In other words, this requires scientific judgment. In a survival study of vertebrates (including most medical studies), the hour at which a death occurs is unlikely to be meaningful, but the year (or for some studies, the month) typically is.

These ideas (left, right, and interval censorship) refer to characteristics of individual data points. Any study may include combinations of these types of data: for example, some data values may be interval censored while others are right censored. Naturally, some researchers feel the need to use special jargon here, and the jargon can be confusing. When you read that a sample is *doubly censored* it usually means that both right and left censorship occurs in the data set. Confusingly, *single censorship* usually refers to something quite different: a singly censored data set is one in which all censored values are the same (e.g., because limits to detection in an instrument were always the same), while a *multiply censored* data set usually means one in which different samples have different censorship levels. Multiple censorship can occur in many ways; for example, visibility conditions or instrument calibration vary over time, or individuals are lost to follow-up in a longitudinal study at different times.

5.1.4 *Censoring in study designs*

It seems clear that the way in which censoring occurs in a study can affect the way one plans to collect data, as well as such critical parts of the statistical analysis as the form of the likelihood function (chapter 2). Thus, censorship should be part of any research design. There are three main types of censoring that apply to designs as a whole:

- *Type I (time) censoring* occurs when censorship levels are, in principle, known in advance. Censoring caused by limits to detection (whether right- or left-censored), is a case of Type I censoring. In time-to-event studies, Type I censoring means that one fixes the date on which data collection ends; any individuals surviving are censored at that ending date. Thus, Type I censoring means that the number of censored data points is a random variable, but the censorship levels are fixed.

  Any study in which the measurements encounter some limit—whether imposed by limits to the resolution of a device (as in remote sensing) or by the choice of the scale of measurement—involves Type I censoring. Consider a behavioral study on the efficacy with which animals locate hidden food, depending on their age. Some study individuals find all of the hidden food items. An ordinary least squares (OLS) regression of the amount of food found as a function of age is inherently problematic, because the food-finding variable is censored. To see this, it is useful to ask whether it is really likely that all individuals finding all of the food actually have the same food-finding abilities. This sort of example was considered in ecological models of plant reproductive effort (Schmid et al. 1994; Gelfand et al. 1997).[1]

  There are two ways that some researchers may be tempted to handle censoring in a case like this. One is to ignore the problem and appeal to the general robustness of OLS as a statistical method. Indeed, sometimes OLS will indeed give a reasonable result in a case like this, but that is not the case in general. Other researchers might be tempted

---

[1] An example from the social sciences may be helpful in providing further illustration. The standardized exam used for admission to many US graduate schools (GRE) is often interpreted as predicting success in graduate school (or sometimes as measuring innate ability). However, everyone getting all answers correct gets the same score, and all students getting all answers wrong also get the same score. In both cases it is unlikely that all of these students really have identical abilities or identical chances of success; the data are censored. Indeed, there is an excess of students getting all answers correct compared with any parametric distribution, suggesting that this is a rather heterogeneous group (see http://www.ats.ucla.edu/stat/sas/dae/tobit.htm for data and a worked analysis in SAS).

to handle this situation by changing their analysis to ask about the fraction of the available food found, using a method like GLM with a binomial link (chapter 6). But there are two issues in doing so: first, a model of the fraction of available food found as a function of age does not really address the same biological questions as a model of the amount of food found as a function of age. Second, the logistic regression model would have a meaningful interpretation only for the particular density of food used in the experiment. It is probably best to recognize that data are censored, and use appropriate methods for analysis.

- *Type II (failure) censoring* refers to the converse case, in which the number of censored data points is determined in advance. The classic example for event times is that one might end data collection after n plants in a sample of N die, so that the censoring levels are random variables but the number of censored data points is fixed. This kind of design may seem a bit unusual to ecologists, but an example familiar to all of us is that of progressive censoring. This refers to the case of planned multiple Type II censoring: one plans that at particular times, fixed numbers of individuals will be sacrificed so that, e.g., one can invasively study physiological state, disease status, or the like.

- *Random censoring* refers to the situation in which individual points are not censored because of their value (Type I) or because there are already some number of recorded data points (Type II), but for some other reason not a deliberate part of the study design. This occurs in many ecological studies, sometimes in addition to either Type I or II censoring. There are at least two sorts of scenarios typical in ecological studies. In one, individuals are removed from the study due to scientist-caused error—for example, a researcher unintentionally disturbs a study plot or harms a study organism. The other scenario involves individuals being removed from a study because of some competing event. For example, a researcher is studying time-to-hatching of bird eggs, but some of the nests are raided by snakes. If the focus is on the ecophysiology of hatching (not on demography), the researcher should consider this a randomly censored value.

All of these cases are partly analogous to data missing at random (chapter 4): censoring at c_i provides no prediction of subsequent values. For example, individuals removed from a time-to-mating study are not any more or less likely to eventually mate. But there is an important difference between censored and missing data: when data are censored, we have information about bounds on the data value, but we have no information about missing values. Some approaches to modeling censored data do involve an imputation step to improve parameter estimates; see section 5.1.6. Regression methods based on maximum likelihood approaches (see section 5.1.7), on the other hand, do not involve imputation but rather use models of the probability of censorship, based on the assumed distribution of the data.

The case corresponding to data missing not at random is called *informative censoring*: there is a correlation between censoring and the true value of the data point. For example, in a survival study one might stop observing an individual because it has fallen ill; the censoring is not independent of the individual survival time. To see why this is problematic, consider a wildlife biologist who decides to stop visiting some bird nests to assess nestling survival because she thinks that her visits themselves are increasing the risk of mortality (e.g., due to nestling stress, or perhaps predator cueing). Aside from the obvious point that this suggests a poor study design, it should be clear that the censoring is not independent of the outcome. There are methods to analyze data in which informative censoring has occurred, but these are highly specialized and outside the scope of this chapter (Kodre and Perme 2013).

5.1.5 *Format of data*

All R libraries that analyze censored data (and other programs that do appropriate statistical analyses, such as SAS) have similar requirements for data formats. Data including censored values are described as a pair (x, c) where x gives the data value and c is an indicator as to whether this is a censored value. For interval-censored data, we need a triplet (x, c_1, c_2) giving the bounds of the interval. These programs typically expect data in which each line contains a single observation, any covariates that will be used in statistical models, and the censorship indicator. For example, data including right censorship for a survival (or other time-to-event) study might be of the form

ID	Treatment	OtherPredictor	EventTime	Event
1	A	17.3	32	1
2	A	5.6	2	0
3	B	12.4	17	1
4	B	3.2	34	0

The default coding for the `survival` library in R is that `Event` is 0/1 for alive/dead (or more generally, for censored at `EventTime` versus true event at `EventTime`). Here, the values with ID 2 and 4 are censored at times 2 and 34, respectively. You can get further information on how to code for cases of interval censorship by `?Surv` or `help(Surv)`. The same type of coding applies to data other than event–time; for example if we had left-censored measurements of counts of bacterial colonies, the data might look like this:

ID	SampleSource	OtherPredictor	Colonies	Censor
1	A	1.4	8	1
2	A	6.2	4	0
3	B	5.3	12	1

Here the point with ID = 2 is left-censored. The `Surv` function in the `survival` library includes a specification for type. If all censored data in a data set are of the same type (say, left-censored) then one can simply specify that in the `Surv` function. Otherwise, one needs an additional column in the data frame to specify whether a point is left- or right-censored. The coding (for example, 0/1 for censored/uncensored) is not universal across all programs that handle censored data; what is universal, though, is the basic assumption that each line contains one observation. Some programs also allow one to specify that there are some number n identical observations, but do not assume that you can enter your data that way unless you have consulted the documentation!

It may seem silly to add the following caution, but I have seen many scientists make this mistake: to include a data point in a study, the data point must actually exist! Many ecologists working with survival data include individuals that have a recorded death at the same time as their birth. Typically the scientist has found a dead newborn individual. Unless this is treated as left-censored, it will generate errors, since one is asking a survival analysis program to consider individuals with a lifespan of 0; it is as though they were never born!

5.1.6 *Estimating means with censored data*

There are several methods that are used widely (especially for environmental data) to estimate the mean (and other aspects of the distribution, including variance and quantiles) of a set of data including left censored values. Here I discuss the most widely used (and well-justified) methods; for a much more extensive discussion, see Helsel (2012).

Two important methods rely on *order statistics*, which are probably most familiar to ecologists from Q–Q plots: a sample is described in terms of the order of values, from smallest to largest. As an aside, rank statistics—the center of what are traditionally called non-parametric statistics—are a transform of order statistics: if one observed the data (2, 1.3, 0.2, 5), the data set would have the order (0.2, 1.3, 2, 5); to use rank statistics, one would replace the original values with the ranks (3, 2, 1, 4).

In any case, a standard statistical result is that, if a probability plot of the data (quantiles for the Normal distribution on the horizontal axis, values on the vertical) is approximately linear, not only is that evidence that the Normal distribution is a good fit; the regression intercept estimates the mean for the distribution, while its slope estimates the standard deviation. As an aside, this result is actually more general than just stated: for any location–scale distribution (see appendix 5A), if the probability plot (using quantiles for that distribution) is linear, the slope of the regression estimates the *scale parameter* while the intercept estimates the *location parameter*. Clearly we can use this method with the uncensored data to estimate the mean and standard deviation for the data. This method is called *regression on order statistics (ROS)*. Obviously a key assumption for use with censored data is that the censored and non-censored values are all sampled from the same distribution.

Somehow this method strikes many people as just being wrong, or possibly even immoral: we are used to calculating means and standard deviations directly from the data, using standard formulae. But the idea is based solidly in statistical theory. Nor should the assumption of a distribution be a problem, for two reasons. First, the method itself provides a check on the assumption: is the plot really close to a straight line? The second reason is that there is a technique that makes ROS much more robust to some departures from the assumed distribution, developed by Helsel (2005). The idea is simple (and closely related to some of the methods in chapter 3): instead of using the slope and intercept of the probability plot to estimate the location and scale parameters, use this regression to impute the values of the censored data, and then calculate the location and scale parameters from that "complete" data set. This is called *robust ROS*, and it is implemented automatically and easily in the NADA library in R.

A distribution-free method uses the Kaplan–Meier (KM) estimator that is the basis of much survival analysis (Fox 2001; Kalbfleisch and Prentice 2002; Klein and Moeschberger 2003). In its more familiar (survival) form, KM is a simple non-parametric way of estimating a cumulative survival curve, and is often called a "life table" by ecologists. The idea is to estimate the fraction of a cohort that survives at time t from a record of times of death; then one just calculates, at each time of death, the chance of not dying at that time (the fraction of those who are "at risk" that do not die at this time). The survival curve itself is just the cumulative product of all of those conditional survival probabilities. It may seem quite strange to use KM for estimating a mean of censored data that does not involve survival, but a moment's thought should tell you that the KM estimator is just an estimate of a cumulative distribution of data. In fact, KM is widely used as a non-parametric estimator of data in such fields as astronomy.

There are only two issues with using it for censored ecological data. First, cumulative survival (in the absence of censorship) starts at 1 and decreases to 0, while a cumulative distribution of other types of data will (without censorship) start at 0 and increase to 1. But KM could just as easily describe the cumulative probability of deaths as of survival, using the same data: one would just use the inverse of survival. The second issue is that KM can only handle right censorship. But inverting the problem takes care of this issue for left-censored data points; they become right-censored. The NADA library's kmfit function handles this all automatically, by "flipping" the left-censored data and then calling the `survival` library's `survfit` function to do standard KM estimation (for an example see appendix 5A). One can use the KM estimator to find quantiles of distributions. The area under the KM curve estimates the mean (though the problem is a bit more subtle with censorship), and there is a large body of survival-oriented literature on methods for estimating the variance of the KM estimator (Klein and Moeschberger 2003).

Finally, we can use censored regression to fit models for censored data. This is a very powerful approach if one's distributional assumptions are reasonable. That sounds like a caution: it is, but the same caution can be made about all statistics that require distributional assumptions. Censorship means that the regression models are not GLMs (chapter 6). On the other hand, since censorship is a central problem in survival analysis, there is a large body of survival-oriented software that can be used to handle censored data problems even when they are not survival problems. Censored regression works especially well with large data sets, and with data having an explanatory structure beyond a single factor.

A simulated example

In appendix 5A, I provide R code to estimate the mean, median, and standard deviation of the simulated data used in figures 5.1–5.3, using each of the ad hoc substitutions discussed previously, as well as the ROS and KM methods. The results are given in table 5.1. In this example, the ROS method reproduces the "true" (uncensored) estimate exceptionally well, and the other methods all do quite good jobs as well. How might you choose which method to use? As always, there are many considerations here, but perhaps the most useful things to recall are these: robust ROS requires a distributional assumption but KM does not. KM cannot work if more than half the sample is censored.

Because it can estimate quantiles, the ROS method can be used to describe a distribution in some detail. Figure 5.4 shows the boxplot for the uncensored data given in appendix 5A, as well as the censored boxplot, drawn with the NADA library, when these data are multiply censored.

Soil phosphorus data

In a study of the impact of the number of crop species raised (agrodiversity) in traditional cropping systems in a rainforest in the state of Veracuz, México, Negrete-Yankelevich et al. (2013) quantified the amount of phosphorus available to the plants, a key limiting factor in this system. Here I analyze a random subset of the data considered in

Table 5.1 Estimates of median, mean, and standard deviation from simulated data, using several ad hoc substitutions for left-censored data, as well as the robust ROS and Kaplan–Meier estimates

Data	Median	Mean	SD
OLS for y^*, uncensored	1.032	1.670	2.073
OLS with censored y –> LOD	1.032	1.711	2.047
OLS with censored y –> LOD/2	1.032	1.658	2.081
OLS with censored y –> 0	1.032	1.604	2.119
ROS	1.032	1.670	2.072
KM	1.021	1.712	2.047

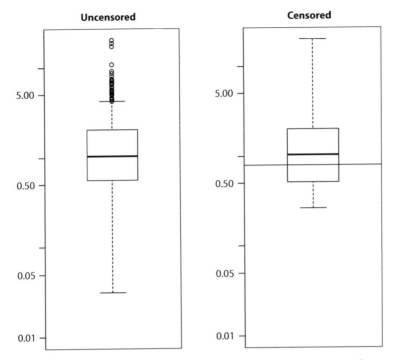

Fig. 5.4 The multiply censored data yield a boxplot for the uncensored data (left) and the censored boxplot estimated with ROS, from the NADA library (right). The horizontal line shows the highest censoring level in the data. The boxes themselves are nearly identical, but the smallest values (left tail of the distribution) are not well described for the censored data.

their paper (Negrete-Yankelevich et al. 2013), as censored data; sample collection and processing methods are given in that paper. R code is in appendix 5A.

The distributions of soil P concentrations are described in figure 5.5, for each of three levels of agrodiversity and nine different *milpas* (small slash-and-burn maize polycultures, normally maintained by different farmers). These censored box plots suggest that much information may be lost by censorship, in this case caused by lower limits to the resolution of the assay used, especially for the milpas with low agrodiversity. There also appears to be a pattern of higher soil P with higher agrodiversity. This suggests that censored regression may prove useful. On the other hand, figure 5.5 also suggests the possibility that the milpas differ substantially in the variance of soil phosphorus. This will prove to be an important issue in formulating a regression model for soil P, as we will see in the following section.

5.1.7 *Regression for censored data*

We want to predict the mean value of an observed variable y by observed explanatory variables x. The difficulty is that some of the observed y's are censored, so OLS regression will yield inconsistent estimates. The central idea behind censored regression is simple: we model both the probability of being censored for different values of x, and the expected value of y given that it is uncensored. By combining these models in a likelihood function (chapter 3) we arrive at estimates that are consistent.

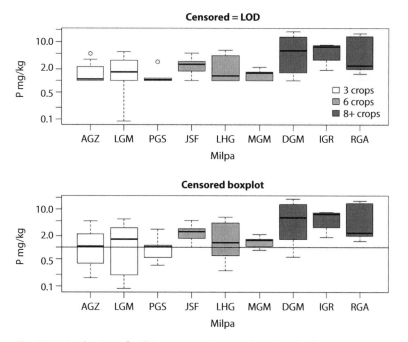

Fig. 5.5 Distribution of soil P concentrations at three levels of agrodiversity, each with three milpas. In the top panel, censored values are set to the limit of detection. The bottom panel was drawn using estimates from robust ROS.

There are a number of models used for censored regression, but by far the most widespread is the same as the model used in parametric failure time models, also called accelerated failure time models (Fox 2001; Kalbfleisch and Prentice 2002; Klein and Moeschberger 2003); the only difference is that the error distribution for censored regression is commonly the Normal distribution. This model is often referred to as tobit regression, (a contraction of the original author's name (Tobin 1958) and probit, a type of categorical model).

Assume for the moment that we have a relatively simple situation: the underlying process generating y as a function of x is linear, and y values are singly censored. For the sake of concreteness, I will assume that they are left-censored. Thus,

$$y_i = \begin{cases} \beta_0 + \beta_1 x_i + \varepsilon_i \text{ if } y_i^* > c \\ c_y \text{ if } y_i^* \leq c \end{cases}. \tag{5.1}$$

Here the βs are regression coefficients and ε_i is a random variate (error term) from a specified distribution (the Normal distribution in this example); following Long (1997), I use c_y to denote the value to which censored data points are set, which is not always c in applications. The expectation of y given $\mathbf{x}$ (which can be multivariate) is

$$E(y_i|x_i) = P(\text{uncensored } |x_i) \times E(y_i |y_i > c, \mathbf{x}_i)$$
$$+ P(\text{censored } |x_i) \times c_y. \tag{5.2}$$

The top panel of figure 5.6 shows the same data as in figure 5.1, but I have superimposed the Normal PDFs for each of three values of x (see also figure 5.2a). We can use this distribution to calculate the probability of a data value being censored. Let $\delta_i = \frac{(\mathbf{x}_i\beta - c)}{\sigma}$, so that δ_i

measures the number of standard deviations that $\mathbf{x_i\beta}$ is above c. Using $\phi()$ to represent the probability density function (PDF) and $\Phi()$ to represent the cumulative distribution function (CDF) for the standard Normal distribution (see book appendix) the probability that a point is uncensored is $\Phi(\delta_i)$, and so by definition the probability that it is censored is $1 - \Phi(\delta_i)$, or (because the Normal distribution is symmetric) simply $\Phi(-\delta_i)$. In the bottom panel of figure 5.6, the curves are the CDFs for y at each of the chosen values of x; the shaded areas under each curve give the probability of y values being censored for that value of x.

Now that we can calculate the probability that a y value is or is not censored, we can also calculate the product term in equation (5.2), as $\Phi(\delta_i) \times$ (the mean of the truncated Normal distribution for parameters μ, σ, and c). This truncated Normal distribution is just what you might guess informally: a Normal distribution with mean μ and standard deviation σ, but instead of the infinite tails of the Normal distribution, it is limited to values greater than (or less than) some limit c. Because of the truncation, though, the mean is not μ and the standard deviation is not σ. The mean of the truncated Normal is (Long 1997)

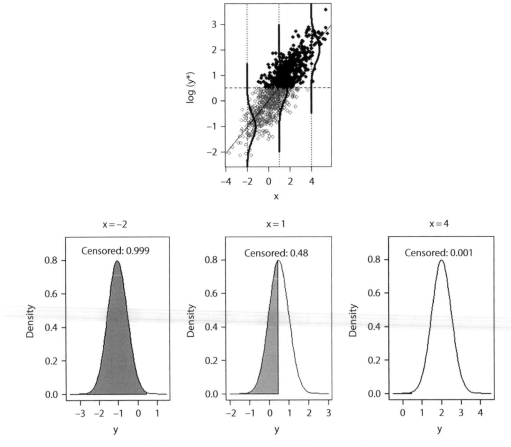

Fig. 5.6 Top: Normal probability density functions (PDFs) for y, each of three x values (−2, 1, and 4). The peak of each PDF is at the regression line, calculated using linear regression treating all data as uncensored. Bottom: Normal cumulative distribution functions (CDFs) for $y|x$ for the three chosen x values. The shaded areas give the probability of a y value being censored given the x value.

$$\mu + \sigma\lambda(\delta_i), \text{ where } \lambda(\delta_i) = \frac{\phi(\delta_i)}{\Phi(\delta_i)}.$$

The function λ is referred to in the statistical literature as both the hazard function and as the inverse Mills' ratio. In any event, the leftmost term in equation (5.2) is, finally, $\Phi(\delta_i)\mathbf{x}_i\beta + \sigma\phi(\delta_i)$, and the rightmost term is $\Phi(-\delta_i)c_y$; we have

$$E(y_i|\mathbf{x}_i) = \Phi(\delta_i)\mathbf{x}_i\beta + \sigma\phi(\delta_i) + \Phi(-\delta_i)c_y. \tag{5.3}$$

The expectation is thus a weighted sum of censored and uncensored values. If there were no censoring, $E(y_i|x_i) = \mathbf{x}_i\beta$; this is a linear function of $\mathbf{x}$, and we can estimate it with OLS. But equation (5.3) is not a linear function of $\mathbf{x}$, because neither the $\Phi()$(cumulative distribution function) nor the $\phi()$ (probability density function) are linear in $\mathbf{x}$. OLS will necessarily give inconsistent results in the relationship between $\mathbf{x}$ and y^* because it is based on this assumption of linearity.

To get a consistent estimate, then, we need a likelihood function that uses information about the censored data. This is actually straightforward: using logic similar to that used to calculate the mean, we weight the censored and uncensored values by their probability. Specifically, if there are N data points, of which k are censored, the log-likelihood is

$$\ln L\left(\beta, \sigma^2|\mathbf{y}, \mathbf{X}\right) = \sum_{i=1}^{N-k} \ln \frac{1}{\sigma}\phi(\delta_i) + \sum_{i=N-k+1}^{N} \ln \Phi(\delta_i), \tag{5.4}$$

with the first sum being over all the uncensored points and the second sum being over the censored points. If there were no censorship, this would reduce to the likelihood for a linear regression model.

The likelihood function is thus built by using probability models for each kind of data point. By doing so it uses all of the data and does not rely on ad hoc fixes to the values of the data. It does not require any great logical leap to extend this kind of likelihood function to include any of the other forms of censored data—such as multiply censored, right-censored, or interval-censored data—even in the same data set. Doing so only requires using the appropriate part of the CDF for points censored in that way. As we will see in the rest of this subsection, models estimated by maximizing likelihoods constructed this way can do an excellent job of estimating relationships.

There are a number of R libraries that will allow this kind of regression. Many, including the NADA library, simply call the `survreg` function in the `survival` library. However, many of them do so without all of the options available from `survreg`. As a result, I recommend using `survreg` directly if possible.

Regression with simulated data

Now that we have a picture of the likelihood function for censored regression, we can fit a model to the data in figure 5.1. Simulations like this are valuable because we know the true relationship ($\ln(y^*) = x/2$). Moreover, we know how well OLS estimation does for the particular simulated sample—quite well, with estimates of the intercept and slope as 0.002 and 0.514, respectively. How well does censored regression do by comparison?

This R code is all that is necessary:

```
library (survival)
sfit <-  survreg (Surv (y_cens_lod, cens == 1, type = "left")
    ~ x, dist = "loggaussian", data = logn_example)
summary (sfit)
```

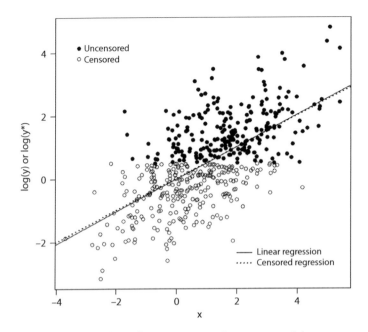

Fig. 5.7 Fits by ordinary least squares on the uncensored data set (y^*) and by censored regression on the censored data set (y).

The `survival` library uses the `Surv` function to indicate survival objects—these are pairs of numbers giving a value and an indicator as to whether this point is censored or not (and if so, what kind of censoring). Part of the output is:

```
               Value      Std. Error      z            p
(Intercept)    0.0257     0.0569          0.453        6.51e-01
x              0.5025     0.0300          16.731       7.83e-63
Log(scale)    -0.0087     0.0353         -0.246        8.05e-01

Scale= 0.991

Log Normal distribution
Loglik(model)= -1039 Loglik(intercept only)= -1158
Chisq= 237.2 on 1 degrees of freedom, p= 0
```

The censored regression on the data set using censorship has estimated the parameters about as well as the linear regression did for the uncensored data (appendix 5A); the two sets of parameter estimates are essentially indistinguishable. The two estimated lines are shown in figure 5.7.

A linear model for the soil phosphorus data

The soil phosphorus data (figure 5.5) has two prominent features: there is a tendency for greater phosphorus concentrations in milpas with greater agrodiversity, and the milpas appear to differ in the variance of soil phosphorus. Fortunately, `survreg` has features that allow us to address both issues. The code for a simple censored regression model (addressing only the pattern in mean concentration) for these data (assuming they have been read in using the code in appendix 5A, and the survival library is loaded) is

```
fit.0 <- survreg (Surv(phosphorus, censor == 1, type = "left") ~
agrodiv.fact, dist = "loggaussian", data = phos.dat)
```

It seems reasonable to regard agrodiversity here as a factor (the last category is not a precise value) and not a continuous variable. Thus, we are modeling the (sometimes censored) P concentration as a function of agrodiversity. The model summary includes this output:

```
                Value  Std.  Error       z        p
(Intercept)    0.0534        0.1778    0.30 7.64e-01
agrodiv.fact6  0.4495        0.2444    1.84 6.58e-02
agrodiv.fact8  1.4573        0.2421    6.02 1.76e-09
Log(scale)    -0.1065        0.0876   -1.22 2.24e-01

Scale= 0.899
```

Comparing this model with an intercept-only model, ΔAIC = 28.8 on 2 df; by any criteria (see chapter 3), the model provides a much better description than the null. The intercept estimates mean soil P at the baseline level of agrodiversity (agrodiv.fact3 = 0), and the terms for the other two levels of agrodiversity should be added to that to estimate each of those means. Thus the mean estimates are 0.0534, 0.5029, and 1.5107. There are several equivalent ways to get confidence intervals for these estimates; I did it the lazy person's way, by using update() to refit fit.0 to eliminate the intercept, and then wrapping confint() around that:

```
R> confint (update (fit.0, ~ . -1))
                 2.5%     97.5%
agrodiv.fact3  -0.29507   0.40186
agrodiv.fact6   0.16974   0.83609
agrodiv.fact8   1.18749   1.83392
```

The 95% confidence intervals for agrodiversity = 3 and agrodiversity = 6 overlap with one another substantially, but both are clearly different from agrodiversity = 8. There is a clear pattern: at least for this study, then, increasing agrodiversity is associated with increasing soil P, and it seems that we can have a good deal of confidence that this pattern is not due to sampling error, at least for the most diverse of the milpas. Had we regarded agrodiversity as a continuous rather than categorical variable (questionable for this data set), we would have found that the single regression parameter was positive with a 95% confidence interval not including 0 (code not shown; you should be able to do this by modifying the preceding line of code).

But does the model fit the data well? In particular, are the data consistent with the distributional assumptions for the tobit model? Like least squares regression, the tobit model assumes that errors are distributed normally, and are homoscedastic. In the linear regression model the principal problem resulting from violations of these assumptions is *inefficiency* in estimating the coefficients (that is, the estimates require a larger sample to achieve the same variance). In the tobit model, violations of these assumptions are more problematic: the maximum likelihood estimates are inconsistent if these assumptions do not hold. Thus model diagnostics are important here!

In a linear regression model, we could plot the residuals against the predictor values, but how do we reasonably calculate an ordinary residual (difference between measured and predicted values) for the censored data? There is no answer to this problem, and we certainly cannot simply ignore the censored data! But we can use the *deviance residuals* (there are actually several more types of residuals for this sort of model; see Therneau and Grambsch (2000)). In a nutshell, the deviance residuals are components of the model deviance. More precisely, for each data point i there is a deviance residual d_i such that

$$\sum_i (d_i)^2 = -2 \ln L,$$

where ln(L) is the log likelihood for the model.

A plot of the deviance residuals for this model is on the left in figure 5.8. The different levels of agrodiversity do appear to differ not only in the mean deviance residual, but also the variance. One might reasonably feel a bit uncomfortable about this model.

Given our suspicion that this model may be problematic, we can consider alternative approaches. In appendix 5A, I fit a so-called frailty model (analogous in some ways to a mixed effects model; see chapter 13) to these data, which treats milpa as a random effect on the mean (intercept). The resulting model is impressive only because of its minuscule improvement in fit (appendix 5A); the frailty model does not address the different variances (figure 5.8) among the milpas. A different approach is to use a stratified model,

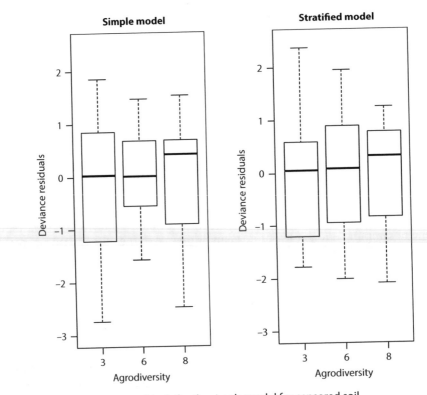

Fig. 5.8 Deviance residuals for the simple model for censored soil phosphorus data as a function of agrodiversity (left), and a stratified model that also allows a different variance for each milpa (right).

which allows both the mean and the variance to differ among milpas. This is quite simple to do, by using the strata() function provided with survreg():

```
R> fit.2 <- survreg
        (Surv(phosphorus,censor==1,type="left") ~agrodiv.fact +
        strata (milpa), dist="loggaussian",data=phos.dat)
```

The summary includes this output:

```
                Value Std.    Error        z          p
(Intercept)     0.0383        0.138    0.279   7.81e-01
agrodiv.fact6   0.4229        0.212    1.991   4.65e-02
agrodiv.fact8   1.5600        0.201    7.754   8.93e-15
milpa=AGZ      -0.1248        0.298   -0.419   6.75e-01
milpa=DGM       0.2827        0.265    1.068   2.86e-01
milpa=IGR      -0.6022        0.230   -2.623   8.72e-03
milpa=JSF      -0.5176        0.291   -1.781   7.50e-02
milpa=LGM       0.3204        0.269    1.193   2.33e-01
milpa=LHG       0.1428        0.306    0.467   6.41e-01
milpa=MGM      -0.8737        0.375   -2.332   1.97e-02
milpa=PGS      -0.7924        0.297   -2.665   7.69e-03
milpa=RGA      -0.0630        0.226   -0.279   7.80e-01


Scale:
milpa=AGZ milpa=DGM milpa=IGR milpa=JSF milpa=LGM milpa=LHG
milpa=MGM milpa=PGS milpa=RGA
     0.883       1.327       0.548       0.596       1.378       1.154
0.417       0.453       0.939


Loglik(model)= -172.5    Loglik(intercept only)= -190.6
        Chisq= 36.1 on 2 degrees of freedom, p= 1.4e-08
n= 90
```

The regression parameter estimates are not terribly different from those we obtained in fit.0, but they now refer to a model that allows the sampling distribution to have different characteristics on the different milpas. The values like milpa=AGZ are for the log scale parameter for each milpa, and the estimated scale parameters (the standard deviations) are provided near the bottom.

Is this a better model? The most obvious way to respond to this is to examine the new deviance residuals, since they were the motivation for fitting this model. They are shown in the right of figure 5.8. The means are more nearly the same (and more nearly 0) in the stratified model, and the variances are also more nearly the same. The stratified model does not appear to suffer from heteroscedasticity. Some readers may take comfort in knowing that a likelihood ratio test also suggests that the stratified model is an improvement: the summary outputs tell us that the log-likelihoods for the two models are –183.1 and –172.5 for the simple and stratified models, respectively; –2 times the difference in their log-likelihoods gives a deviance of 21.23. With 8 df (since the stratified model uses 12 parameters while the simple model uses only 4), the P-value (from the χ^2 distribution) is 0.0066—suggesting that the better fit of the stratified model is not likely to be due to sampling error. Finally, we can compare AICs (see chapter 3); ΔAIC = 5.22 on 8 df. If we blindly relied on AIC alone we might conclude that the stratified model was not justified.

But since it appears to better meet the distributional assumptions of our model (an issue not addressed by the AIC statistic, and a crucial issue for tobit models), it would be a mistake to draw this conclusion. It should be clear that the stratified model is the one we should use to draw conclusions.

Censored regression (including models with categorical explanatory variables) is a powerful and fairly straightforward method that can be used for many ecological data sets. Obviously one can use an ad hoc substitution for handling the censored data values and then use some method like a linear mixed model on the new data set. Sometimes this will lead to values similar to those obtained from censored regression—but there is generally no way to tell when this is the case, and generally no reason to use an approach of this sort. Doing so always imposes pattern on the data that is not present in the actual data set; analyses of such false patterns are inherently suspect. It is much better to use a method actually designed for the problem at hand. Tobit models provide one such method. They are easy to use, but sensitive to violations of their assumptions. Beyond the method used here (including heteroscedasticity in the model) other approaches include modifying the model to include omitted variables, using different distributions, or use of other modeling approaches such as censored quantile regression (Powell 1986).

I have focused on Normal (or lognormal) data rather than on event–time data, because there is already a substantial literature on survival statistics in ecology (Polakow and Dunne 1999; Fox 2001; Ibañez 2002; Keith 2002; Dungan et al. 2003; Moritz 2003; Silvertown and Bullock 2003; Nur et al. 2004; Beckerman et al. 2006; Fox et al. 2006; Schwartz et al. 2006; Sim et al. 2011; Saino et al. 2012). Studies of event times almost invariably involve data censorship, and by definition the assumption of normality must generally be quite wrong in event–time studies.

There is, nevertheless, a continuing tradition of ecological and evolutionary studies that examine event–time data using tools like analysis of variance. So far as I can tell, there are two reasons for this: it is a tool that most ecologists know, and it lends itself to partitioning of variance. However, fitting a wildly wrong model to data may well lead to estimates, *P*-values, and variance partitions, but none of these will be reliable, and may not even be meaningful.

5.2 Truncated data

5.2.1 *Introduction to truncated data*

Truncation is quite different from censorship, but sometimes the distinction appears to be quite subtle. Truncation is a characteristic of an entire data set: censoring means that for some data points we know only an inequality, but truncation means that we do not even know about data points with values over some range. For example, a plant ecologist might want to estimate the densities or diversity of seeds in the soil, but has at his disposal only a sieve with relatively large pores; small seeds will be missed. The data are truncated: the ecologist will never have any information (explanatory or response) about the small seeds, and if he ignores this fact, his estimates will be inconsistent. As another example, consider a study of the spread of an invasive plant, using satellite imagery in which each pixel is 30m in diameter. Infestations of the plant smaller than 30m (or perhaps, with cover less than, say, 50% of a 30m patch) will not be detected. Maps, estimated rates of spread, and many hypothesis tests are likely to suffer severely from this data truncation. Still another common sort of example occurs in longitudinal studies: an ecologist interested in estimating lifetime reproductive success censuses her study organisms every two weeks;

by doing so she is guaranteed to miss mortality among newborns, and thus misestimate the chance of survival to adulthood. Finally, consider estimating the density of a plant population from aerial photos. Some individuals cannot be counted because they are so close to others that we cannot distinguish them, some individuals are shaded by others, and still others are too small to be observed using this technique. The true density of plants is at least as large as what we estimate naïvely.

Fortunately, there are methods for coping with truncated data. They are related to, but not as well-developed as, methods for censored data. On the other hand, the ideas are somewhat simpler, and can be discussed more briefly.

5.2.2 *Sweeping the issue under the rug*

For truncated data, there is really only one commonly used ad hoc method: ignoring truncation. This may be done consciously—one can rationalize that since some data are lacking, (a) one cannot do anything about it, and can only make inferences about data that are detectable, and besides, (b) OLS approaches should be robust. I suspect that it is much more commonly unconscious, because few ecologists are aware of potential problems with data truncation.

Whether done deliberately or unconsciously, truncation can be quite problematic. Figure 5.9 uses the same data as in figure 5.1, but now we compare the relationship between the full, untruncated data set (OLS regression of y^* on x) with the naïve regression of y on x, using only the observed data points. The naïve regression does a poor job of estimating the relationship between y^* and x. As we shall see, it does not provide a very good description of the relationship between y and x either. However, there is a method for truncated regression, which provides a very good model.

The difference is not simply one of sample size. I used a random subset of the full data set to get the OLS estimate of the regression of y^* on x, to guarantee that the data sets were of identical size. The differences between these models are mainly due to the nature of truncation. Indeed, an examination of Figure 5.9 suggests that, as x becomes small (and thus truncation is more likely), the naïve regression will be dominated by the untruncated values—which in this case are all above the regression line. The naïve regression will certainly give poor estimates if there is much truncation.

Does this imply that, if there is little truncation, one can use OLS? As always, the answer is "maybe," which is best qualified by adding "sort of; it depends." As with censorship, it is a matter of scale. With a large data set and little truncation, the violations of the assumptions of OLS (or other sorts of regression models) will be mild. With relatively more truncation, the problems will grow. The point is not to be a statistical fundamentalist, waving a holy book of rules; the point is to use models that are both biologically and statistically sensible for our data. We have to use our judgment; my argument is that we should embrace methods for truncated (and censored) data, not fear them.

5.2.3 *Estimation*

Estimation is somewhat simpler with a truncated data set than with one that includes censored data. After all, the very definition of truncation is that we lack information on even the existence of data points smaller (or larger) than some value. Thus there are no analogs to the methods discussed in section 5.1.6 for estimating means with censored data.

By definition, a truncated data set is not a random sample of the population. If we only measure units that are large (small) enough, or old (young) enough, at best we have a

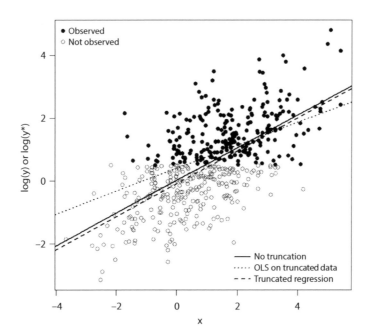

Fig. 5.9 Regression with the data set now truncated at c—points below c are now unknown to the researcher. Figure shows the "true" relation (OLS regression of y^* on x), the OLS regression of y on x, and the truncated regression of y on x.

random sample of the population conditioned on measurement being greater (less) than truncation limit. Estimation focuses on including this conditionality.

For a regression model, deriving the likelihood is similar to, but simpler than is the case for censored regression. Again, assume that the underlying process generating y as a function of $\mathbf{x}$ is linear, and y values are truncated at c. Thus, the relationship between x and y is the same as for uncensored values in equation (5.1):

$$y_i = \beta_0 + \beta_1 x_i + \varepsilon_i \text{ if } y_i^* > c. \tag{5.5}$$

Using the same reasoning as for censored regression, the expectation of y given $\mathbf{x}$ (which can be multivariate) is just

$$E(y_i|\mathbf{x}_i) = \mathbf{x}_i\boldsymbol{\beta} + \sigma\lambda(\delta_i). \tag{5.6}$$

Again, this is not a linear function in $\mathbf{x}$, so we cannot use OLS. The contribution to the likelihood is the same as for uncensored data points, but we must adjust for truncation by the area of the Normal distribution that has been truncated.

To get some insight on this problem, consider an informal discussion based on figure 5.10. The probability density for y^* (for a standard Normal distribution), $\phi(y)$, is given by the solid curve, and the density for y is given by the dashed curve. If the data are truncated at c, then the area under $\phi(y)$ and to the left of c is the cumulative distribution of y at that point, $\Phi(y|y = c)$. Since the total area under a probability density function is 1, and for the truncated distribution the quantity $\Phi(y|y = c)$ has been truncated, the density function for the truncated distribution is necessarily more peaked, and the additional area under the truncated PDF is exactly $\Phi(y|y = c)$,

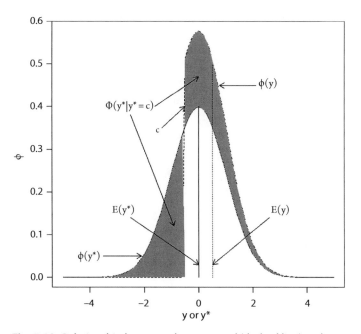

Fig. 5.10 Relationship between the truncated (dashed line) and untruncated (solid line) PDFs. The shaded region to the left of *c* is truncated (and is equal to $\Phi(c)$). A region of the same area augments the PDF for the truncated distribution. The mean of the truncated distribution is shifted away from the truncation. Although not directly illustrated, it should make sense that the variance of the untruncated distribution is smaller than that of the full distribution.

$$f(y_i) = \frac{\frac{1}{\sigma}\phi(\delta_i)}{\Phi(\delta_i)}, \qquad (5.7)$$

and the log likelihood for the entire data set is

$$\ln L = \sum_{i=1}^{N} \ln[f(y_i)]. \qquad (5.8)$$

5.2.4 *Regression for truncated data*

A Normal example

As an introduction to truncated regression, consider the data we used previously (figure 5.1), but now instead of treating the points for $y \leq c$ as censored, I will remove them from the data set entirely; we now have a truncated data set.

```
R> lgn <- subset (logn_example, y > lod)
```

In order to make all comparisons with the full data set reasonable, I will use a random sample of the full data frame (called `subs_logn`; see R code in appendix 5A), that is the same size as the truncated data frame `lgn`.

Table 5.2 Parameter estimates for regressions for the data in figure 5.1, but truncating at c

Model	Intercept (SE)	Slope (SE)
Untruncated	0.0155 (0.0574)	0.5215 (0.0318)
Naïve	0.4188 (0.0571)	0.3669 (0.0292)
Truncated	−0.0880 (0.1239)	0.5262 (0.0486)

Now I compare truncated regression with both OLS regression on the sample of the untruncated data set and with OLS regression on the truncated data:

```
R> lm_untruncated <- lm (log(y) ~ x, data = subs_logn)
R> lm_naive <- lm (log (y) ~ x, data = lgn)
R> treg <- truncreg (log (y) ~ x, lgn, point = log (lod),
direction = "left")
```

The truncated regression is estimated using the truncreg function from the library of the same name. The model is specified as usual, but point specifies where the data are truncated, and direction specifies whether the truncation is to the left (below) or right (above) that value.

Recall that the model generating the data is $\log(y) = x/2$, so we would hope that the estimated intercepts would be near 0 and the slopes near 1/2. The estimates are given in table 5.2 and the regression lines in figure 5.9, which suggests that the estimates for the untruncated and truncated models are close to the correct values, but those for the naïve model are considerably further away. Bootstrapped confidence intervals support this conclusion strongly (see appendix 5A).

A survival example

Time-to-event data are likely to involve censorship, but it is also fairly common that they involve truncation. Few, if any, published analyses of ecological time-to-event data account for truncation. I simulate a data set here to underline its importance for future studies.

The simulation (see appendix 5A for R code) assumes that the time-to-event is distributed as a Weibull distribution, with two predictive factors. The Weibull distribution is very commonly used in models and statistics for survival analysis. Under a Weibull distribution, the hazard either increases or decreases continually with time, depending on whether the Weibull shape parameter is larger or smaller than 1 (Fox 2001). If the shape parameter is exactly 1, the Weibull distribution reduces to the exponential (constant risk) distribution. As with the preceding (Normal) example, I compare three models: one that accounts for truncation, one that ignores truncation, and one that is a random subset of the untruncated dataframe, but of the same length as the truncated data. The parameter values used to generate the data were Weibull shape = 1.2, log(scale) = 0.03, and for the two predictive factors, $x_1 = 2$, $x_2 = -1$.

To get the estimates for the truncated data, I used the aftreg function from the eha library. This function uses the Surv function from the survival library to allow one to specify truncation and censorship. The survreg function in survival could have handled the untruncated and naïve models (but not the truncated model), but to keep things comparable, I used aftreg to fit all three models.

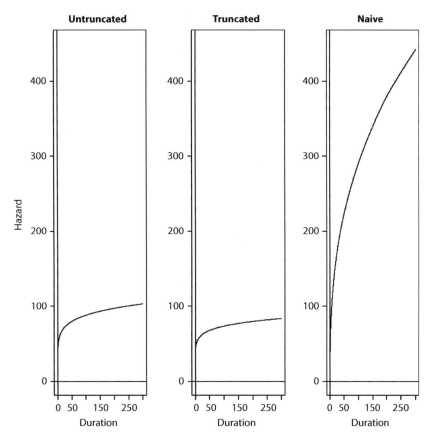

Fig. 5.11 Hazard functions estimated by the untruncated, truncated, and naïve Weibull models.

The estimates from the three models (see appendix 5A for details) are:

	True	Untruncated	Naive	Truncated
x1	2.00	2.01168	1.85494	2.04962
x2	-1.00	-0.98537	-0.87885	-0.99654
scale	0.03	0.03969	0.07463	0.03819
shape	1.20	1.14300	1.38230	1.11738

Again, the untruncated and truncated models both estimate the true parameters well. The naïve model does a considerably worse job. Most notably, it estimates the Weibull scale parameter as more than twice its true value. This has considerable importance: figure 5.11 shows the estimates from all three models for the hazard function (the instantaneous probability of an event, given that it has not yet occurred; sometimes called the force of mortality). The hazard estimates from the naïve model are dramatically different than the other two estimates.

5.3 Discussion

Both censorship and truncation are probably much more common in ecological data than is usually recognized. These issues may arise in almost any study, but they are especially likely to occur in studies that are longitudinal in time or space, or in which limits of resolution for measurements are likely to be encountered. Both issues can have large impacts on parameter estimates and on the statistical inferences we make about data.

Data censorship is well known to be an issue in survival and other time-to-event problems. Among ecologists, it is much more poorly appreciated as an issue when there are limits to measurement. The analyses discussed in this chapter—for soil phosphorus and for simulated data—should help to persuade readers of two simple points: (1) ignoring data censorship can severely compromise our data analyses, and (2) there are reasonable approaches to analyzing data that may be censored, and these methods are accessible to any ecologist with a bit of statistical training. There is a very large literature on censored data; Helsel's (2005) book provides a rich source of information for ecologists.

While truncation may occur because of decisions made in the data analysis phase (e.g., to delete censored data), I suspect it is more often caused in ecology by choices made about sampling design. For example, many ecological studies concern non-random subsets of larger statistical populations, such as survival among adults or dispersion among individuals moving more than some minimal distance. These kinds of choices mean that one is working with truncated data. The examples in this chapter should help to convince readers that using statistics designed for random samples on non-random subsets of the data is generally a poor idea, as quantities estimated this way cannot be trusted. As with censorship, there is a body of statistical techniques that allow us to account for the truncation, and one need not be a statistics wizard to use these techniques.

Must you use these techniques in analyzing data involving censorship and/or truncation? No, not necessarily. Statistics is not a religion, with commandments to be followed by its adherents; it is a mathematical discipline that provides quantitative tools that aid researchers in modeling and understanding their data. One must make assumptions and compromises in using any statistical model. This means that there are times when a reasonable person might choose not to use, say, censored (or truncated) regression. The theory used for censored (or truncated) regression is not as well developed as for some other problems. For example, there are issues addressed by generalized linear mixed models (see chapter 13), or by spatial regressions (chapter 10), for which there is not yet any theoretical basis for coping with data limitations like censorship or truncation. In such cases, one must judge what the best way is to analyze the data. The key thing is not to assume that the data limitations are unimportant!

The goal of data analysis is to learn something about the natural systems under study (see Introduction). Data analysis may often require attention to issues like censorship and truncation—but we pay attention to them because we would like to have confidence that we have made reasonable estimates about the things we study. Limitations on the data sets we use can change those estimates, but the good news is that there is already a toolkit that allows us to account for these types of data limitations. As more ecologists begin to use these tools, they will become a part of our culture, and, I suspect, our field may help to expand these toolkits.

Acknowledgments

I thank Simoneta Negrete-Yankelevich for her careful reading and criticism of earlier versions of this chapter. Jody Harwood introduced me to the problem of non-detects. This work was supported by grant number DEB-1120330 from the U.S. National Science Foundation.

## CHAPTER 6

Generalized linear models

Yvonne M. Buckley

6.1 Introduction to generalized linear models

Ecologists in the past have often been taught individual statistical techniques in piecemeal fashion without the unifying conceptual model that lies behind them. This is changing rapidly and the increasing use of *generalized linear models* (GLM) is leading to a unification of modeling concepts in the minds of users. The beauty of generalized linear modeling is that it provides strong and stable "hooks" on which to hang additional knowledge, enables a re-interpretation of previously learned ANOVA and regression techniques, and integrates well with more advanced modeling techniques introduced in chapters 12, 13, and 14.

Generalized linear modeling unifies several statistical and modeling techniques. It is much easier to remember the structure and use of a GLM than to recall several disparate tests and rules to appropriately use ANOVA, ANCOVA, regression, multiple regression, logistic regression, and logit regression. Any of these models, and more, can be described using an appropriate combination of just three properties of the GLM: the *linear predictor*, *error distribution*, and *link function*, which together enable the modeling of simple linear relationships with Normal errors as well as more complex non-linear relationships with alternative error distributions. To make life even easier the linear predictor is always linear, just like a standard regression, so we need only be concerned with variations in the error distribution and link functions used. The error distribution describes how the variation, which is not explained as part of the linear predictor, is distributed. The link function provides a link between the linear predictor and the response variable, as the response variable may not be directly linearly related to the explanatory variables in the linear predictor.

Classical linear models assume that errors, also called residuals or deviations from the fitted model, are normally distributed. This means that the errors take continuous values and are distributed according to the Normal or Gaussian probability distribution (see online appendix 6A.1). In a statistical modeling framework, the Normal distribution is the familiar bell-shaped curve with a high frequency of values close to zero (many small errors) and the frequency of larger positive or negative errors declining toward the tails of the distribution. The width or spread of the distribution is determined by the variance of the errors. Because the Normal distribution of the errors around a fitted line has its mean at zero, the only parameter of interest is the error variance around those fitted values. There is a very persistent myth that when we assess normality we

Ecological Statistics: Contemporary Theory and Application. First Edition. Edited by Gordon A. Fox, Simoneta Negrete-Yankelevich, and Vinicio J. Sosa. © Oxford University Press 2015. Published in 2015 by Oxford University Press.

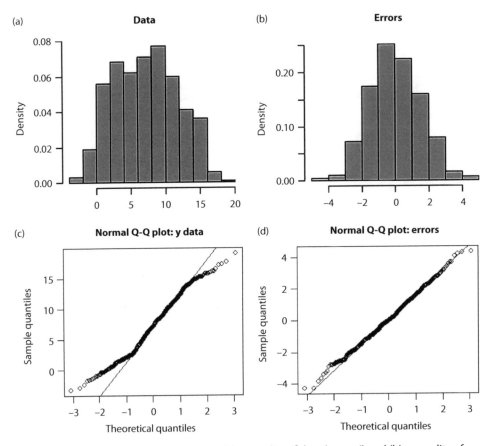

Fig. 6.1 The difference between assessing (a) normality of data (wrong!) and (b) normality of errors (right!) in linear modeling is shown here. Errors are the residuals from the fitted relationship. These are histograms of simulated data drawn from a Normal distribution with a specified relationship between x and y (intercept = 0, slope = 5) and a standard deviation of 1.5. As can be seen from the Normal quantile–quantile plots the original data are not necessarily normally distributed (c), but the errors are normally distributed (d). See appendix 6A.1.

are referring to the distribution of the original data. This is wrong. When we assess normality we do so with reference to the residuals or errors (see chapter 1, figure 6.1, appendix 6A.1).

Prior to the development of GLMs, researchers often applied transformations to data where the errors were not normally distributed, e.g., counts or proportions. This practice has continued among ecologists despite the wide availability of statistical software enabling "off-the-shelf" GLM construction and testing. Warton and Hui (2010) and O'Hara and Kotze (2010) provide useful commentary, contrasting transformation approaches with GLM in an ecological context (O'Hara and Kotze 2010; Warton and Hui 2010), and both strongly advocate the use of GLMs rather than transformation for count, proportion, and other data where Normal errors may not be appropriate. While O'Hara and Kotze use simulated data, real data sets can be more complex, especially if there are unmodeled or unknown sources of variance. In some cases the transformed data with Normal errors may lead to a better fitting, or simpler, model than a GLM (see appendix 6A.2 for an example using Normal, Poisson, Quasi-Poisson, and Negative Binomial error structures). Model comparison and criticism will enable you to determine what the best course of action is in a particular circumstance.

Generalized linear models entered the ecological statistician's toolkit with the publication of McCullagh and Nelder's landmark book in 1989 (McCullagh and Nelder 1989) and slowly gained in popularity among ecologists as statistical modeling programs were developed to implement GLM. GLM is now routinely taught to ecologists at undergraduate and graduate level. There are several texts that explain GLMs from different angles and with varying ecological, statistical, and mathematical flavors (e.g., Crawley 2007; Gelman and Hill 2007; Bolker 2008), and I refer readers to these for complementary material. While students who have had experience with GLMs generally have a good grasp of the "how" to undertake a GLM analysis, in my experience they tend to lack an intuitive understanding of the model structure and properties of data that make some error distributions and link functions more appropriate than others. There is also confusion around how best to evaluate a GLM in terms of how appropriate it is for the data, how to plot data and models, and what statistics to report. Here I aim to augment existing books, book chapters, and papers on GLMs by focusing on the data/GLM interface.

Much of the data that ecologists commonly collect lead to the assumption of normality of errors being contravened. Count data such as number of species, number of cones on pine trees, and number of offspring wolves produce are discrete, constrained to be positive, and may be clustered due to a biological process that is not directly modeled. Binary data (e.g., male/female, alive/dead) are discrete and bounded at 0 and 1; for example, a tree cannot be more dead than dead (0) or more alive than alive (1). Proportion data come in two main flavors: proportion data with a meaningful denominator—for example, the proportion of seeds in a known sample size that germinate, or the proportion of frogs in a sample succumbing to chytrid fungus disease—and proportion data without a meaningful denominator, for example, the % sand content of a soil sample or % cover of grass in a quadrat. Both types of proportion data are bounded at 0 and 1. The bounded and/or discrete nature of much biological data leads to errors which are not normally distributed around a fitted relationship and distributions such as the Binomial, Poisson or Negative Binomial (see book appendix) might be more appropriate for modeling the errors (figure 6.2).

If we handle count, binary, and proportion data improperly, our results can be biased, underpowered, or difficult to interpret (Warton and Hui 2010). Maximum Likelihood estimates of the parameters of a fitted relationship depend on the specified error distribution (see chapter 3); therefore, misspecified error distributions can bias the parameter estimates, standard errors, and subsequent statistical tests. I show below in an example about pine cones that different error distributions change the significance of the explanatory variables (appendix 6A.4). If we are to move beyond the aim of determining statistical significance alone, to appropriately modeling a biological process, it makes sense to use a model that captures more of the properties of the process we are trying to understand with our model. Often we are not only interested in the fitted relationship or mean response, but also the unexplained variation that surrounds the response. Variation in biological processes is commonly important; for example, variation in population growth rate can influence extinction risk. Thus, it is necessary to describe the nature and magnitude of this variation appropriately, including how variation is distributed around the fitted values, as it may not be homogeneous. In addition, appropriate modeling of error variance allows uncertainty in model outcomes to be accounted for.

GLMs are commonly used to parameterize population dynamics models as several of the vital rates determining population growth rate are best modeled using non-Normal

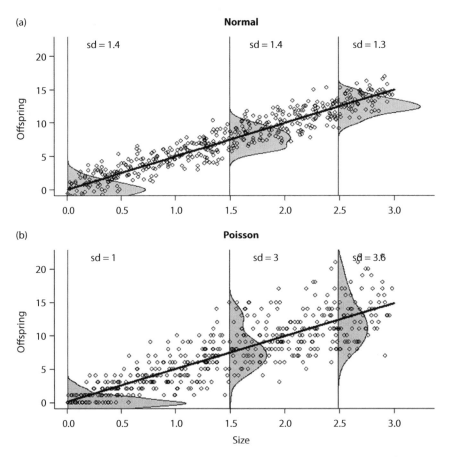

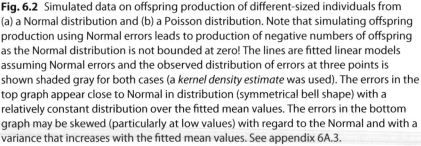

Fig. 6.2 Simulated data on offspring production of different-sized individuals from (a) a Normal distribution and (b) a Poisson distribution. Note that simulating offspring production using Normal errors leads to production of negative numbers of offspring as the Normal distribution is not bounded at zero! The lines are fitted linear models assuming Normal errors and the observed distribution of errors at three points is shown shaded gray for both cases (a *kernel density estimate* was used). The errors in the top graph appear close to Normal in distribution (symmetrical bell shape) with a relatively constant distribution over the fitted mean values. The errors in the bottom graph may be skewed (particularly at low values) with regard to the Normal and with a variance that increases with the fitted mean values. See appendix 6A.3.

error distributions (e.g., survival and number of offspring produced) (Merow et al. 2014). Testing different model structures informs on the ecological process and knowing the ecological process informs the appropriate model structure (Coutts et al. 2012). The resulting population models are used to make predictions about population growth rates or times to extinction. Scientists make recommendations for management of pests, or species of conservation concern, based on statistical models of biological processes, therefore incorrectly modeled populations may lead to wasted management resources or adverse outcomes for the species concerned.

Proportion, binary, and count data by their very nature tend to have fundamentally non-linear relationships with explanatory variables. For example, low numbers of seeds may be produced on a tree until a threshold tree size is achieved, leading to a log-linear relationship between seed production and size. Another example is that survival may be low at small sizes and high at large sizes leading to an S-shaped, or logistic, relationship. GLMs enable the inclusion of inherent non-linearity through the link function. The link function relates the linear predictor, η_i, which is not bounded and has homogeneous variance, to the expected values of the data, which may be bounded and have heterogeneous variance. Link functions enable fitted values from the model of the linear predictor to be converted to non-linear predictions on the same scale that the data were collected; for example, in a GLM of survival data the appropriate link function (e.g., logit) enables fitting and predictions of probabilities bounded at 0 and 1.

Plot your data before starting the modeling process in order to determine what the likely relationships are, and to explore the structure of variance in your data. While we should have mental models of what we expect our data to look like, there are often surprises in store once it is collected. There is no substitute for thorough knowledge of your data. Throughout the modeling process, confront your models with data by looking at fitted relationships in contrast to your data. This is not always trivial, particularly for GLMs where data and fitted relationships can be plotted at the observed scale or at the scale of the linear predictor. Throughout the chapter, I give several examples of plotting techniques and associated code.

6.2 Structure of a GLM

So, you suspect or know that you will require a GLM. Perhaps you have undertaken a classical linear model with Normal errors and are unhappy with the standard diagnostic plots (heteroscedasticity of errors, non-linearity of the relationship, poor performance on a Q-Q normality plot—see appendix 6A.4 for an example of what these might look like). Perhaps due to the nature of your data: count, proportion, binary, or time-to-event, you suspect that a GLM would be appropriate. How do you choose an appropriate GLM structure? Let's look first at the structural options for a GLM.

6.2.1 *The linear predictor*

There are few exotic options here; the linear predictor is equivalent to a classical linear regression, ANOVA or ANCOVA model with one response variable and some combination of categorical and/or continuous explanatory variables as main effects and/or in interaction. Here your linear predictor η_i, for each data point i, is a linear function of some explanatory variables (also called "covariates" if they are continuous or "factors" if they are categorical) X_{pi}, with an intercept, α, and slope or additional intercept parameters, β_p, corresponding to the 1 to p explanatory variables.

$$\eta_i = \alpha + \beta_1 X_{1i} + \cdots + \beta_p X_{pi}.$$

Note that there is no error specified in the linear predictor, it just contains the systematic part of the model.

As with classical regression, you can specify polynomial functions of continuous explanatory variables to deal with non-linear relationships; with polynomial models, the linearity is retained in the parameters. To illustrate this: specifying a model as $a + bx + cx^2$, to capture non-linearity in the response variable y in relation to variation in the x variable, is equivalent to calculating a new variable $z = x^2$ and fitting $a + bx + cz$ where y clearly has linear relationships with both x and z. For an example of fitting and testing a quadratic term, see appendix 6A.2.

A useful extension of the classical linear model is the potential for specifying *offsets* in the model. An offset is an a priori value for a parameter in the model; you can then test the significance of the estimated difference between the given offset parameter and the freely estimated parameter. For example you might have a good reason to suspect that the slope of a relationship with a particular explanatory variable would take a particular value, e.g., slope = 1. You set the value of that parameter in the model as an offset that means that instead of estimating the slope directly you get an estimate of the difference between a slope equal to the offset value and the estimated slope without the offset. You can then easily assess significance of the difference between the offset and the estimated parameter; this difference would be zero if the offset parameter and the estimated parameter were similar. See appendix 6A.4 for an example.

6.2.2 *The error structure*

Once an initial model for the linear predictor is constructed, the next step is to decide on an appropriate error structure. There are good a priori reasons for choosing particular error distributions for particular types of data. The first distinction is between continuous and discrete error distributions. If your response variable contains continuous data (e.g., size, weight, gene expression) you will need a continuous error distribution. The most familiar distribution in this class is the Normal or Gaussian error distribution, which has two parameters, the mean and the variance.

For a Normal error distribution the error distribution model would be:

$$\mathbf{Y} \sim \mathrm{N}(\boldsymbol{\mu}, \sigma^2).$$

Predictions of the vector of $\mathbf{Y}$ values are estimated to come from a Normal distribution with a mean $\boldsymbol{\mu}$ and a constant variance σ^2, hence $\mathrm{N}(\boldsymbol{\mu}, \sigma^2)$. The mean, $\boldsymbol{\mu}$, is a vector of fitted values describing the relationship between the response variable and the explanatory variables. The value of the vector $\boldsymbol{\mu}$ therefore changes depending on which values of the explanatory variables you look at. There is just one value for the variance σ^2, which does not change depending on the fitted values. For the case of the Normal distribution the link function between the linear predictor and the predicted values is the identity link which means $\eta = \boldsymbol{\mu}$. Other error structures are presented in section 6.3.

6.2.3 *The link function*

The final part of a GLM structure is the link function: this important element links the linear predictor (linear model whose fitted values can take any negative or positive value) to the original response variable via a function. Where we use Normal errors, the link function is the identity link, which means the linear predictor automatically predicts on the scale of the original response variable. For counts, however, the fitted values should only take values greater than zero, so we need a function that maps the linear predictor (any value) to only positive values.

In the case of the Normal distribution of errors the appropriate link function which connects the linear predictor with the mean of the Normal distribution is the identity link, $\eta = \boldsymbol{\mu}$. In general, $\eta = \mathrm{g}(\boldsymbol{\mu})$, with g() called the link function. Usually we have the linear predictor η (fitted values on the scale of the linear predictor) and we want to convert these to the predicted means $\boldsymbol{\mu}$ (equivalent to the fitted values on the scale of the response variable $\mathbf{Y}$). To do this we apply the inverse of the link function $\mathrm{g}()^{-1}$ and, for the Normal error case, the inverse of the identity link is still the identity link ($\eta = \boldsymbol{\mu}$ and $\boldsymbol{\mu} = \eta$).

Table 6.1 Appropriate error distributions, canonical link functions, and alternative link functions for commonly modeled response variables (for details on distribution properties see book appendix)

Response variable	Error distribution	Canonical link function	Alternative link functions
Continuous positive and negative values	Gaussian/Normal	Identity	Log, Inverse
Counts	Poisson	Log	Identity, Sqrt
Counts with over-dispersion	Negative Binomial, Quasi-Poisson	Log Log	As per Poisson
Proportions (no. successes/total trials)	Binomial	Logit	Probit, Cauchit, Log, Complementary Log-Log
Binary (male/female, alive/dead)	Binomial (Bernoulli)	Logit	As per Binomial
Proportions or binary with overdispersion	Quasi-Binomial	logit	As per Binomial
Time to event (germination, death)	Gamma	Inverse	Inverse, Identity, Log

The expectation of the response variable $\mathbf{Y}$ is therefore equivalent to the fitted values at both the linear predictor and response scales, $E(\mathbf{Y}) = \boldsymbol{\mu} = \boldsymbol{\eta}$. Other link functions are presented in table 6.1.

A particular value of the response variable Y_i is modeled using a Normal distribution with mean coming from the linear predictor, η_i, and error variance σ^2 with a random number from that error variance given by $\varepsilon_i \cdot Y_i = \eta_i + \varepsilon_i$. Note that negative fitted values could be predicted if η_i is predicted to be negative for some values of the explanatory variables, or if negative errors reduce Y_i sufficiently. This is important to note when modeling a process that is bounded inherently at zero; negative values may make no biological sense for variables such as size (chapter 5). The full model including the linear predictor, error distribution, and link function is therefore:

$$\mathbf{Y} \sim N(g(\boldsymbol{\eta})^{-1}, \sigma^2).$$

6.3 Which error distribution and link function are suitable for my data?

In general, GLMs allow for the error distribution to come from an exponential family other than the Normal and the link function may be any *monotonic differentiable function* (continuously increasing or decreasing function with a gradient calculable using differentiation). *Canonical links* are specified for each family of error distributions; they are statistically convenient but are not necessarily always the best fit, in which case alternative link functions can be used.

How do we choose an appropriate link function? The canonical link function is a good starting point. See table 6.1 for a list of error distributions and associated link functions. However, there may be several different link functions that could be applied to your model and it makes sense to try a few and compare model fits to determine the most appropriate link. In R you can get a list of potential links for each distribution by typing `?family`.

The Gamma, Beta, Exponential, and LogNormal distributions are alternative continuous error distributions that are used for different kinds of continuous response variables.

For example, the Beta distribution can be useful for modeling proportion data that is not from success/failure trials, such as percent cover data for plant community assessment. The Gamma distribution is useful for "time to failure" data such as time to germination or time to death (chapter 5). See chapter 3, book appendix, and Bolker (2008) for a discussion of distributions and their properties.

If your response variable contains discrete data, you will need to use a discrete probability distribution function. Discrete response variables include counts (which are constrained to be greater than zero and constrained to integer values), binary events (such as success/failure, survived/died, male/female), and proportion data which were collected as number of successes from a series of trials. The most commonly employed error distributions for ecological data include the Binomial for binary and proportion data, and the Poisson for count data. Below I give two examples for ecological data that are analyzed appropriately using the Binomial error distribution and the Poisson error distribution, respectively. I have also provided worked examples of Binomial (appendix 6A.5) and Poisson, quasi-Poisson, and Negative Binomial (appendix 6A.2 and 6A.4)

6.3.1 *Binomial distribution*

The probability of success in any trial is p, for example the probability of a tossed coin landing heads up (a *success*) is 0.5, $p = 0.5$. You might undertake a number of trials, n, resulting in a number of successes, k. We can use the Binomial distribution to estimate, for example, the probabilities of getting 4 heads ($k = 4$) in a series of 5 trials ($n = 5$):

$$\Pr(X = k)\binom{n}{k}p^k(1-p)^{n-k} \quad \text{for } k = 0, 1, 2, \ldots, n.$$

In this way you can work out the probability of any outcome from the n trials giving the distribution of all possible outcomes (e.g., 1/5, 2/5, . . . , 5/5 successes). The mean of the Binomial distribution is np (number of trials × probability of success in one trial) which for our coin example is 2.5; the most likely outcome from a trial of 5 coin tosses is either that you get 2 heads or 3 heads (you can't get exactly 2.5). The variance of the Binomial distribution is $np(1-p)$ which for the coin example is 5*0.5*0.5 = 1.25. You can see from figure 6.3 that if the probability of success (equivalent to survival in that example) is either very high or very low, the variance is low and variance is maximized at $p = 0.5$, which is where success and failure are both equally likely, as is the case of the tossed coin.

Trees are generally long-lived and tree death is difficult to observe without long-term repeated measures. However, we can take advantage of mass mortality events, such as during droughts, to look at individual vulnerability to mortality. Data were collected on tree mortality, a binary variable, as well as tree size (diameter at breast height) and other potential influences of mortality, such as neighborhood density and location (appendix 6A.5). Here I use this example to analyze the effects of tree size and location on determining vulnerability to mortality during a drought episode; for further analyses of these data see Dwyer et al. (2010).

Survival is a binary variable, taking only two values, alive or dead. The errors may therefore be distributed according to the Bernoulli distribution, which is a special case of the Binomial distribution with just one trial. Thus, each tree can be viewed as a trial with a possible outcome of alive or dead. As tree size increases we would expect survival to increase as adults are more deeply rooted, potentially accessing water reserves unavailable to seedlings. At small sizes, perhaps most trees die so variance in mortality is again

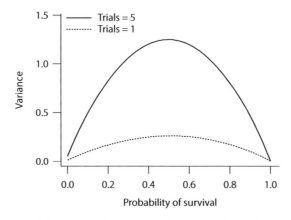

Fig. 6.3 Variance from a Binomial distribution with number of trials $n = 5$ or the special case of the Bernoulli distribution where $n = 1$. Variance is at its lowest (0) at probability of survival = 0, where all individuals die, and probability of survival = 1 where all individuals live, and is at its maximum at probability of survival = 0.5 where survival and mortality are both equally likely. See appendix 6A.6.

low. Somewhere in the middle trees might be equally likely to die or survive leading to high variance at intermediate sizes. In this situation, variance is obviously not homogeneous across the range of the explanatory variable (size), and is likely to be distributed according to the Binomial/Bernoulli distribution around the mean, with the variance of the distribution changing with the mean (figure 6.3).

Binary data can be difficult to visualize in relation to explanatory variables. For the survival example, binary alive or dead is the response variable (figure 6.4) with a continuous tree diameter as the explanatory variable. Below I show four ways of visualizing the relationship between these variables. A simple plot of the binary variable can mask the density of points as points are overlaid on each other (figure 6.4a). Introducing a jitter, a random displacement of small magnitude to the binary points, enables better visualization of the density of points along the x-axis (figure 6.4b). A more intuitive understanding of the relationship with the explanatory variable can be gained by *binning* binary values along intervals and taking the means for each interval along the x-axis. This can be done by taking the mean of the binary points at even spacing along the x-axis or by creating bins with equal numbers of observations. However the patterns observed vary depending on the intervals used for finding the means (figure 6.4c,d).

The linear predictor in this example is given by:

$$\eta_i = \alpha + \beta_1 X_{1i},$$

where X_1 is tree diameter, and α and β are the intercept and slope, respectively. The linear predictor η is a Real Number and can be positive or negative depending on the values of the parameters and the explanatory variable. However, survival can only lie between 0 and 1 so we need a relationship between survival and the linear predictor that is bounded at 0 and 1. The logit link function can achieve this:

$$\eta = \text{logit}\,(\boldsymbol{p}) = \log\left(\frac{\boldsymbol{p}}{1-\boldsymbol{p}}\right).$$

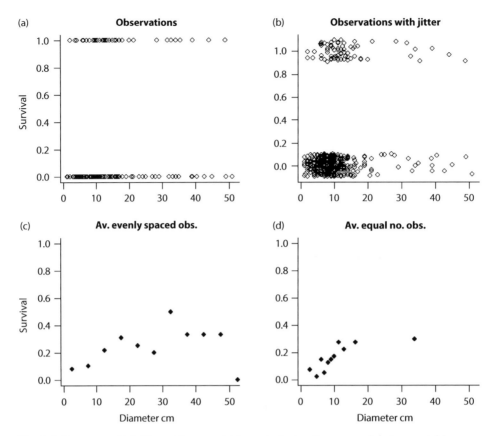

Fig. 6.4 Four ways of plotting a binary variable against a continuous explanatory variable. Observations (a) were of trees that were either alive (1) or dead (0). Observations with jitter (b) are the observations of 0's and 1's with a small uniformly distributed random variable added to separate the points vertically to enable better visualization of point density along the x-axis. The average of the evenly spaced observations (c) was found by taking the means of survival in 11 × 5 cm diameter categories. The average (Av.) of the equal number of observations (d) was found by taking the means of survival in 11 categories with approx. the same number of observations in each ($n = 40$). These data were taken from site 1 of the *Eucalyptus melanophloia* survival data set (appendix 6A.5).

The inverse link function used to rescale the linear predictor back to the 0–1 probability scale is:

$$\mathbf{p} = \text{logit}^{-1}(\eta) = \frac{\exp(\eta)}{\exp(\eta) + 1},$$

(6.1)

and the full error model for the Binomial (*B*) error distribution is:

$$\mathbf{Y} \sim B(\mathbf{n},\mathbf{p}),$$

where *n* is the number of trials (in the Bernoulli case $n = 1$ but in the general Binomial case **n** can be a vector of trials with, potentially, a different number of trials for each data point) and **p** is the inverse logit of the linear predictor as in equation 6.1.

I like to explore and present my data as close to how they were collected as possible so would prefer to see the original data points if possible on figures. Figure 6.5 attempts

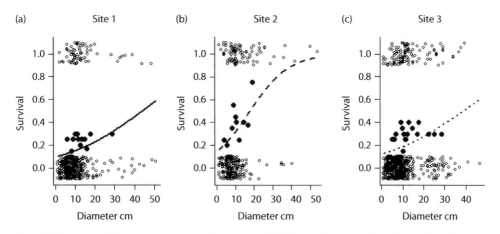

Fig. 6.5 Survival of *Eucalyptus melanophloia* trees of different diameters from three sites. The unfilled circles are the observed values (0 and 1 values with a vertical jitter added to aid visualization), the filled circles are averages of survival in 11 categories with approximately equal numbers of observations in each category ($n = 20$) and the line represents the predicted probabilities from a GLM with Binomial errors and a logit link function. The predictions are not straight lines and are bounded at 0 and 1 as they have been rescaled from the linear predictor to the survival probability scale using the inverse link function (see appendix 6A.5).

to reconcile this preference with showing the fit of the data to the model (it shows a combination of figures 6.4b and d with fitted lines from a GLM for three sites).

The denominator is a very important component of modeling proportion data. The denominator is your friend. Are you more confident of an estimate of 0.5 for the probability of observing heads from a sample of 4 or 100 coin tosses? In general, we have more confidence in probability estimates that come from larger samples. By calculating proportions and throwing away information about the denominator of the proportion (the sample size) we are essentially weighting all proportions equally. Knowing the denominator allows correct weighting of data toward the larger sample sizes (appendix 6A.6). This is important even in a well-designed experiment with an equal number of trials across all sampling events, as organisms can die, petri dishes can fall off benches, and infections can compromise cultures, resulting in uneven numbers of trials and variation in the denominator of your proportion. Always monitor and collect the sample size of a series of trials that make up your proportion data. You model proportion data in a GLM as a two-column response variable of (success, failure) where the number of failures is calculated as the number of trials minus the number of successes. Use the `cbind()` function in R to combine the (success, failure) data and treat it like a single response variable. See appendix 6A.6 for an example of a proportion calculated as the mean of eight individual proportions ignoring the denominator, contrasted with a proportion calculated using the binomial denominator, which weights estimates toward the larger denominators.

6.3.2 *Poisson distribution*

Count data have a number of properties that we need to account for in models: many zeros, bounded at zero, non-linearity, and non-homogeneity of variance. The occurrence of many zeros may call for zero-inflated error distribution methods (chapter 12; see appendix 6A.4 for a discussion of zeros in the pine cone example). Because data are bounded at zero, the relationship with covariates (e.g., tree age, year, and pine tree density) is often non-linear and the variance often increases with the mean (figure 6.2). A pervasive recommendation, to normalize errors and linearize the relationship with covariates, is to log-transform the response variable and use classical regression methods with Normal errors. However, this is potentially problematic for a number of reasons, the most serious of which in my view is the addition of an arbitrary amount (usually 1 or smaller) to the response, because one cannot take a log of zero. See O'Hara and Kotze (2010) for a thorough discussion of this issue from an ecological viewpoint. The use of a GLM may solve these problems; by modeling errors using alternative distributions it may be possible to find a distribution that adequately captures the increase in variance with the mean, and an appropriate link function can linearize the relationship for the linear predictor.

Pine cones contain seeds which determine the reproductive output of individuals and the population. A simple linear model of how tree size predicts number of cones with Normal errors is inappropriate for a number of reasons, including that the number of cones on trees is a discrete count variable, bounded at zero (figure 6.6).

There is only one parameter in the Poisson (P) distribution, the mean μ_i, and the variance is assumed to increase in direct proportion to the mean with a dispersion parameter of 1 (figure 6. 7):

$$\mathbf{Y} \sim P(\mathbf{\mu}).$$

The response variable $\mathbf{Y}$ is therefore Poisson (P) distributed with mean μ and variance μ, where μ is determined by the linear predictor and link function of the values of the explanatory variables in the model. If errors are Poisson distributed the fitted values on the original count scale will have an increasing spread of errors around them as μ increases. At large count means, and if the range of values you are modeling is not large relative to the mean, a Normal error distribution may be an adequate approximation, particularly in combination with a log transformation of the response (figure 6.7).

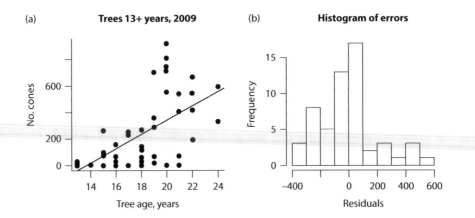

Fig. 6.6 The number of cones on individual pine trees of different ages (a) demonstrates several common issues with count data: many zeros which inflate residual variance, bounding at zero which decreases negative residual variance at low fitted values, non-linearity of the response to the explanatory variable, and variance which increases with the explanatory variable. The line in (a) is an inappropriate linear model fitted to the data: (b) shows that the distribution of errors resulting from the linear fit is obviously not Normal (appendix 6A.4). Data from Caplat et al. (2012); Coutts et al. (2012).

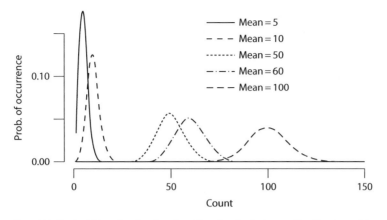

Fig. 6.7 Probability distribution function for the Poisson distribution with means of 5, 10, 50, 60, and 100 counts. The variance or errors increase as the mean increases, however; at large mean counts the distribution becomes symmetrical and a Normal distribution may be a reasonable approximation (appendix 6A.7).

A commonly used link function for Poisson models of count data is the log link, $\eta = \log(\mu)$ with the inverse link as $\exp(\eta) = \mu$ to convert the fitted values from the linear predictor back to positive values that can be compared directly with the observed count data.

Zeros can be informative about biological processes, so do not ignore zeros or exclude them without some thought about how they could be used. Zero counts can be directly incorporated into Poisson models; however, many zeros may cause the variance to increase faster than the mean (section 6.3.3). In these cases you can split models, so model one process (e.g., probability of reproducing, probability of detection) as a Binomial model of 0's and 1's and a separate model (e.g., number of offspring produced if you do reproduce, number of individuals observed if detected) as a Poisson model of counts. You could also simultaneously model both processes in a mixture model (chapter 12). If zero counts arise as a separate process from other counts, then consider collecting data on additional explanatory variables (which could be determinants of zeros), as these data will greatly improve the resulting two-stage or mixture model (e.g., organism age or maturity, or characteristics of observer, appendix 6A.4).

6.3.3 *Overdispersion*

Overdispersion is a relatively common problem encountered when using Poisson and Binomial error distributions. There is assumed to be a strong (for Poisson, exact) relationship between the mean and the variance in both distributions. Quite often with biological data, there is more variance in the errors than can be explained by the Binomial or Poisson distribution—see chapter 12 for a thorough treatment of overdispersion. This can be assessed by dividing the residual deviance (of the GLM model) by the residual degrees of freedom—if this quotient is substantially greater than unity (1) then overdispersion is a problem (for examples of overdispersion see appendix 6A.2 and 6A.4). It is difficult to give a threshold at which overdispersion becomes *substantial*. Minor overdispersion can be dealt with if the underlying process can be adequately modeled with a known distribution such as the Binomial or Poisson. You should explore the consequences of overdispersion for your model inference and use that as a guide to whether it is problematic or not. For example, you could take some of the steps suggested below and see if they change your conclusions. More serious overdispersion may be a result of a *misspecified model*.

There are some quick and dirty fixes for mild overdispersion including switching from using chi-squared to more conservative F-tests to assess significance, and the use of quasi-distributions. Which methods you should use depend largely on the purpose of your model; if you are looking to assess significance of explanatory variables alone and are not interested in parameter estimates then switching to F-tests might suffice. However, this is not a philosophy with which I am particularly comfortable; I prefer to model to get insight into process rather than just model for significance tests. Quasi-methods do not use a specified error distribution. Rather, they estimate an additional parameter to relate the mean and the variance, but are still based on either the Binomial or Poisson distributions, hence quasi-Poisson or quasi-Binomial. Quasi-methods are therefore not necessarily self-contained error distributions that capture characteristic behavior of a process; they can be viewed as *hacks* of underlying distributions. If your errors are not really a result of Binomial or Poisson processes, then quasi-models are also not appropriate. They should only be used to account for mild overdispersion resulting from minor deviation from the assumptions of Binomial or Poisson distributions. The additional parameter in a quasi-model is named the overdispersion or scale parameter. Quasi-models are parameterized using quasi-likelihood methods (chapter 3). You will notice that if you compare a quasi-model with a specified distribution model (Poisson or Binomial), the standard errors of the parameter estimates become larger using the quasi-model (appendix 6A.2 and 6A.4). Thus, you can easily come across situations where the use of the Poisson or Binomial family can lead to conclusions of significance but accounting for overdispersion using a quasi-model can change those conclusions. It is therefore very important to be aware of overdispersion and to know how to deal with it.

If you have serious overdispersion, you may be misspecifying your model, and rather than switching immediately to quasi approaches, it may be more appropriate to consider alternative model constructions and error distributions. For count data, a negative binomial error distribution can sometimes give a better fit than a Poisson model. Figure 6.8 shows Poisson and Negative Binomial models fitted to the same count data for the pine cones example; in particular the negative binomial model seems to fit better at low tree ages, better capturing low numbers of cones produced (appendix 6A.4). Another alternative is to use mixed-effects models and include a random effect at the level of the observations (chapter 13). Generalized mixed-effects models have improved considerably in computational tractability making them a useful alternative to quasi approaches.

Zeros can arise due to underlying environmental heterogeneity or individual differences in quality which cannot be readily observed, causing overdispersion. Overdispersion can arise from unknown or unmodeled sources of variation that cause heterogeneity in the response variable at different scales. Overdispersion may be eliminated, or substantially reduced, by collecting information on the right covariates. Consider collecting spatial coordinates/locations or individual identities and using these as random effects within mixed-effects models—inclusion of random effects at the individual level can be very effective at dealing with overdispersion (chapters 10 and 13); also see chapter 12 for the use of mixture models to deal with overdispersion.

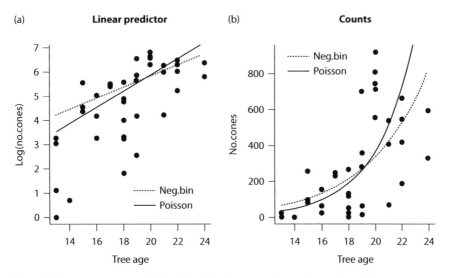

Fig. 6.8 Here we compare (a) the fit of the linear predictor when using Poisson and Negative Binomial error distributions and (b) the same models and data plotted on the original count scale. This figure illustrates that the error distribution you choose affects the parameter estimates, and also demonstrates the difference between the fitted values from the linear predictor and the fitted values from the linear predictor with the inverse link function applied to convert them back to the count scale (appendix 6A.4).

6.4 Model fit and inference

Parameters for GLMs can be estimated using maximum likelihood, quasi-likelihood or Bayesian approaches. GLMs with specified error distributions (e.g., Binomial, Poisson, Negative Binomial etc.) are fit using Maximum Likelihood (chapter 3) or Bayesian approaches (chapter 1). Quasi-distribution models are fit using quasi-likelihood, which uses the variance function where the variance is a function of the mean with the relationship specified by the overdispersion or scale parameter. Typically, the variance function of a quasi-model collapses down to the Poisson or Binomial when the scale parameter is set to one, hence quasi-Poisson or quasi-Binomial.

Often we would like to know whether our experimental treatment or covariate makes a *significant* difference to our observations, i.e., whether the differences observed between treatments are likely to have arisen by chance or not. An appropriate statistical model will adequately model the *by chance* component of this, i.e., the error distribution. Without an appropriate model of the errors, we risk finding false significance or not finding significance when we should (type I and II errors, chapter 2).

Model fit for GLMs is assessed using deviance. The (residual) *deviance* is the GLM equivalent of the residual sum of squares. Technically, it is the difference between the log-likelihood of the *saturated model* and the log-likelihood for the fitted model (chapter 3). In other words deviance is the difference between the *logs* of the probability of the data given the fitted values, and the probability of the data given a fully saturated model (multiplied by –2; chapter 1). The probability of the data in the saturated model will be large (close to 1, so its logarithm will be close to 0) because you are explaining all the variance in the data by using as many parameters as there are data points. The probability of the data in the alternative fitted model will be smaller (less than 1, so its logarithm will be negative). Subtracting 0 from a negative number gives a negative number; multiplying it by –2 gives a positive number. Thus the deviance will be larger, the worse the fitted model is.

Chapter 3 presents a thorough treatment of model selection processes. I provide a brief summary with some GLM specific issues below. If your fitted model is a good description of the data you would expect a small deviance from the saturated model, if the deviance is small you would accept the simpler model (all models which are nested within the saturated model will be simpler!). Operationally we are interested in whether one model (which is a simplification of a more complicated model) is better. In this case we can use deviance to compare the likelihoods of the more complex model and the nested simpler one (Likelihood ratio [LR] test; chapter 3). We use a chi-squared distribution to assess whether the resulting deviance is small or large. We therefore compare the difference in deviance between the two models to a chi-squared distribution with $p1 - p2$ parameters, which is the difference between the numbers of parameters in the two models. A low P value indicates that the deviance is large and that the models are different; we would infer that you lose important information using the simpler model and may therefore choose to retain the more complex model. A high P value indicates that the deviance is small, there is little difference in explanatory power between the two models, and we would therefore choose the simpler model. We report the LR_{df} or χ^2_{df} as our test statistic and its associated P value.

For small data sets, t-tests on the parameters and deviance (LR) tests may give different answers but deviance tests are more reliable. For large data sets, t-tests and LR tests should be similar assuming that there is not serious collinearity (correlations among explanatory variables). For overdispersion, a different kind of deviance test is used: $\frac{D_2 - D_1}{\rho(p-q)} \sim F$, where the test statistic is assumed to follow the F distribution rather than the chi-squared distribution. ρ is the dispersion parameter and $p + 1$ and $q + 1$ are the number of parameters in models 1 and 2, respectively. D_1 is the deviance of the more complex model and D_2 of the simpler nested model. The LR test statistic follows an F-distribution with $p - q$ and $n - p$ degrees of freedom so we report $LR_{p-q,n-p}$ or $F_{p-q,n-p}$ and its resulting p-value.

As GLMs with specified error distributions have an associated log-likelihood an Akaike Information Criterion (AIC) can be calculated and multi-model inference using information criteria can be carried out easily (chapters 1 and 3). Quasi-models have a quasi-likelihood and Quasi-Akaike Information Criterion (QAIC) can be used for inference. There are methods to extract QAIC from these models, but note that R does not give you QAIC as part of the model summary automatically. See Richards (2008) for a discussion and ecological examples of the use of QAIC for model selection.

6.5 Computational methods and convergence

Sometimes, despite our best efforts models can fail. Models can fail in a number of different ways; one of the most common is non-*convergence*. The default method for fitting GLMs in R is iteratively reweighted least squares (IWLS). This algorithm uses the maximum likelihood function to determine iteratively *better* estimates of the parameters, ideally converging on a set of parameter estimates where the likelihood of the data is maximized. A model is determined to have converged when successive iterations of the algorithm no longer improve the fit, or improve it only a very small amount. The shape of the *likelihood surface* determines how quickly and how well this process of convergence happens. Imagine two explanatory variables for which we are seeking to find best-fit parameter values in a model; all possible parameter values lie on the x and y axes of a graph. Now imagine a 3-D surface which is the likelihood of the data given a particular set of parameter values x and y. The likelihood surface is like a landscape determined by the likelihood values of all possible sets of parameter values. If the likelihood surface is very flat, i.e., the data are relatively equally likely to occur under a large number of combinations of parameter values, there can be difficulty with convergence.

Non-convergence can occur for a number of reasons:

- your original mental model of the process may be wrong—perhaps you are using non-informative explanatory variables
- the process is inherently non-linear and you have not appropriately modeled the non-linearity
- insufficient data were collected
- use of an incorrect likelihood function (i.e., the wrong error distribution), and/or link function.

Appendix 6A.4 shows an example of non-convergence when modeling data with lots of zeros using a negative-binomial error distribution. In this case, an appropriate single distribution could not be found, so alternative models such as mixture models should be explored (chapter 12).

For binomial GLMs the warning message "fitted probabilities numerically 0 or 1 occurred" comes up relatively often. This means that for values of certain variables or combinations of variables the fitted values were very close to 0 or 1, called *complete separation*. For example, all of your small trees died and all of your large trees lived. This is a particular problem as you increase the number of covariates, because you increase the chance that for a particular combination of values there will be no variation in the probabilities of success (see Venables and Ripley (2002), pp. 197–8). This warning message may result in large model coefficients for the variable and correspondingly large standard errors (the Hauck–Donner effect), making it difficult to assess significance. Use of likelihood ratio tests on models with and without the offending variable can help to determine if the variable is a useful predictor. Sometimes it may be because that variable truly is a great predictor or it may be that you have too few data points for a particular variable or combination of variables. If it is a sampling problem then you should increase your sample size, or try a different (probably simpler) model.

Warning messages about non-convergence should not be ignored; instead, you should investigate the reason/s for non-convergence and be aware that the parameter estimates reported from a non-converged model may be nonsense. If you can't find a structural problem with your data or model causing the non-convergence then you can try to increase the number of iterations and/or specify some appropriate starting values. You should consider specifying starting values if you are using a quasi-model or an unusual combination of error distribution and link function.

6.6 Discussion

GLMs free us to collect more biologically meaningful data as we are not constrained to collection of just continuous response variables or response variables that can be transformed readily to meet the assumptions of linear models. There are, however, limits to what GLMs can do. Moving from classical linear models to GLMs means the replacement of assumptions of linearity and normality of errors with alternative assumptions about linearity and alternative error distributions. These new assumptions may not be appropriate. Errors from count data may not be Poisson distributed if the variance increases more rapidly than the mean (overdispersion) or there is clustering in the data. Errors from binary or proportion data may not be Binomial if there is overdispersion in the data. Models may not converge. The assumptions and predictions of GLMs need to be investigated just as carefully as the assumptions of linearity and normality in classical linear models.

Despite these limits, GLMs are foundational units of many more complex modeling techniques. The skills and knowledge you attain by fitting and critiquing GLMs will pay dividends when learning more complex modeling techniques. It is important to note that real world data are messy; sometimes, despite a well-designed experiment and careful data collection, the assumptions of GLMs just cannot be met. Perhaps an appropriate error distribution cannot be found, the models all fit poorly, or models do not converge. There are many alternative ways of modeling data including (but not restricted to): nonlinear models, non-parametric methods (particularly useful for error distributions which do not conform to common distributional assumptions), mixture models (chapter 12), and mixed-effects models (chapter 13). If you think something biologically interesting is going on with your data set, you may want to invest in reading about and learning appropriate alternative techniques, starting with those in this book and moving beyond.

It's the ecology, stupid! We pay attention to these statistical issues so we can have some confidence in the answers we get. With our models, we look for insight into the ecological processes producing the patterns around us. The right error distribution and link function are not just statistical rules to be followed blindly. Informed and appropriate modeling of errors and link functions can enable more general predictions to be made as we strive to capture the ecological processes important to the generation of observed patterns.

## CHAPTER 7

A statistical symphony: instrumental variables reveal causality and control measurement error

Bruce E. Kendall

7.1 Introduction to instrumental variables

The goal of much ecological research is to determine causality, whether in test of ecological theory or to develop predictive power for making management decisions in the face of environmental change. In its simplest form, we want to test whether, and how strongly, x causes y—for example, does lizard density determine the density of spiders, one of the lizard's prey items (Spiller and Schoener 1998)? Although it is straightforward to demonstrate whether there is correlation between x and y, that alone does not tell us whether the observed relationship is actually an epiphenomenon of both variables being caused by a third factor, z (e.g., moisture; Spiller and Schoener 1995). The reason that experiments are so powerful is that, when done well, they rule out alternate causal explanations of the observed relationships. Randomization ensures that variation across sample units in z (and p, and q, and however many other factors might be relevant) is assigned randomly with respect to our experimental variable, x. Although we still can't attribute the response of a single sample unit to its value of x (it may, after all, have an extreme value of z), this ensures that our estimate of the statistical relationship between y and x has the desirable property of *consistency*. In brief, a consistent statistical estimator converges to the quantity being estimated—a clear prerequisite to valid statistical inference. Inconsistency feels a lot like bias in that it leads to systematically incorrect parameter estimates and p-values. However, unlike bias, which attenuates with sufficiently large sample size, inconsistency is structural, and cannot be remedied simply by collecting more data.

However, by their very nature experiments are limited in spatial and temporal extent, and abstract away much of the complexity of real ecological systems. To demonstrate causality in the field, ecologists have made use of natural experiments, in which natural variability in x across sites (e.g., differences in lizard density across islands) is used as a substitute for an experimental treatment. The problem, of course, is that a lot of other stuff besides x may be varying across sites. The usual prescription is to make sure that

Ecological Statistics: Contemporary Theory and Application. First Edition. Edited by Gordon A. Fox, Simoneta Negrete-Yankelevich, and Vinicio J. Sosa. © Oxford University Press 2015. Published in 2015 by Oxford University Press.

the sites are as similar as possible in every aspect other than x, but this is limited to the variables we can measure, and will be difficult to maintain across more than a few variables anyway. Furthermore, this isn't the right analogy to controlled experiments. True experiments don't "hold everything else equal;" they hold some things equal, but randomize to account for everything else. Natural experiments, even if carefully matched across a set of potential confounding variables, lack this randomization, so we cannot be sure that the observed variation in y is caused by x or by an unobserved z. We have seemingly lost our ability to make strong causal conclusions, which leads many ecologists to discount natural experiments, chanting the mantra "correlation does not imply causation." This view was reinforced by a number of influential books in the latter part of the 20th century (e.g., Underwood 1997), and the standard biometry text books focus heavily on analysis of experimental data, maintaining an artificial distinction between ANOVA and linear regression (see Introduction chapter).

In many cases we want to do more than simply determine the presence or absence of a causal relationship (the usual outcome of an ANOVA), and estimate the magnitude and direction of the relationship (information provided by the parameter estimate in a regression). If the unobserved variable z is correlated with x, the predictor variable of interest, then, even if there is a strong causal relationship between x and y, this relationship *cannot* be estimated from the simple statistical association between x and z (figure 7.1).

Economists have had to face this problem head-on, because most of the questions they are interested in are not amenable to controlled experiments, and they aim to inform decisions in the real world outside the academy. Thus most of their tests of causality have to make use of natural experiments, and they have developed techniques that allow them to estimate the causal relationship between x and y in the face of uncontrolled variation in z. One approach that is general and powerful, and almost completely missing from the ecologist's toolbox (Armsworth et al. 2009), is the method of *instrumental variables* (IV), which is the subject of this chapter. An instrumental variable (or more commonly,

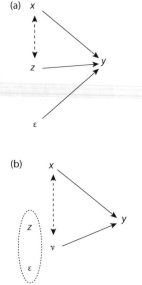

Fig. 7.1 The omitted variable problem. (a) Suppose that y is causally influenced by both x and z (solid arrows), as well as a host of small effects that are collected into the residuals ε. The dominant causal variables x and z are correlated (dashed arrow); this could result from direct or indirect causation in one or both directions, or from statistical non-independence in a finite sample (e.g., x and z have similar patterns of spatial autocorrelation). As long as the residuals are uncorrelated with x and z, then ordinary least squares (OLS) regression (e.g., `lm(y ~ x + z)`) will provide consistent estimates of the coefficients relating each of the predictor variables to the response variable. Each coefficient will be a partial regression coefficient, which accounts for the effects of the other predictor variable—this accurately represents the direct causal effect of each variable on its own. The only effect of the correlation between x and z is to (possibly) inflate the uncertainty around the parameter estimates. (b) If z is unobserved, then the residuals from the *causal* relationship between x and y (not to be confused with the *statistical* relationship that we see in a bidirectional scatterplot) are $v = z + \varepsilon$, which are correlated with x. By construction, OLS regression finds parameter estimates that eliminate any such correlation; these estimates may be consistently larger or smaller than the causal relationship, and may not even have the correct sign!

a collection of instrumental variables) can effectively randomize x with respect to z, allowing us to infer the causal effect of x on y. The actual statistical implementation of this approach is only marginally more complex than least squares regression, and the whole thing can feel a bit magical. The "trick" is that instrumental variables need to meet some very rigorous requirements, so that finding a measurable IV (or even worse, identifying an IV from among a set of measurements that have already been taken) is challenging and may not always be possible.

Before proceeding further, let's take a look at the impact such an unmeasured variable has on a regression equation.

7.2 Endogeneity and its consequences

From a formal statistical perspective, *endogeneity* is the presence of a non-zero covariance between one or more predictor variables and the residuals of the model (the name comes from the fact that one cause of such covariance is bidirectional causality, in which the predictor variable is not independent of the response variable). To see how this can arise, suppose that spider density (y) depends on lizard density (x) and moisture (z): $y \sim x + z$. Furthermore, suppose that moisture also affects lizard density, so that $\text{cov}(x, z) \neq 0$. The full model that we want to estimate is

$$y_i = \beta_0 + \beta_1 x_i + \beta_2 z_i + \varepsilon_i, \tag{7.1}$$

where β_1 is the marginal effect of x on y—that is, a unit increase in x, holding all else equal, will increase y by β_1. This is illustrated by hypothetical data in figure 7.2, where, for ease of visualization, "moisture" (z) takes on just two values. The two predictor variables have opposite effects on y, and are correlated with each other. If we have measurements of all three variables, we can use multiple regression to get consistent parameter estimates.

However, if we haven't measured z, we are left with trying to estimate a short-form model,

$$y_i = \beta_0 + \beta_1 x_i + v_i, \tag{7.2}$$

where $v_i = \beta_2 z_i + \varepsilon_i$; this is the solid line in figure 7.2 (this makes clear that the "error" term in a regression includes the causal effects of unmodeled variables). The problem is that the residuals from this model are not independent from x: $\text{cov}(x, v) = \beta_2 \text{cov}(x, z)$. In figure 7.2, the v_i are the residuals from the solid line to the points; they are mostly negative for small values of x and mostly positive for large values of x. If we ignore this and fit equation (7.2) using OLS, the estimated coefficient of x will be a combination of the direct effects of x on y and the indirect effects via z, giving a slope estimate that bears little relationship to the underlying causal effect (figure 7.2, dotted line). This error is not simply bias, which could be reduced by increasing the sample size. Even in the limit of infinite data, this estimate will not converge on β_1: in the presence of correlations between the residuals and the predictors, OLS parameter estimates are inconsistent. For the data in figure 7.1, the short-form regression gives the wrong sign for the effect of x on y; this will not always be the case, as it depends on the correlation structure among the variables. Nevertheless, even if by good luck the regression estimates the correct sign of the relationship, the confidence intervals and p-values will be meaningless.

In general, we will not be able to identify the existence of the missing variable z by looking at diagnostic plots of the regression of y on x. In the illustrated example, with a single predictor variable and a single omitted variable, it appears that the skew of the

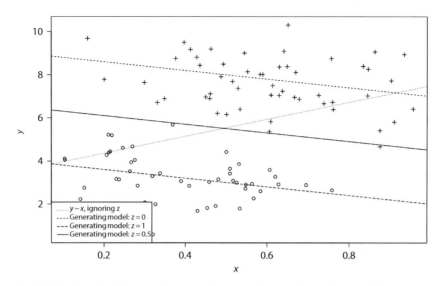

Fig. 7.2 Omitted variables lead to inconsistent estimates. These simulated data were generated by $y_i = 4 - 2x_i + 5z_i + \varepsilon_i$, where z takes on values of zero (circles) or one (plusses) and is correlated with x. The dashed lines show the generating model with z taking on values of zero and one; the solid line is the marginal effect of x evaluated at the "average" value of z (which is ultimately the effect that we are trying to recover). However, regressing y on x gives a slope estimate of 3.93 (dotted line), which is not just the wrong sign but very convincingly so: $P = 0.001$ for the t test against a null hypothesis of zero slope. Code for generating and analyzing these data is in appendix 7.A.

estimated residuals varies as a function of x, which might be revealed as a trend in the conditional median (not mean) of the residuals. However, this is unlikely to generalize to the more typical situation of multiple predictors and omitted variables. Thus, we must use our theoretical understanding of the system to generate lists of putative variables that might causally influence the response variable and could be correlated with the predictor variable. If we have measured these variables then we can add them to the model; but if we are not so fortunate, then we cannot know whether our analysis is science or science fiction.

7.2.1 *Sources of endogeneity*

The endogeneity in the example of the previous section is generated by an omitted variable (confusingly, the phenomenon is conventionally called "omitted variable bias," even though the problem is inconsistency rather than bias). Given the complexity of ecological causality, it is likely that this is a common phenomenon in ecological data. Without measurements of the variable in question, or experimental approaches that randomize the observations with respect to the omitted variable, it is difficult to imagine how to address this problem. We shall see in the next section that instrumental variables, if they can be found, can control for this effect, allowing us to draw causal conclusions.

There are two additional common causes of endogeneity: measurement error and bi-directional causality. It is straightforward to demonstrate that measurement error in a predictor variable generates endogeneity. Suppose that

$$y_i = \beta_0 + \beta_1 x_i + \varepsilon_i, \tag{7.3}$$

but that we have an imperfect observation of x,

$$x_i^* = x_i + v_i. \tag{7.4}$$

If we apply OLS to this imperfect data,

$$y_i = \beta_0 + \beta_1^* x_i^* + \varepsilon_i^*, \tag{7.5}$$

then we are estimating a model with rather different residuals: $\varepsilon_i^* = \varepsilon_i - \beta_1 v_i$.

Typically we expect the measurement error v_i to be uncorrelated with the true value x_i. However, this means that, assuming that the expected value of v is zero,

$$\begin{aligned}
\mathrm{cov}(\varepsilon^*, x^*) &= \mathrm{cov}(\varepsilon - \beta_1 v, x + v) \\
&= \mathrm{cov}(\varepsilon, x) + \mathrm{cov}(\varepsilon, v) - \beta_1 \mathrm{cov}(v, x) - \beta_1 \mathrm{cov}(v, v) \\
&= 0 + 0 - 0 - \beta_1 \mathrm{var}(v).
\end{aligned} \tag{7.6}$$

Thus the residuals between the actual causal relationship and the observed data are correlated with the observed predictor x^*, with sign opposite to the sign of the underlying relationship β_1. The former generates inconsistent estimates; the latter ensures that the inconsistency is in the direction of zero. Because this is inconsistency rather than bias, it cannot be reduced simply by increasing the number of data points in the regression (once again, the conventional terminology—"measurement error bias"—muddies this distinction). This is why precise measurement of the predictor variable is one of the key assumptions of OLS. However, this assumption is rarely discussed in ecological studies that use such models.

This effect is not entirely unknown to ecologists (e.g., Freckleton 2011; Solow 1998). In general, estimating and correcting the magnitude of this "attenuation" requires information about the measurement error variance. If the estimates of x come from a well-characterized statistical sampling process, this variance is available and there are standard techniques for correcting the attenuation (Buonaccorsi 2010). Absent such information, however, there is little that can be done to eliminate the endogeneity directly (although there are specialized tools for certain circumstances, such as using state–space models to estimate parameters of autoregressive processes). However, recognizing that measurement error creates this inconsistency by inducing endogeneity shows that we can use apply tools that address endogeneity directly.

The third major source of endogeneity is bidirectional causality (often called "simultaneity" in the economics literature), in which two variables causally affect each other. For example, Bonds et al. (2012) sought to analyze whether a country's biodiversity affects the disease burden on its human inhabitants. An important control variable for this analysis is the country's per capita income: all else equal, we expect richer countries to have lower disease burdens, because of investment in public health and medical facilities. However, a high disease burden tends to reduce per capita income, by reducing productivity and life expectancy. Clearly, then, per capita income cannot be treated as "independent" in a regression in which disease burden is the dependent variable. The solution is to use a "simultaneous equation model" (not to be confused with a structural equation model; chapter 8), in which two equations, one with disease burden as the predictor variable and the other with per capita income as the dependent variable, are solved simultaneously. There is still endogeneity within each equation, however, which we will address below.

Interestingly, endogeneity can arise even in randomized experiments, if the putative causal variable isn't under direct experimental control. For example, suppose you are using microcosms to study the effect of phytoplankton density on *Daphnia* growth. You manipulate phytoplankton density indirectly by applying one of a number of different nutrient supplies to each microcosm. You know that temperatures are uneven in your growth chamber, so you randomize microcosm placement to ensure that treatments are random with respect to these position effects, and then you regress the observed *Daphnia* growth rate on the observed phytoplankton density (measured as the opacity of the water in the container). However, there may be a problem: the phytoplankton in the microcosms in the "warm" part of the chamber will probably grow faster than average and achieve a higher than average density, given the nutrient supply level; and the *Daphnia* may also grow faster than average, given the phytoplankton density. Thus *Daphnia* in a warm location have a double advantage: they get more food than average, given the experimental treatment; and they grow faster than average, given the amount of food. Despite the experimental randomization with respect to the treatment, the unobserved temperature variation has correlated effects on the causal and response variables, creating endogeneity in spite of the careful experimental randomization. In this case, with a positive relationship between *Daphnia* growth and phytoplankton density, and a positive correlation between the conditional responses of phytoplankton and *Daphnia* to random variation, the regression will overestimate the magnitude (and probably, the statistical significance) of the relationship. You could analyze the response of *Daphnia* to the level of nutrient addition; but while this gives a statistically consistent result, it probably will not be biologically useful.

7.2.2 *Effects of endogeneity propagate to other variables*

By now it should be clear that endogeneity leads to serious problems in identifying and estimating causal relationships in bivariate data. Even more distressing, however, is that when performing multiple regression (more than one predictor variable), a single endogenous variable can cause the estimated coefficients of the other predictor variables to be inconsistent as well. This will occur in the common situation where the predictor variables are correlated with one another. Not surprisingly, the effect becomes stronger as the correlations among the variables becomes stronger. Thus, one badly endogenous variable can spoil a whole regression! In particular, because measurement error is a source of endogeneity, this means that one should think very carefully before introducing a poorly measured variable into a model otherwise comprising accurately measured quantities.

7.3 The solution: instrumental variable regression

Instrumental variable regression solves the endogeneity problem by using a suitable new variable to "instrument"[1] the endogenous predictor variable, controlling for the correlation between the predictor variable and the residuals. Mathematically, this solution is disarmingly simple; the challenge is finding a variable that meets the conditions to be an instrument.

[1] The conceptual origin of the term "instrumental variable" is not recorded; the term was introduced without comment in a 1945 dissertation (Morgan 1990). However, insofar as the instrumental variable is (imprecisely) measuring the variation in the regressor that is not correlated with the error term, the term may have it roots in the idea of a scientific instrument as a measurement device. Of interest to biologists, the technique itself was first developed 20 years previously by Sewall Wright's father, probably with Sewall's input (Stock and Trebbi 2003).

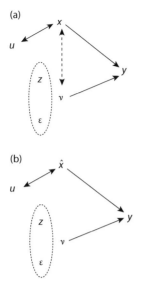

Fig. 7.3 The instrumental variable solution to the omitted variable problem of figure 7.1. (a) The instrumental variable, u, has a causal or structural relationship with the predictor variable x, which creates a correlation between the two variables. In addition, u has no relationship with the response variable, y, except through the pathway involving x; in particular, this means that u is uncorrelated with z, ε, and v. In essence, u captures (some of) the variation in x that is uncorrelated with z and v. (b) Thus, the predicted value of x given u, $\hat{x}$, is uncorrelated with v and hence does not suffer from endogeneity. As a consequence, the OLS regression of y on $\hat{x}$ gives a consistent estimate of the causal relationship between x and y; the cost is greater uncertainty in the parameter estimates.

An instrumental variable must satisfy two conditions (figure 7.3):

(1) *Relevance:* It must be correlated with the endogenous regressor variable; the stronger the correlation, the more effective it will be.
(2) *Exogeneity:* It must have no direct causal effect on the response variable, and be correlated with the response variable *only* because of its relationship to the endogenous regressor variable. This is often called the "exclusion condition".

Finding a variable that satisfies the first criterion is usually easy. The second condition, which has the effect that the instrumental variable is uncorrelated with the residuals of the regression, is the bottleneck. In economics, uncovering a new instrument for an important class of problems can launch a successful academic career. I discuss possible examples of instrumental variables in ecology below; first, let us examine how instrumental variables solve the endogeneity problem.

The general idea is that the instrument represents a treatment that, while not experimentally randomized, is effectively random with respect to the process that introduces endogeneity. As a concrete example, a central feature of life history theory is the notion of a trade-off between current and future reproduction (de Jong and van Noordwijk 1992). Consideration of energetics suggests that, for a given individual, a greater investment in reproduction today will reduce the individual's resources for survival and future reproduction; evolved reproductive strategies are thought to take this into account. When trying to measure this trade-off in the field, it is tempting to simply look for a relationship between a given year's reproduction and some measure of subsequent reproduction (survival, reproductive success in the following year, etc.). However, doing this as simple regression often results in positive estimates of the coefficient for the effect of current reproduction on future success. This occurs because the analysis looks across individuals instead of within individuals, and any variation in individual quality (i.e., some individuals, whether because of good genes, good parents, or good environments, are better than average at both current and future reproduction) tends to swamp the effects of within-individual trade-offs. This unmeasured quality variable is like the omitted variable in figure 7.1. Formally, what we are interested in is the coefficient β_1 in

$$F_i = \beta_0 + \beta_1 C_i + \beta_2 Q_i + \varepsilon_i, \tag{7.7}$$

where F is future reproductive success, C is current reproductive success, and Q is quality, which in addition to affecting future reproductive success as shown in the equation, is causally related to current reproductive success,

$$C_i = \gamma_0 + \gamma_1 Q_i + \delta_i. \tag{7.8}$$

If Q is not observed, then running the regression $F \sim C$ will be analogous to fitting equation (7.2), and will clearly give the wrong estimate for β_1.

What we need is a way to manipulate current reproduction independently of quality. Thinking specifically of birds, we can imagine that nest predators might do this for us. We have to think carefully about this, because it might be that high-quality individuals (from the perspective of potential reproductive success) are also better at defending against predators, but in at least some cases we could be justified in concluding that this correlation is weak at best (a rigorous argument would entail a detailed consideration of antipredator behavior, nest siting, etc.). Now consider circumstances in which predators consume all eggs before they hatch, and the pair doesn't renest. At this point the male's energetic investment is modest, and we might expect that males who have had their eggs depredated should have, on average, more resources for future reproduction than those who haven't. If, on average, birds whose nests were depredated have the same quality as those whose nests were not depredated, then a regression of future success on depredation status should not suffer from endogeneity, and so the resulting parameter estimate is consistent.

However, the relationship between depredation status and future success is not what we are interested in. Define depredation status to be instrumental variable, Z. As an IV, it must be correlated with C but have no direct causal relationship with F. Another way of saying the latter is that adding Z to the full model does not improve the fit of the model. A third way of saying it is that Z is uncorrelated with both Q and with ε ; this implies that Z is also uncorrelated with the residuals of $F \sim C$.

To get an unbiased estimate of the parameter describing the effect of current reproduction on future success, conditioned on quality, we must combine the relationship between depredation and future success with the relationship between current reproduction and depredation. First we estimate the relationship between the predictor variable (current reproduction) and the instrumental variable (depredation status):

$$C_i = \alpha_0 + \alpha_1 Z_i + \omega_i.$$

This is called the "first-stage regression." Then we estimate the model we conceptually described when we introduced the depredation variable,

$$F_i = a + bZ_i + e_i. \tag{7.9}$$

This is the "reduced form regression." Through some rather tedious statistical theory, it can be shown that $\rho = b/\alpha_1$ is a consistent estimate of β_1 (e.g., Greene 2008).

Although the above calculation is straightforward for that simple example, it is rather more complex when there are multiple predictor and/or instrumental variables. Instead, this analysis is commonly done using a procedure called *two-stage least squares* (2SLS), which scales to models of arbitrary complexity. It starts with the first-stage regression, but then creates a synthetic variable that is, in effect, the predictor variable scrubbed of its endogeneity. First, estimate the coefficients from the first-stage regression, equation (7.9), which we will call $\hat{\alpha}_0$ and $\hat{\alpha}_1$. Then, use these to estimate the expected value of C given Z:

$$\tilde{C}_i = \hat{\alpha}_0 + \hat{\alpha}_1 Z_i. \tag{7.10}$$

where $\tilde{C}$ is the conditional expected value of current reproduction, conditioned on depredation status. Then regress F on $\tilde{C}$ (the "second-stage regression"); the resulting regression coefficient for $\tilde{C}$ is also a consistent estimate of β_1 (for the simple model here, it is identical to ρ).

To develop a graphical intuition, we return to the simulated data of figure 7.2. Figure 7.4(a) shows the first-stage regression, the relationship between the predictor variable (x) and an instrumental variable (u). Notice that, whereas u is uncorrelated with z, the residuals from the first-stage regression are almost perfectly correlated with the omitted variable, z (positive residuals for $z = 1$; negative residuals for $z = 0$). Thus, the projection of x onto the regression line (which is the conditional expected value of x, $\hat{x}$) is uncorrelated with z. Panels (b) and (c) show that, while x is on average lower for $z = 0$ than for $z = 1$, $\hat{x}$ has a very similar distribution for the two values of z. Finally, figure 7.4(d) shows that the second-stage regression is very close to the generating model evaluated at the mean value of z, which is the relationship that we are actually trying to recover. Contrast this with the simple regression of y on x shown in figure 7.2.

Getting correct standard errors of the parameter estimate requires some more work, so in practice we use software that does this automatically. Traditionally economists use Stata, which has lots of tools for instrumental variable regression. In R, the two functions

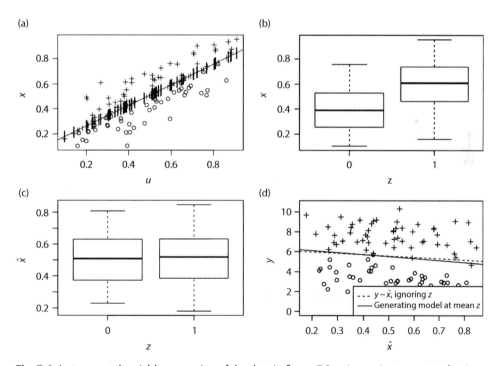

Fig. 7.4 Instrumental variable regression of the data in figure 7.2, using an instrument u that is positively correlated with x but uncorrelated with z. (a) Relationship between x and u, showing the data (circles and plusses) and the fitted values ($\hat{x}$; vertical hashes on the diagonal). Plusses: $z = 1$; circles: $z = 0$. (b) On average, x is larger when $z = 1$ than when $z = 0$. (c) In contrast, $\hat{x}$ is uncorrelated with z. (d) The regression of y on $\hat{x}$ is very similar to the generating model evaluated at the mean value of z.

Table 7.1 Results of fitting the data in figure 7.2 using OLS with the generating model and without the omitted variable, and using two-stage least squares with an instrumental variable. The "Generating model" column shows the values of the coefficients used to generate the simulated data. Standard errors of parameter estimates in parentheses

Coefficient	Generating model	OLS y~x + z	OLS y~x	2SLS y~x | u
Intercept	4	3.99 (0.29)	3.57 (0.61)	6.24 (0.82)
x	−2	−1.87 (0.61)	3.93 (1.13)	−1.35 (1.55)
z	5	4.72 (0.25)	–	–

that have been developed are `ivreg()` in the AER library (Kleiber and Zeileis 2008) and `tsls()` in the sem library (Fox et al. 2013). The former (which I use in this chapter) has a slightly simpler syntax; for basic regressions it works just like `lm()` except that the formula includes a vertical bar followed by the instrumental variables; e.g., `ivreg(y ~ x | u)`. The fitted model can be viewed with `summary()`, generating output very similar to that for `lm()`. Parameter estimates and standard errors for the full OLS regression, the short-form regression, and the 2SLS regression are shown in table 7.1. Although the standard error on the 2SLS slope estimate is large, the estimate itself is clearly better than the short-form regression (the error in the intercept might be improved by centering the data—see chapter 4).

7.3.1 *Simultaneous equation models*

Bidirectional causality is best solved through a system of simultaneous equation models. For example, in examining the effects of disease on economic outcomes, Bonds et al. (2012) write down two equations, one for disease burden as a function of income and some other variables, and the other for income as a function of disease burden and some other variables. As discussed above, there is endogeneity between disease burden and income, so these equations cannot be estimated via ordinary least squares. Instead, instruments are identified for both income and disease burden; and in the equation for disease burden, for example, the income variable is replaced with its predicted value from the first-stage regression. The same is done for the other equation.

This model can be estimated using two-stage least squares, as for the single-equation model. However, another technique, called three-stage least squares, is more efficient; but it requires additional restrictions on the instrumental variables (notably, that they be uncorrelated with the residuals from *all* of the equations, not just the equation to which they are being applied).

The easiest way to fit these types of models in R is with the `systemfit` library. The documentation for this library (Henningsen and Hamann 2007) has a good overview of the various statistical approaches to fitting these models.

7.4 Life-history trade-offs in Florida scrub-jays

As a concrete example of the life-history trade-off problem, let's look at some data from the population of Florida scrub-jays (*Aphelocoma coerulescens*) that has been studied at Archbold Biological Station since the early 1970s (Woolfenden and Fitzpatrick 1984). The focus here is on the trade-off between current-year reproductive success and survival over the succeeding winter. For reasons that will become clear in a bit, I focus on reproductive males.

The file `JayData.csv` has the following variables:

- `fledge_number`: the number of offspring that a breeding pair fledged in a given year. This is our measure of reproductive success.
- `dad_survival`: a zero/one variable indicating whether the male of the breeding pair survived to the subsequent breeding season. This is our measure of subsequent fitness.
- `hatching_success`: a zero/one variable indicating whether the breeding pair hatched any chicks in the year. This will be our instrumental variable.

First let's look at a simple OLS regression.

```
> jays <- read.csv("JayData.csv")
> jay_lm <- lm(dad_survival ~ fledge_number, data=jays)
> summary(jay_lm)
Call:
lm(formula = dad_survival ~ fledge_number, data = jays)
Residuals:
    Min         1Q      Median        3Q         Max
-0.7881     0.2232      0.2288     0.2458      0.2458
Coefficients:
               Estimate Std. Error t value Pr(>|t|)
(Intercept)    0.754193   0.019067  39.555   <2e-16 ***
fledge_number  0.005659   0.008024   0.705    0.481
---
Signif. codes:   0 *** 0.001 ** 0.01 * 0.05 . 0.1   1
Residual standard error: 0.4246 on 1204 degrees of freedom
Multiple R-squared: 0.0004128, Adjusted R-squared: -0.0004174
F-statistic: 0.4973 on 1 and 1204 DF, p-value: 0.4808
```

Thus, subsequent survival is positively related to fledgling number, although the coefficient is not significantly different from zero. Given that the response variable is binary, one might reasonably argue that logistic regression is more appropriate for these data; however, the consistency guarantee of IV regression only applies to least squares models. Just to check that the least squares assumption isn't too egregious, let's look at a logistic regression; the result is qualitatively similar (the magnitudes of the coefficients are different, because of the transformation of the response variable; but the p-value for the effect of `fledge_number` is identical):

```
> jay_glm <- glm(dad_survival ~ fledge_number, data=jays,
        family=binomial)
> summary(jay_glm)
Call:
glm(formula = dad_survival ~ fledge_number, family = binomial,
    data = jays)
Deviance Residuals:
    Min       1Q    Median       3Q      Max
-1.7599   0.7109    0.7208   0.7512   0.7512
Coefficients:
               Estimate Std. Error z value Pr(>|z|)
```

```
(Intercept)     1.12077    0.10465   10.709    <2e-16 ***
fledge_number   0.03149    0.04463    0.705     0.481
---
Signif. codes:  0  ***  0.001  **  0.01 *  0.05 .  0.1   1
(Dispersion parameter for binomial family taken to be 1)
    Null deviance: 1316.5 on 1205 degrees of freedom
Residual deviance: 1316.0 on 1204 degrees of freedom
AIC: 1320
Number of Fisher Scoring iterations: 4
```

So does this mean that there is no trade-off? Well, as described above, there might be differences in individual quality that affect both fecundity and survival, but in this data set we have no measures of quality that we could include in the regression. So now let's look for an instrumental variable: a quantity that is correlated with fledgling number, but has no direct causal effect on survival, and is not correlated with individual quality.

The Florida scrub-jay is a social breeder, and the presence of helpers is known to increase reproductive success (Woolfenden and Fitzpatrick 1984). It is at least plausible that helpers have little effect on parental survival, and thus their presence or number could serve as an instrument. However, helpers are the offspring of those same breeders from previous years of successful reproduction, so that breeders that, because of quality, have higher than average fledgling numbers, are also more likely to have helpers. Thus the presence of helpers is probably correlated with breeder quality, disqualifying them as an instrument.

Another candidate variable is nest failure, that is, the loss of all eggs in the nest prior to hatching, whether because of predation or because of the nest falling to the ground. Certainly we would expect this variable to be correlated with fledgling number, as the latter will always be zero when no eggs hatch (this species will sometimes renest; here I use a "hatching success" variable that is zero only if all nests fail to hatch any chicks). Female breeders will have expended some energy laying these eggs, so their reproductive investment is not zero; but male breeders whose nests fail to hatch have expended almost no reproductive energy. If hatching failure is effectively random across male breeders, then comparing the survival of those who hatched chicks and those who did not should reveal the average cost of raising an average-sized brood (conditioned on hatching).

The question is whether hatching success is correlated with male quality. It is probably not correlated with energetic or physiological quality; but predator defense ability is a form of quality that might affect both fledgling number and individual survival. In addition, helpers may help defend the nest from predators, and as mentioned above, the presence of helpers is probably associated with breeder quality. Thus, hatching success is at best an imperfect instrument; but it is probably less correlated with the omitted quality variable than is raw fledgling number. Thus we will try using it as an instrumental variable in a two-stage least squares.

```
> library(AER)
> jay_2sls <- ivreg(dad_survival ~ fledge_number | hatching_success,
          data=jays)
> summary(jay_2sls)
Call:
ivreg(formula = dad_survival ~ fledge_number | hatching_success,
    data = jays)
```

```
Residuals:
    Min         1Q      Median         3Q          Max
-0.8209     0.1791     0.2100      0.2719       0.4265
Coefficients:
               Estimate Std. Error t value Pr(>|t|)
(Intercept)     0.82090    0.03699  22.191   <2e-16 ***
fledge_number  -0.03092    0.01913  -1.617    0.106
---
Signif. codes: 0  ***  0.001  **  0.01  *  0.05 . 0.1   1
Residual standard error: 0.4282 on 1204 degrees of freedom
Multiple R-Squared: -0.01684,    Adjusted R-squared: -0.01769
Wald test: 2.614 on 1 and 1204 DF, p-value: 0.1062
```

The result is suggestive of a trade-off: every extra fledgling reduces male survival probability by 0.031, or 3.1 percentage points. The standard errors are relatively large, and under ecological conventions for α levels we cannot reject the null hypothesis of no effect. Nevertheless, we have relatively high confidence that the effect is not negative, and theory says that this estimate is more reliable than the one from the OLS regression.

7.5 Other issues with instrumental variable regression

The mathematical proofs that IV regression gives consistent parameter estimates depend both on the linearity of the response and the use of least squares, so IV cannot be used with generalized linear regression or non-linear regression. This is why I used least squares in the scrub-jay analysis, even though the response is a dichotomous variable. With some care, this is not as much of a shortcoming as it might seem. For example, as long as the fitted probabilities are not too close to zero or one, the assumption of linearity on an untransformed scale is not too bad (and for other types of GLM, an appropriate transformation can often be applied to the response variable to linearize the relationship). The only effect of applying least squares when the theoretical residuals are not Normally distributed is to generate biased standard errors—the parameter estimates are consistent and unbiased—and this effect gets small as sample size gets large.

Although the IV estimates are consistent (as long as the instrument truly meets the criteria outlined above), they can still be biased with a finite sample size. In general, for a given sample size, this bias is lower the better the instrument is at predicting the endogenous variable. A commonly cited rule of thumb is that the F statistic for the first-stage regression should be at least 10 for the instrument to be useful. If the F statistic is lower than that, then it is called a "weak instrument," and the estimates from the IV regression are likely to be no more reliable than the inconsistent estimates from the OLS regression. If there are multiple instruments, and some of them are very weak, then retaining them in the IV regression may tend to increase the bias. In that case, it may be beneficial to prune variables from the first-stage regression in much the same way one would do for model selection in ordinary multiple regression (with the caveat that any exogenous variables must be left in).

We run this test simply by explicitly running the first-stage regression using OLS. Here is the scrub jay example:

```
> jay_1st <- lm(fledge_number ~ hatching_success, data=jays)
> summary(jay_1st)
Call:
lm(formula = fledge_number ~ hatching_success, data = jays)
Residuals:
    Min      1Q    Median      3Q       Max
 -2.0513  -1.0513   0.0000   0.9487   5.9487
Coefficients:
                   Estimate Std. Error t value Pr(>|t|)
(Intercept)      -8.184e-15  1.194e-01    0.0        1
hatching_success  2.051e+00  1.266e-01   16.2   <2e-16 ***
---
Signif. codes: 0 *** 0.001 ** 0.01 * 0.05 . 0.1    1
Residual standard error: 1.382 on 1204 degrees of freedom
Multiple R-squared: 0.179, Adjusted R-squared: 0.1784
F-statistic: 262.6 on 1 and 1204 DF, p-value: < 2.2e-16
```

We see that $F_{1,1204} = 262$, so bias will not be an issue here.

Starting with version 1.2, the AER library added a helpful diagnostics argument to the summary() function for IV regression objects:

```
> if (packageVersion("AER") >= 1.2)
      summary(jay_2sls, diagnostics=TRUE)
Call:
ivreg(formula = dad_survival ~ fledge_number | hatching_success,
    data = jays)
Residuals:
    Min      1Q    Median      3Q       Max
 -0.8209   0.1791   0.2100   0.2719   0.4265
Coefficients:
              Estimate Std. Error t value Pr(>|t|)
(Intercept)    0.82090    0.03699  22.191   <2e-16 ***
fledge_number -0.03092    0.01913  -1.617    0.106
Diagnostic tests:
                  df1   df2   statistic  p-value
Weak instruments   1   1204    262.577   <2e-16 ***
Wu-Hausman         1   1203      4.546    0.0332 *
Sargan             0    NA        NA       NA
---
Signif. codes: 0 '***' 0.001 '**' 0.01 '*' 0.05 '.' 0.1 ' ' 1
Residual standard error: 0.4282 on 1204 degrees of freedom
Multiple R-Squared: -0.01684, Adjusted R-squared: -0.01769
Wald test: 2.614 on 1 and 1204 DF, p-value: 0.1062
```

The thing to look at here is the statistic reported for the "weak instruments" test—it is the F from the first-stage regression that we calculated above (there are a variety of ways of calculating p-values for the weak instruments test; the precise meaning of the p-value reported here is not documented). The "Wu–Hausman" and "Sargan" tests are discussed below.

The examples so far have had a single predictor variable, which was endogenous. More generally, we will want to perform regressions with multiple predictor variables, some of which may be exogenous and others endogenous. To do instrumental variable regression in this case, we need at least as many instruments as there are endogenous variables; and the exogenous variables need to be included in both the first-stage and second-stage regressions. Thus, for example, if x_1 and x_2 were exogenous, x_3 and x_4 were endogenous, and u_1, u_2, and u_3 were each instruments for one or the other (or both) endogenous variables, then the call to `ivreg` would be `ivreg(y ~ x1 + x2 + x3 + x4 | x1 + x2 + u1 + u2 + u3)`. My economics colleagues tell me that it is relatively rare to instrument more than one endogenous variable in a single equation, however, as it is quite difficult to interpret the result unless the problem is set up very carefully (e.g., Angrist 2010).

It is often the case that there are more instruments than endogenous variables. This is fine; subject to the caveats about very weak instruments above, to the extent that multiple instruments improve the explanatory power of the first-stage regression, they will improve the final model. Having multiple instruments also allows one to conduct "overidentification" tests, the most common of which is the Sargan test reported by the AER library. Under the assumption that at least one of the instruments is exogenous, then the Sargan test tests the null hypothesis that all of the other instruments are also exogenous. A small p-value from this test suggests that at least one of the instruments is endogenous and fails the exclusion condition; in contrast, a large p-value can be evidence of valid instruments. However, without a priori confidence that at least one of the instruments is valid, the outcome of this test is meaningless.

7.6 Deciding to use instrumental variable regression

Although the computational mechanics of running an instrumental variable regression are quite simple, the identification of observed variables that meet the assumptions required to serve as instruments is often quite challenging. Thus IV regression should not be undertaken lightly. Unfortunately, there is no simple diagnostic test of your original data that will tell you whether your predictor variables are endogenous. Endogeneity means that the predictor variable is correlated with the actual residuals, but the latter are not observable; OLS regression, by its very design, produces estimated residuals that have zero linear correlation with the predictor variables. As I suggested above, one might imagine that a single omitted variable with a single predictor variable might produce heteroscedasticity in the skew of the estimated residuals; but I know of no statistical theory or practice to support that conjecture, and even if it applies, it will not help with other sources of endogeneity.

The only solution is to think hard and critically about potential relationships among your variables, just as we do when designing field experiments: Are the quadrats truly randomly distributed with respect to the statistical population that I want to draw inferences about? How will the raptor perches created by my exclosures affect the ecological processes I am interested in? Here are some questions that you can ask about your data:

• What is the likely magnitude of measurement error in my predictor variables? Can I estimate it, and can I apply one of the other techniques for correcting measurement error bias?

- What are all of the processes that might causally affect my response variable? Of the ones whose effects are not guaranteed to be small, how might they be correlated with the predictor variables that I actually want to draw inferences about?
- Are there any feedback loops in my system? If so, are there plausible pathways whereby my response variable might directly or indirectly affect my predictor variable?

If these questions lead to the conclusion that endogeneity is likely in your model, then you should try to identify one or more instrumental variables (see section 7.7).

Once you have identified some instrumental variables, and you are confident that they are good instruments, then you can use them to test whether your predictor variable really is endogenous. This is important to do because if the predictor variable is actually exogenous, then the OLS regression is already consistent and the IV regression is both less efficient (confidence intervals will be wider) and more prone to bias than the OLS regression.

The standard test for endogeneity is the Wu–Hausman test. The basic idea of this test is that if the predictor variable (x) is exogenous, then the residuals of the first-stage regression describing the relationship between the instruments and the predictor variable ($x \sim u$) will be uncorrelated with the response variable (y). In practice the test is performed by taking the estimated residuals from the first-stage regression and adding them as an additional variable in the short-form equation; then using OLS regression to test the null hypothesis that the coefficient of these residuals is zero. Failing to reject this null hypothesis leads to the conclusion that the predictor variable is exogenous; rejection leads to the conclusion that the predictor variable is endogenous, and IV regression is required.

Here is the procedure for the scrub-jay example:

```
> jay_1st <- lm(fledge_number ~ hatching_success, data=jays)
> jay_wh <- lm (dad_survival ~ fledge_number + resid(jay_1st),
              data=jays)
> summary(jay_wh)
Call:
lm(formula = dad_survival ~ fledge_number + resid(jay_1st),
   data = jays)
Residuals:
    Min       1Q     Median       3Q       Max
-0.8209    0.1791    0.2296    0.2432    0.2705
Coefficients:
                Estimate Std. Error t value Pr(>|t|)
(Intercept)      0.82090    0.03662   22.415   <2e-16 ***
fledge_number   -0.03092    0.01894   -1.633   0.1027
resid(jay_1st)   0.04456    0.02090    2.132   0.0332 *
---
Signif. codes: 0  ***  0.001 ** 0.01 * 0.05  .  0.1     1
Residual standard error: 0.4239 on 1203 degrees of freedom
Multiple R-squared: 0.004176, Adjusted R-squared: 0.00252
F-statistic: 2.522 on 2 and 1203 DF, p-value: 0.0807
```

The p-value for the first-stage residuals is 0.033, so we reject the null hypothesis of exogeneity, and conclude that the IV regression is the appropriate analysis for this model. Notice that this p-value is reported by diagnostics=TRUE argument to the summary() function for IV regression objects (see section 7.5).

Of course, if the instruments are not appropriate—they are poor predictors of the predictor variable (which can be seen in the first-stage regression) or are not themselves exogenous (which can only be evaluated through reasoning)—then the results of the Wu–Hausman test are meaningless.

7.7 Choosing instrumental variables

How does one find a good instrument? It is relatively easy to determine whether a candidate instrument is strongly correlated with the endogenous variable—that is a simple statistical test. However, the second criterion, that the instrument affects the response variable *only* through its effect on the endogenous variable, requires careful thought. The case of endogeneity arising from measurement error is probably the easiest, as it is relatively easy to find an instrument that is uncorrelated with the measurement error process. However, even here it is critical to ensure that there are no other pathways by which the instrumental variable can influence the response variable.

One fairly straightforward situation is when addressing endogeneity in an experiment, where the experimental treatment is not the predictor variable (e.g., the hypothetical phytoplankton–*Daphnia* experiment, where phytoplankton density was the predictor variable, but it was manipulated indirectly using variable nutrient input). Here, the experimental treatment itself is a clear instrumental variable: it is correlated with the predictor variable (if it was not, then the experiment would have failed), and has no direct effect on the response variable (if it does, then the whole rationale for the experiment breaks down). Basically, you are predicting the response variable using the *expected* value of the predictor variable, conditioned on the experimental treatment, rather than the observed value of the predictor variable. This removes the potential correlation between the responses of the predictor variable and the dependent variable to the uncontrolled variation among experimental units.

In a time series context, *lagged* variables can often fit the bill. This was used by Creel and Creel (2009) in estimating the population dynamics of Rocky Mountain elk. The authors fit a density-dependent Gompertz model,

$$r_t = \log(N_{t+1}/N_t) = a + b\log(N_t) + \beta E_t + \varepsilon_t, \tag{7.11}$$

where E is a vector of environmental variables such as snow depth and wolf density. The problem is that population abundance, N_t, is measured with substantial uncertainty, and this measurement error will tend to bias the density dependence parameter (b) toward zero. The authors addressed this with two techniques, instrumental variable regression and state-space modeling, and found that both gave similar results. As instruments they used lagged values of the environmental variables, E_{t-1} and E_{t-2}; the authors do not explicitly state their reasoning, but it would typically be that such variables may affect the "true" value of N_t (through their effects on past population dynamics) but have no direct effect on either the current population growth rate or the measurement error. As is common practice in economics when instruments are generic rather than carefully crafted to emulate an experiment, squares and interactions of the lagged environmental variables were included as well. The authors did not report the F statistic of the first-stage regression, but noted that $R^2 = 0.20$, which they considered to be "large considering the overall noise in the data."

The only other example of instrumental variables in the ecological literature at the time of writing draws heavily on economic reasoning. Hanley et al. (2008) evaluated the effects of grazing on plant biodiversity over the past 400 years in Scotland. Although historical plant diversity could be reconstructed from pollen records, there were not data on stocking rates over the entire study period. The authors used meat price as a proxy variable, arguing that no single farmer influenced price and so would increase stocking rate when prices were high without endogenously affecting price. However, there are times when price is high because of a regional collapse of stock, perhaps due to an unusually harsh winter or a disease outbreak. Thus, the authors sought instrumental variables that would indicate an unusual spike in demand or trough in supply: "the English population (a measure of market demand); the presence or absence of garrisons in a particular region (which increases local demand); the passing of the Act of Union between England and Scotland (this reduced trade barriers); the advent of refrigerated transport from the New World (this reduced demand for UK meat); and grain prices, as a substitute for meat in consumption." These were used as instruments for meat price in the first-stage regression, such that the fitted value of price from the first-stage model can be treated as an exogenous proxy for stocking rates in the second stage. Note that dealing with this endogeneity between supply, demand, and price is the bread and butter of econometrics.

An important use of instrumental variables in the social sciences is in program evaluation: e.g., determining whether a public health program targeted at poor people results in improved health outcomes. The challenge comes from potential selection bias: the service is more likely to be used by people in particularly poor health, who may still be sicker than average after treatment. The solution is to use *program eligibility* to instrument program participation, essentially comparing outcomes for those who were eligible, regardless of participation, with those who were not. This works as long as there is some source of arbitrariness in the eligibility rule. If the rule is issued based on the same factors that may affect your outcome of interest, then it is not a good instrument. Of course, such analyses need to control for a variety of other variables as well. Similar analyses have been done on the effects of higher education on subsequent earning, using access to education as an instrument for actual enrollment to address the endogeneity arising from the fact that individuals who are ambitious enough to go to college may be predisposed to success in the workplace.

Similar approaches may be useful to evaluate the ecological effects of various human activities. For example, the effects of marine protected areas (MPAs) on biodiversity and species abundance are usually assessed by comparing measurements inside an MPA with a nearby site that is not protected. However, within a planning area, the MPAs are not sited at random, and may be consistently higher or lower in recovery potential than sites that are not protected. One can imagine instrumenting protection status of a site with "eligibility to be protected," i.e., is the site inside the planning area that was being considered when setting up the MPAs? Of course, this requires that there be sites outside the planning area that are broadly ecologically comparable to those in the planning area, but in many cases that may be possible.

The biggest challenging in determining whether a variable is a suitable instrument is verifying the exogeneity criterion. Just as in the initial assessment of potential sources of endogeneity, this requires a critical analysis of potential causal pathways, based on a scientific understanding of the system. For example, in the elk model, one can imagine that the previous year's precipitation could affect vegetation structure and greenness in the current year, which could, in turn, affect the detectability of elk (and hence the measurement error residual). The plausibility of such a scenario would have to be evaluated using prior knowledge about the effects of rainfall on vegetation and about the effects of vegetation on detectability. An ideal analysis would examine such potential critiques with as much quantitative and qualitative data as is available.

7.8 Conclusion

In summary, instrumental variable regression may provide a way to develop robust causal interpretations of patterns in observational ecological data sets. The challenge in this technique is not technical, but in the conceptual identification of suitably independent instruments. Until we ecologists become more practiced at thinking through all of the ways that putative instruments might have hidden correlations with the factors causing endogeneity, it may be best to discuss any such analyses with economist colleagues (as was done, for example, by Creel and Creel 2009; Hanley et al. 2008). For those who want to delve more deeply into the subject, some useful econometrics textbooks with extensive treatments of instrumental variables include Angrist and Pischke (2009) and Wooldridge (2010).

A general checklist for doing instrumental variable regression is as follows (relevant chapter sections are in parentheses):

(1) Determine plausible or known sources of endogeneity in your predictor variables, considering measurement error, omitted variables, and bidirectional causality (7.2.1).
(2) Identify instrumental variables (7.7).
(3) Run the first-stage regression to determine if the instruments are strong enough to be useful ($F > 10$) (7.5).
(4) If you have multiple instrumental variables, use the Sargan test to identify any that might be exogenous (and remove them!) (7.5).
(5) Run the Wu–Hausman test of endogeneity (7.6).
(6) If the instruments are sufficiently strong, and the Wu–Hausman test confirms endogeneity, then run the IV regression. Otherwise, stick with your original OLS regression (7.3).

In this chapter, I have focused on consistent estimates of regression parameters. However, endogeneity also renders p-values meaningless, precluding hypothesis tests and qualitative assessments of causality. Instrumental variable regression solves this problem: although the power can be substantially reduced, the p-value is accurate and can be used in statistical hypothesis tests. Furthermore, by effectively randomizing with respect to the potential sources of endogeneity, a well-constructed instrumental variable provides as robust a demonstration of causality as does a well-designed experiment.

Acknowledgments

I thank Gordon Fox, Simoneta Negrete, and Paulina Oliva for feedback and encouragement on the text. The Florida scrub-jay data appear courtesy of Glen Woolfenden, John Fitzpatrick, Reed Bowman, and the Archbold Biological Station. This work was funded by NSF grant number DEB-1120865.

CHAPTER 8

Structural equation modeling: building and evaluating causal models

James B. Grace, Samuel M. Scheiner,
and Donald R. Schoolmaster, Jr.

8.1 Introduction to causal hypotheses

Scientists frequently wish to study hypotheses about *causal* relationships, but how should we approach this ambitious task? In this chapter we describe structural equation modeling (SEM), a general modeling framework for the study of causal hypotheses. Our goals will be to (a) concisely describe the methodology, (b) illustrate its utility for investigating ecological systems, and (c) provide guidance for its application. Throughout our presentation, we rely on a study of the effects of human activities on wetland ecosystems to make our description of methodology more tangible. We begin by presenting the fundamental principles of SEM, including both its distinguishing characteristics and the requirements for modeling hypotheses about causal networks. We then illustrate SEM procedures and offer guidelines for conducting SEM analyses. Our focus in this presentation is on basic modeling objectives and core techniques. Pointers to additional modeling options are also given.

8.1.1 *The need for SEM*

Consider a task faced by the US National Park Service (NPS), the monitoring of natural resources. For documenting conditions, they can use conventional statistical methods to quantify properties of the parks' ecosystems and track changes over time. However, the NPS is also charged with protecting and restoring natural resources. This second task requires understanding cause–effect relationships such as ascribing changes in the conditions of natural resources to particular human activities. Causal understanding is central to the prevention of future impacts by, for example, halting certain human activities. Active restoration through effective intervention carries with it strong causal assumptions. These fundamental scientific aspirations—understanding how systems work, predicting future behaviors, intervening on current circumstances—all involve causal modeling, which is most comprehensively conducted using SEM.

Ecological Statistics: Contemporary Theory and Application. First Edition. Edited by Gordon A. Fox, Simoneta Negrete-Yankelevich, and Vinicio J. Sosa. © Oxford University Press 2015. Published in 2015 by Oxford University Press.

Modeling cause–effect relationships requires additional caveats beyond those involved in the characterization of statistical associations. For the evaluation of causal hypotheses, biologists have historically relied on experimental studies. SEM allows us to utilize experimental and observational data for evaluating causal hypotheses, adding value to the analysis of both types of information (e.g., Grace et al. 2009). While experimental studies provide the greatest rigor for testing individual cause–effect assumptions, in a great many situations experiments that match the phenomena of interest are not practical. Under these conditions, it is possible to rely on reasonable assumptions built on prior knowledge to propose models that represent causal hypotheses. SEM procedures can then be used to judge model–data consistency and rule out models whose testable implications do not match the patterns in the data. Whether one is relying on experimental or non-experimental data, SEM provides a comprehensive approach to studying complex causal hypotheses.

Learning about cause–effect relationships, as central as it is to science, brings with it some big challenges. We emphasize that confidence in causal understanding generally requires a sequential process that develops and tests ideas. SEM, through both its philosophy and procedures, is designed for such a sequential learning process.

Beyond testing causal hypotheses, we are also interested in estimating the magnitudes of causal effects. Just because we have properly captured causal relationships qualitatively (A does indeed affect C through B) does not guarantee arriving at unbiased and usable estimates of the magnitudes of causal effects. Several conditions can contribute to bias, including imperfect temporal consistency, partial confounding, and measurement error. The consequences of such biases depend on their context. In many ecological studies, SEM analyses are aimed at discovering the significant connections in a hypothesized network. In such studies, the relative strengths of paths are the basis for scientific conclusions about network structure. In other fields such as medicine or epidemiology, often the focus of a causal analysis may be on a single functional relationship that will be used to establish regulatory guidelines or recommended treatments (e.g., isolating the magnitude of causal effect of a drug on the progress of an illness). The general requirements for SEM are the same in both situations, but the priorities for suitable data for analyses and levels of acceptable bias may differ. Investigators should be aware of these distinctions and strive to obtain data suitable to their study priorities.

8.1.2 *An ecological example*

In this chapter, we use data from Acadia National Park, located on the coast of Maine (USA) for illustration. At Acadia, wetlands are one of the priority ecosystem types designated for protection. For these ecosystems, both resource managers and scientists wish to know how things work, what kinds of changes (favorable or unfavorable) they can anticipate, and what active steps might prove useful to protect or restore the wetlands.

Acadia National Park is located on Mount Desert Island, a 24,000 ha granite bedrock island. As a consequence of its mountainous topography, wetlands on the island are in numerous small watersheds. The soils are shallow in the uplands while the wetlands are often peat-forming. Additionally, they receive their water largely from acidic and low-nutrient inputs of rain and surface runoff, making them weakly-buffered systems (Kahl et al. 2000). For our illustrations, we use data from 37 nonforested wetlands recently studied by Little et al. (2010) and Grace et al. (2012) who examined the effects of human development on biological characteristics, hydrology, and water quality.

The studies measured various types of historical human activities in each wetland catchment area: (1) the intensity of human development in a watershed, (2) the degree of hydrologic alteration, (3) human intrusion into the buffer zone around wetlands, and (4) soil disturbance adjacent to wetlands. A human disturbance index (HDI) of the summed component measures was used to identify biological characteristics of plant communities that serve as bioindicators of human disturbance (figure 8.1a; Schoolmaster et al. 2012). Altered environmental conditions were also recorded, including water conductivity (as an indicator of nutrient loading) and hydroperiod (daily water depth). A subset of key biological characteristics was chosen to represent components of biotic integrity (Grace et al. 2012). We focus here on one biological property, native plant species richness, and its relationship with human disturbance as shown in figure 8.1b. For this example, we want to know how specific components of the human disturbance index might lower species richness and what might be done to reduce such impacts.

(a)

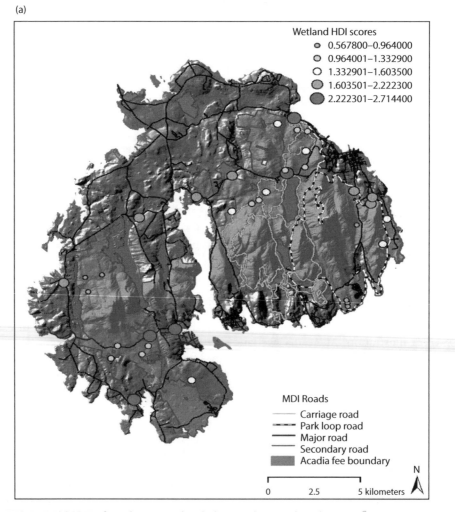

Fig. 8.1 (a) Map of Acadia National Park showing human disturbance index scores. (b) Native richness (species per plot) against land-use intensity scores.

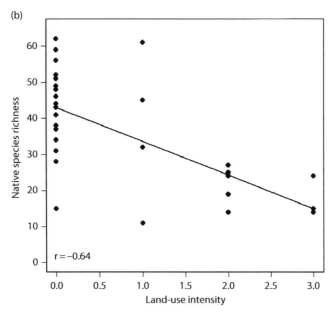

Fig. 8.1 (*continued*)

8.1.3 *A structural equation modeling perspective*

In our example, information from previous studies allows us to propose a hypothesis about how human activities and environmental alterations can lead to a loss of native species. In figure 8.2a we first represent our ideas in the form of a *causal diagram* (Pearl 2009) that ignores statistical details and focuses on hypothesized causal relationships. The purpose of a causal diagram is to (a) allow explicit consideration of causal reasoning, and (b) guide the development and interpretation of SE models. What we include in such a diagram is a function of our knowledge and the level of detail we wish to examine. Causal diagrams, distinct from structural equation models, are not limited by the available data.

Several causal assumptions implied in figure 8.2a are represented by directional arrows. The causal interpretation of *directed relationships* is that if we were to sufficiently manipulate a variable at the origin of an arrow, the variable at the head of the arrow would respond. In quantitative modeling, a relationship such as $Y = f(X)$ is assumed to be causal if an induced variation in X could lead to changes in the value of Y (see also the book Introduction). Generally, we must be able to defend a model's causal assumptions against the alternative that the direction of causation is the opposite of what is proposed or that relationships between variables are due to some additional variable(s) affecting both and producing a spurious correlation.

In this example, the assumptions expressed are: (1) Increasing land use in a watershed leads to more physical structures (ditches and dams) that control or alter hydrology. (2) Physical structures that influence hydrology can lead to changes in water-level variations (flooding duration). (3) Reduced water-level fluctuations (e.g., resulting from impoundment of wetlands) would create a plant community made up of the few highly flood-tolerant species. Collectively, these fundamental assumptions are only partially testable with observational data, since actual responses to physical manipulations are required to demonstrate causality unequivocally.

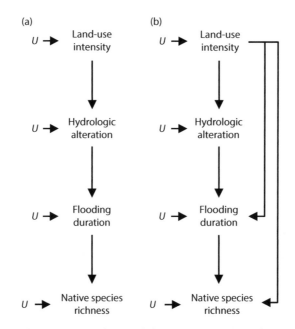

Fig. 8.2 (a) Simple causal diagram representing the hypothesis that there is a causal chain connecting land use to richness reduction through hydrologic alteration and subsequent impacts on flooding duration. The letter U refers to other unspecified forces. (b) Alternative diagram including additional mechanisms/links.

There are several ways data could be inconsistent with the general hypothesis in figure 8.2a. The direct effects encoded in the model might not be detectable. Also, the omitted linkages implied in the model might be inconsistent with the relations in the data. It is entirely possible that land use leads to changes in flooding duration, community flood tolerance, or native richness in ways not captured by the observed hydrologic alterations. Such additional omitted mechanisms (e.g., figure 8.2b) would result in residual correlations among variables not connected by a direct path. As explained later, these alternative possibilities are testable (i.e., these are "testable implications" of a model).

Generally, it is natural to think of cause–effect connections in systems as component parts of *causal networks* that represent the interconnected workings of those systems. *Structural equations* are those that estimate causal effects and a *structural equation model* is a collection of such equations used to represent a network (or portion thereof). Defined in this way, we think of causal networks as properties of systems, causal diagrams as general hypotheses about those networks, and structural equation models as a means of quantifying and evaluating hypotheses about networks. SEM originated as path analysis (Wright 1921); however, it has now evolved well beyond those original roots.

SEM represents an endeavor to learn about causal networks by posing hypotheses in the form of structural equation models and then evaluating those models against appropriate data. It is a process that involves both testing model structures and estimating model parameters. Thus, it is different from statistical procedures that assume a model structure is correct and only engage in estimating parameters. A key element of SEM is the use of graphical models to represent the causal logic implied by the equations (e.g., figure 8.2a).

SEM is a very general methodology that can be applied to nearly any type of natural (or human) system. At the end of the chapter we list and provide references for a few of the types of ecological problems that have been examined using SEM.

8.2 Background to structural equation modeling

The history of SEM and its mathematical details are beyond the limited space of this chapter, though a brief description of the equational underpinning of SE models is given in appendix 8.1. References to both general and specific topics related to this background are presented in the Discussion (section 8.4). Here we focus on fundamental principles related to the development and testing of causal modeling.

8.2.1 *Causal modeling and causal hypotheses*

Achieving a confident causal understanding of a system requires a series of studies that challenge and build on each other (e.g., Grace 2006, chapter 10). Any SEM application will have some assumptions that will not be explicitly tested in that analysis. Thus, SEM results will support or falsify some of the proposed ideas, while implying predictions that are in need of further testing for some of the other ideas. *SEM results should not be taken as proof of causal claims, but instead as evaluations or tests of models representing causal hypotheses.* With that qualifying statement in place, we can now ask, "What are the requirements for a causal/structural analysis?"

Structural equations are designed to estimate causal effects. We say "designed" to connote the fact that when we construct a SE model, we should be thinking in terms of cause–effect connections. More strictly, we are thinking about probabilistic dependencies as our means of representing causal connections. Careful causal thinking can be aided by first developing conceptual models and/or causal diagrams that focus on processes rather than just thinking about the available variables in hand. Each directed relationship in a causal model can be interpreted as an implied experiment. Each undirected relationship (e.g., double-headed arrow) connecting two variables is thought to be caused by some unmeasured entity affecting both variables. Further, in causal diagrams, the unspecified factors (U) (figure 8.2) that contribute to residual variances can be thought of as additional unmeasured causal agents (although they may also represent true, stochastic variation). Ultimately, our intent is to craft models that match, in some fashion, cause–effect relations. This is a more serious enterprise than simply searching for regression predictors. By our very intention of seeking causal relations, the onus is placed on scientists to justify causal assumptions. A strength of SEM is its requirement that we make these assumptions explicit.

The phrase "no causes in, no causes out" encapsulates the fact that there are certain assumptions embedded in our models that cannot be tested with the data being used to test the model. These untested assumptions include the directionality of causation. Such assumptions have to be defended based on theoretical knowledge; sometimes that is easy, sometimes it is more challenging. While links are not tested for directionality, we can still evaluate consistency in proposed direct, and *indirect effects*, as well as statements of *conditional independence*.

One point that is sometimes overlooked (and is commonly treated as implicit) is the assumption that causes precede effects. We should, for example (figure 8.2a), recognize that the plant diversity of today has been influenced by the flooding duration during some prior time period. Similarly, the flooding duration this year is influenced by hydrologic alterations made in prior years. It is not uncommon that the data may fail to strictly meet the precedence requirements desired for causal effect estimation. When proper temporal precedence does not hold, we must assume temporal consistency, meaning that current values of a predictor are correlated with values when the effect was generated. For example, we might only have data on flooding duration for a single year and have to assume that the variation among sites in duration was similar in past years. Such assumptions are not always reasonable. In such cases, one needs to develop dynamic SE models using time-course data (e.g., Larson and Grace 2004).

8.2.2 *Mediators, indirect effects, and conditional independence*

Arguably the most fundamental operation in SEM is the test of *mediation* (MacKinnon 2008). In this test we hypothesize that the reason one system property influences another can be explained by a third lying along the causal path. In our example (figure 8.2a), we hypothesize that one reason plant species richness is lower in areas with greater human land use is because of a series of processes involving hydrology that mediate/convey an effect. The ability to express causal chains and indirect effects is a distinguishing attribute of structural equation models (appendix 8.1). When we specify that flooding duration is influenced by land-use intensity through hydrologic alterations, we are making a causal proposition representing causal hypotheses that can be tested with observational data for model-data consistency. Tests of mediation are most powerful when a SEM analysis leads an investigator to conduct a follow-up study or to obtain additional measurements that permit possible mediators to be included in models.

Model–data consistency is critical for obtaining proper parameter estimates. First and foremost, variables not directly connected by a single or double-headed arrow in a model are presumed to exhibit conditional independence—that is, having no significant residual associations. Finding residual associations can suggest either an omitted direct, causal relationship or some unmeasured joint influence. Depending on model architecture, omitted links may result in biased estimates for some parameters. In figure 8.2a we pose the hypothesis that the effects of land use on flooding duration are due to hydrologic alterations. If we find that our data indicate that land use and flooding duration are not conditionally independent once we know the hydrologic alterations, either the intensity of land use influences flooding duration in ways unrelated to observable hydrologic alterations or an unmeasured process is causing the association. If land use and flooding duration are causally connected through two pathways (direct and indirect), both need to be included in the model to obtain unbiased estimates of effects along the causal chain (e.g., figure 8.2b).

Typically, after discovering a residual relationship (e.g., a significant correlation among residuals), we would revise our model either to include additional linkages or alter the structure of the model so as to resolve model–data discrepancies. It is critical that model revision be based on theoretical thinking and not simply by tinkering to improve model fit, otherwise our modeling enterprise is just a descriptive exercise rather than a test of a hypothesis or theory.

It can be helpful to know the minimum set of variables needed to be measured and modeled so as to properly specify a model, especially if one is working from a causal diagram. A general graphical-modeling solution to this problem, the *d-separation criterion*, has been developed by Pearl (1988). We omit describing this somewhat intricate concept and instead refer the reader to a more complete treatment in Pearl (2009).

8.2.3 *A key causal assumption: lack of confounding*

A classic problem in causal modeling is to avoid *confounding* (see chapter 7). Confounding occurs when variables in a model are jointly influenced by variables omitted from the model. Identifying and including the omitted variables can solve this problem, as represented in figure 8.3a. Here there is some factor U′ that jointly influences both intensity of land use and hydrologic alterations. If we are unaware of such an influence and estimate effects using a SE model that treats the two variables as independent, then, the directionality of linkages may still be causally correct, but our parameter estimate linking the two will be biased. An extended discussion of how confounding affects causal estimates can be found in Schoolmaster et al. (2013). Here, we consider only a single illustration.

Let us imagine a case where there is a planning process that determines which watersheds to develop and how many hydrologic alterations to install to support those potential developments. If planners assessed the topographic suitability for both land development and modifications of hydrology, we would need to include some measured variable to represent this decision process in our model if we are to avoid bias in our parameters. Figure 8.3b illustrates how we resolve this problem. By including a measured variable representing a planner's perceived topographic favorability in our SE model, we block the "back-door connection" between intensity of land use and hydrologic alterations (for more on the back-door criterion see Pearl 2009).

8.2.4 *Statistical specifications*

There is a relationship between how models are specified and how their parameters can be estimated. Options for statistical specification (response distributions and link forms;

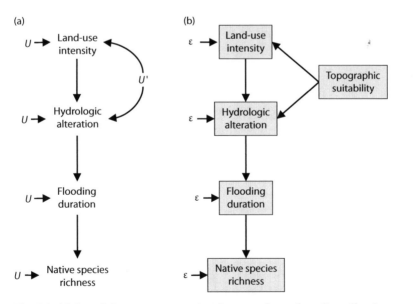

Fig. 8.3 (a) Causal diagram representing the case of a confounding effect by U′, an unmeasured factor that influences both ends of a causal chain, the effect of intensity of land use on hydrologic alteration. (b) A structural equation model that resolves the potential confounding by including measurements of the factor creating the confounding in (a).

e.g., Poisson responses with log-linear linkage) are well covered in conventional statistics textbooks and other chapters in this volume (see chapters 5, 6, 7, 13, and 14). In this chapter we provide a few examples of various response specifications, present some guidelines for the order in which specification choices might be considered (e.g., figure 8.5), and mention some of the criteria that may be used.

In any statistical model, including SE models, we must choose a probability distribution for each response (*endogenous*) variable and the form of the equation for relating predictors to responses. It is common to assume linear relationships with Gaussian-distributed independent errors, but we are not restricted to this assumption and a SE model can include any form for a particular causal relationship, including logistic, quadratic, and binary. Of course, for any functional form one must be cognizant of the statistical assumptions involved related to the data, model specifications, and estimation methods. The choices of model specification and estimation methods will depend on both the form of the data and the questions being asked. Each method has its array of specific assumptions and potential hazards and limitations, a topic too vast to cover in this chapter. We urge the reader to be cautious with any analysis, but especially when using unfamiliar procedures.

8.2.5 *Estimation options: global and local approaches*

There are two general approaches to parameter estimation in SEM, a single global approach that optimizes solutions across the entire model and a local-estimation approach that separately optimizes the solutions for each endogenous variable as a function of its predictors (figure 8.4). Much of the focus in SEM in the past few decades has been on *global estimation*, where data–model relationships for the entire model are summarized in terms of variance–covariance matrices (upper analysis route in figure 8.4). Maximum likelihood procedures (see chapter 3) are typically used to arrive at parameter estimates by minimizing the total deviation between observed and model-implied covariances in the whole model. Sometimes, alternatives to maximum likelihood, such as two-stage least squares, are used (Lei and Wu 2012).

Maximum likelihood global estimation typically relies on fitting functions such as

$$F_{ML} = \log |\hat{\Sigma}| + tr(S\hat{\Sigma}^{-1}) - \log |S| - (p+q). \tag{8.1}$$

Here, F_{ML} is the maximum likelihood fitting function, $\hat{\Sigma}$ is the model-implied covariance matrix, S is the observed covariance matrix, while $(p+q)$ represents the sum of the exogenous and endogenous variables. For a discussion of the statistical assumptions associated with estimation methods used in SEM, refer to Kline (2012). Global analyses have historically not used the original data, but instead only the means, variances, and covariances that summarize those data. This simplification allows for the estimation of a tremendous variety of types of models, including those involving latent (unmeasured) variables, correlated errors, and nonrecursive relations such as causal loops and feedbacks (Jöreskog 1973). (In appendix 8.2 we illustrate a simple application of this type of analysis.)

The alternative to global estimation is *local estimation*, estimating parameters for separable pieces of a model (figure 8.4, lower analysis route). Modern approaches to local estimation are implemented under a graphical modeling perspective (Grace et al. 2012). Consider the causal diagram in figure 8.2b. In graphical models we often talk about nodes and their links. The graph here has four nodes and five links; nodes that are directly linked are said to be adjacent. Causal relations within the graph can be described using familial terminology; adjacent nodes have a *parent-child relationship*, with parents having causal

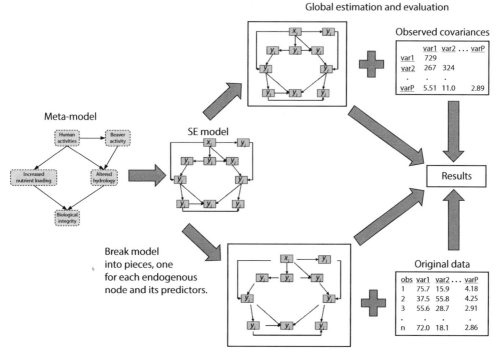

Fig. 8.4 Comparison of global- to local-estimation procedures. Starting with a meta-model based on a priori ideas, an SE model is defined. SE models can be analyzed either under a global-solution framework or through piecewise estimation of local relationships. While analytical procedures differ, both approaches represent implementations of the SEM paradigm.

effects on children. Three of the four nodes are endogenous because they have parents within the diagram; land-use intensity is *exogenous* because it does not have a parent. The node for hydrologic alteration has one parent (land-use intensity), while the nodes for flooding duration and native species richness both have two parents. While hydrologic alteration is an ancestor of native species richness, it is not a parent because there is no direct linkage. So, for the SE model, we have four equations representing the four parent-child relationships. These equations are of the form, $y_i = f(pa_1 + pa_2 + \cdots + \varepsilon_i)$, with one equation for each child node in the diagram and where y_i is any response variable, pa_i refers to the parent variables for each response variable, and ε_i is the residual variation. A local solution approach involves estimating the parameters for each of those four equations separately. Once that is done, there needs to be a separate analysis (and confirmation) of the conditional independence assumptions before the estimates are to be trusted.

Local estimation is a useful alternative because it permits great latitude for the inclusion of complex specifications of responses and linkages. It is also potentially advantageous because it avoids propagating errors from one part of a model to another, which can happen with global-estimation methods. Further, Bayesian estimation procedures optimize parameters locally in most cases, seeking optimal solutions for individual equations rather than the overall model. Bayesian estimation of SE models is increasingly popular (Lee 2007; Congdon 2010; Song and Lee 2012; for ecological applications see Arhonditsis et al. 2006; Grace et al. 2011, 2012). Here we present the local-estimation approach as an umbrella that permits a wide variety of statistical estimation philosophies, including Bayesian, likelihood, and frequentist methods (see chapter 1).

Despite philosophical preferences one may have for global-estimation versus local-estimation approaches, practical considerations are of overwhelming importance when considering the options, as we illustrate in section 8.3. Without question, the capabilities of available software are an important consideration and both software and instructional materials supporting SEM are continuously evolving. In the next section we provide further guidance for the choice of estimation method and how it relates to both model specification and modeling objectives.

8.2.6 *Model evaluation, comparison, and selection*

Few problems in statistics have received more attention than the issue of how models are critiqued, evaluated, and compared. This can ultimately be viewed in the context of a decision problem. The question is, "What variables should I leave in my model?" or, alternatively, "Which of the possible models should I select, based on the data available?" For models that represent *causal networks*, the question is a bit different. Here we wish to know, "Are the linkages in a structural equation model consistent with the linkages in the causal network that generated the data?" In this situation there should be theoretical support for any model that is considered, as we are not shopping for some *parsimonious* set of predictors; instead, we are seeking models representing realistic causal interpretations.

An important consideration in causal modeling is that it combines theoretical a priori knowledge with the statistical analysis of data. We bring some context to this enterprise by distinguishing between SEM applications that are model-generating versus model-comparing versus confirmatory. These applications represent a gradient from situations where we have relatively weak confidence in our a priori preference for a particular model to situations where we have great confidence in our a priori model choice. The companion ingredient is our degree of confidence in the data. For example, if we have a very large and robust sample, we must give the data strong priority over our initial theoretical ideas. Conversely some data sets are relatively weak, and we may have greater confidence in our views about the underlying mechanisms. The extreme example of this theory-weighted case is in system simulation models where data are used only to estimate parameters, not to critique model structure. Therefore, when dealing with models containing causal content, context and judgment matter in arriving at final conclusions about processes. Of course, it is important that one clearly notes any difference between statistical results and any final conclusions derived from other considerations. There is a parallel here to the issue of using informed priors in Bayesian estimation. In some cases it is appropriate to weight posterior estimates based on prior information, but we must make clear what the new data say when we arrive at conclusions. (This theme, that one needs to use judgment and not blindly rely on statistical procedures, is also touched upon in the introductory chapter and chapters 2, 3, 5, and 7.)

While model performance and model support have different nuances—assessments of explanatory power versus relative likelihoods, respectively—in this treatment we do not emphasize this distinction. When evaluating network models, there are always both implicit and explicit comparisons. Further, evaluating model predictions, explanations, or residuals informs us about both our specification choices and also whether we have overspecified our models. Ultimately, in causal modeling there are many different kinds of examinations (including comparisons to previous or subsequent analyses) that contribute to model selection and the inferences drawn.

The classical approach to evaluating SE models using global-estimation methods is based on the function shown in equation (8.1) (see section 8.2.5). That function goes to zero when there is a perfect match between observed and model-implied covariances. In contrast, when evaluating SE models using local-estimation methods, one evaluates individual linkages. This process of testing the topology of the model, while compatible with global-estimation methods, is more general, applies to any network-type model, and is essential with local-estimation methods. The first step in local estimation is usually to determine whether there are missing connections in the model, such as testing for conditional independence. Each unlinked pair of variables can be tested for a significant omitted connection (Shipley 2013). Information-theoretic methods, such as the Akaike Information Criterion, are commonly used for model comparisons, both for the global-estimation and local-estimation cases. chapter 3 covers the theory behind AIC methods.

8.3 Illustration of structural equation modeling

8.3.1 *Overview of the modeling process*

In this section we provide general, practical guidelines for SEM and illustrate core techniques and their application using our ecological example. Grace et al. (2012) present an updated set of guidelines for the SEM process (figure 8.5), which we briefly describe here. First, be clear on the goals of your study (step 1). The specific goals and the focus of an

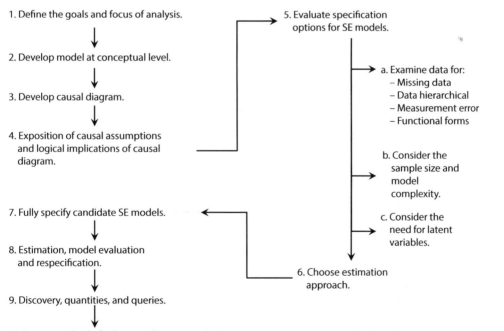

Fig. 8.5 Steps in the modeling process (from Grace et al. 2012).

analysis influence the data needed, model specifications, and estimation choices. Explicitly articulating the conceptual model (step 2), both verbally and graphically, is critical for conveying the logic that translates concepts into variables and ideas about processes into models made up of those variables (Grace et al. 2010). These goals can then be used to consider what is needed for drawing particular causal inferences as well as evaluating the testable *causal propositions* (steps 3 and 4).

A number of things need to be considered when developing a fully-specified model (steps 5–7). The characteristics of the data must be evaluated, both for the purpose of attending to data issues (Are there missing data? Do variables need transformation?) and for informing decisions about the equational representations (Are data hierarchical? Are non-linear relationships anticipated?). One must decide how complex to make the model. Model complexity is influenced by many factors, including objectives, hypothesis complexity, available measurements, number of samples, and the need for latent variables to represent important unmeasured factors. All of these choices influence the choice of estimation method, based on the criteria previously discussed comparing global versus local approaches. See Grace et al. (2010) for more background on model building.

For the next step (step 8), the processes of model estimation and model evaluation/comparison, it is ideal if there is a candidate set of models to compare. However, in SEM the issue of the overall fit of the data and model is of paramount importance. An omitted link is a claim of conditional independence between two unconnected variables, a claim that can be tested against the data. It is possible that all of the initially considered models are inconsistent with the data. In that case, you need to reconsider the theory underpinning the models and develop a revised hypothesis about the system, which can be subsequently evaluated. Once no missing links are indicated, the question of retaining all included links can be addressed. This is inherently a model comparison process. Only when a final suitable model is obtained are parameter estimates to be trusted. At that point parameter estimates, computed quantities, and queries of interest are summarized (step 9) and used to arrive at final interpretations (step 10).

8.3.2 *Conceptual models and causal diagrams*

The conceptual model for our ecological example (figure 8.6a) represents a general theoretical understanding of the major ways human activities impact wetland communities in this system. The conceptual model, termed a *structural equation meta-model* (SEMM; Grace and Bollen 2008; Grace et al. 2010), represents general expected dependencies among theoretical concepts. The SEMM provides a formal bridge between general knowledge and specific structural equation models, serving both as a guide for SE model development and as a basis for generalizing back from SEM results to our general understanding.

In the example, our general hypothesis (figure 8.6a) is that human activities primarily affect wetlands through changes in hydrology and water chemistry, especially elevated nutrient levels (Grace et al. 2012). Here we focus on two biological responses, cattail (*Typha*) abundance and native species richness (figure 8.6b). Cattails are invasive in this system and known to dominate under high-nutrient conditions. Thus, high cattail abundance is an undesirable state while high native species richness is a desirable state.

The causal diagram (figure 8.6b) is a statement of the processes operating behind the scene. This particular model does not portray the dynamic behavior of the system. Instead, it represents a static set of expectations appropriate to the data being analyzed. Given that simplification, we need to carefully consider the unmeasured (*U*) variables and associated processes if we are to avoid confounding (see section 8.2.3).

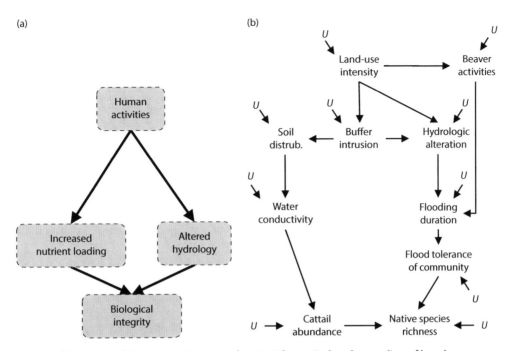

Fig. 8.6 (a) Meta-model representing general a priori theoretical understanding of how human activities most commonly affect wetland communities in cases such as this one. (b) Causal diagram representing a family of possible hypotheses about how specific activities might affect one particular component of integrity, native plant richness.

In this study, several environmental covariates were considered as possible confounders of relationships (distance from the coast, watershed size); ultimately none were considered to be sufficiently important for inclusion.

Causal diagrams can include variables that we did not measure such as hypothesized processes for which we have no direct measures (termed *latent variables*). A strength of SEM is the ability to include latent variables and evaluate their effects on observed quantities (Grace et al. 2010). Another alternative at our disposal is to absorb the effects of some variables in a *reduced-form model*. For example, we hypothesize that human activities may influence beaver populations and that the species pool for plants may be limited to flood tolerant species (figure 8.6b); however, we chose to not include beavers in our SE model (figure 8.7) because we lack appropriate data. Instead, the model has a direct link from intensity of land use to flooding duration to represent that process (thus, absorbing the node for beavers in the causal diagram). We also omit the variable "flood tolerance of community" even though data exist, because our sample size is small and we wish to keep our SE model as simple as possible. As all of this illustrates, the complexity of our SE models may be constrained for a variety of practical reasons.

8.3.3 *Classic global-estimation modeling*

In this section we demonstrate global-estimation approaches to model specification, estimation, and evaluation, including different implementations of SEM. We begin with a popular R library (lavaan; <u>la</u>tent <u>va</u>riable <u>an</u>alysis) that implements SEM using maximum likelihood methods that seek global solutions. Further information about

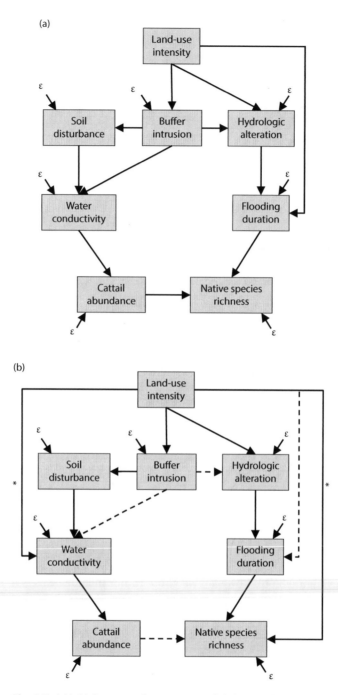

Fig. 8.7 (a) Initial structural equation model. (b) Revised model based on global analysis using `lavaan`. Paths with asterisks were added based on the discovery of non-independence. Paths represented by dashed lines were not supported (i.e., non-significant) based on the sample data.

lavaan can be found in Rosseel (2012, 2014). For simplicity of presentation, we assumed linear Gaussian relations throughout the model, the default setting of lavaan. We know that assuming Gaussian residuals is not appropriate for some of the variables in this model. For example, the variables representing human activities are all ordered categorical measurements. The lavaan library has an option for declaring ordered categorical variables that permits a more appropriate specification, though we do not use it in this demonstration for simplicity of presentation.

Once we begin the estimation process, a first task is to determine whether there are missing links that should be included for model–data consistency. When evaluating model fit, one should be aware that perfect fit automatically occurs when a model is saturated, i.e., there are as many parameters estimated as there are variances plus covariances. This is usually the situation when all possible links in a model are included. For any given model being evaluated, observed discrepancy is compared to a saturated model using a chi-squared statistic. The *model degrees of freedom* is the difference between the number of known covariances and the number of parameters estimated. The subsequent p-value represents the asymptotic probability that the data are consistent with the candidate model. Because in SEM our "default model" is our a priori theoretical model, not a null model, we use a logic that is the reverse from that used in null hypothesis testing. The hypothesized model is interpreted as being consistent with the data unless the *p*-value is small; in the case of a small *p*-value, we conclude that the data obtained are very unlikely given the model in hand. A chi-squared test is commonly used for evaluating overall fit and when comparing models differing by only a single link. However, when evaluating SE models we do more than use $p < 0.05$ for model rejection; instead, there are a number of different model fit assessment criteria. The literature relating to ways of assessing fit (and comparing SE models) is voluminous and well beyond what we can cover in this chapter; see Schermelleh-Engel et al. (2003) for further background. Ultimately, our goal is to detect and remedy any omitted associations, as their absence can substantially alter parameter estimates. In contrast, if a model includes unneeded links, their impacts on parameter estimates is generally small.

The lavaan code and some basic fit statistics for the first phase of analysis are shown in box 8.1, with the code presented in Part A. In this example, the very low *p*-value for the chi-squared test in Part B indicates a lack of fit. This lack of fit is reflected in the large residual covariances (differences between observed and model-implied covariances) shown in Part C. These discrepancies in turn are used to produce a set of *modification indices* that suggests ways of adding links to our model that would improve the fit (part D). These suggestions should not be used blindly, as some may make no scientific sense. The investigator must consider what plausible alternative hypotheses are worthy of consideration before re-estimating a new hypothesis.

In our model, several possible omitted linkages are suggested by the modification indices. Developing a revised model based on this kind of information can involve a bit of trial and error because modification indices are not perfect predictors of actual changes in model fit. So, some of the suggested modifications will reduce model–data discrepancy, but others will not. The reason for this paradoxical situation is that the raw material for the modification indices is the residual covariance matrix and large residuals can be created for a variety of indirect reasons. In our example, it appears that we should add a link between buffer intrusion and native species richness because that implied path had the largest modification index. However, there is no reason to think that buffer intrusion would have a direct effect. Ultimately, one must select theoretically supportable modifications and then make changes that seem most reasonable, continuing until a set of

Box 8.1 EXAMINING OVERALL GOODNESS OF FIT AND LOOKING FOR
OMITTED LINKS IN INITIAL MODEL: R CODE AND SELECT RESULTS

```
# PART A: LAVAAN CODE
# creating data object for the analysis
semdat_8 <- data.frame ( landuse, buffer, hydro, flooding,
    richness, soil, cond, cattails)
# specify model
mod_8a <- 'buffer ~ landuse
           hydro ~ buffer + landuse
           flooding ~ hydro + landuse
           soil ~ buffer
           cond ~ soil + buffer
           cattails ~ cond
           richness ~ flooding + cattails'
fit_8a <- sem ( mod_8a,             # estimate model
    data = semdat_8)
summary (fit_8a, rsq = T,           # select results in Part B
    fit.measures = TRUE)
resid ( fit_8a,                     # select results in Part C
    type = "standardized")
modindices ( fit_8a)                # select results in Part D


# PART B: INITIAL MODEL FIT RESULTS
lavaan (0.5-11) converged normally after 60 iterations
  Number of observations                        37
  Estimator                                     ML
  Minimum Function Test Statistic           36.776
  Degrees of freedom                            17
  P-value (Chi-square)                       0.004

# PART C: STANDARDIZED RESIDUAL COVARIANCES
          buffer hydro flooding soil cond cattails richness
landuse
buffer     0.000
hydro      0.000  0.000
flooding  -2.037    NA   NA
soil        NA    0.466 -0.028   NA
cond        NA    0.863 -0.066   NA    NA
cattails   1.232  0.111  0.759  1.485  NA      NA
richness  -1.502 -1.497 -0.303 -0.934 -1.214 -0.732   0.336
landuse    0.000  0.000   NA    0.211 1.311   1.237  -2.034  0.000
```

```
# PART D: SELECT MODIFICATION INDICES
Variable Pair            Implied Path            Modification Index
richness ~~ landuse      richness <-> landuse     8.941
richness ~ buffer        richness <- buffer      10.450
richness ~ landuse       richness <- landuse      8.941
richness ~ cond          richness <- cond         6.232
cond ~ landuse           cond <- landuse          5.513
```

defensible changes is obtained. Working through this we find that including links from land use to native richness and to conductivity (box 8.2, Part E) reduce model discrepancy to generally acceptable levels based on the chi-squared *p*-value (box 8.2, Part F), and in the process the other suggested modifications are resolved.

Model simplification, asking whether our model is parsimonious, is the next phase of evaluation. It turns out our revised model includes some links that may not be supported by the data (box 8.2, Part G *p*-values). One method for deciding whether a link actually

Box 8.2 ANALYSIS OF REVISED MODEL WITH LINKS ADDED: R CODE AND SELECT RESULTS

```
# PART E: LAVAAN CODE FOR MODEL WITH LINKS ADDED (added component
in bold)
mod_8b2 <- 'buffer ~ landuse
           hydro ~ buffer + landuse
           flooding ~ hydro + landuse
           soil ~ buffer
           cond ~ soil + buffer + landuse
           cattails ~ cond
           rich ~ flooding + cattails + landuse'
fit_8b2 <- sem ( mod_8b2, data = semdat_8)       # estimate model
# select results in Parts F and G
summary ( fit_8b2, rsq = T, fit.measures = TRUE)

# PART F: REVISED MODEL FIT
lavaan (0.5-11) converged normally after  67 iterations
  Number of observations                         37
  Estimator                                      ML
  Minimum Function Test Statistic            18.076
  Degrees of freedom                             15
  P-value (Chi-square)                        0.259
  RMSEA                                       0.074
  90 Percent Confidence Interval       0.000 0.180
  P-value RMSEA <= 0.05                       0.348
```

(continued)

Box 8.2 (*continued*)

```
# PART G: PARAMETER ESTIMATES
                    Estimate    Std.err    Z-value    P(>|z|)
buffer ~
    landuse          1.048       0.089      11.743     0.000
  hydro ~
    buffer           0.427       0.396       1.078     0.281
    landuse          0.966       0.468       2.065     0.039
  flooding ~
    hydro           29.854      11.784       2.533     0.011
    landuse         15.263      22.861       0.668     0.504
  soil ~
    buffer           0.211       0.040       5.289     0.000
  cond ~
    soil             0.179       0.102       1.748     0.081
    buffer           0.070       0.058       1.202     0.229
    landuse          0.163       0.064       2.549     0.011
  cattails ~
    cond             1.038       0.145       7.139     0.000
  rich ~
    flooding        -0.082       0.011      -7.571     0.000
    cattails         4.812       3.084       1.560     0.119
    landuse         -6.228       1.546      -4.029     0.000
```

can be removed is the *single-degree-of-freedom chi-squared test,* which is computed for two models that differ by only a single link/parameter. Either standard frequentist or likelihood ratio tests can be used to compare models (see chapter 1). A different approach that is preferred when comparing several alternative models is the use of information theory measures such as the Akaike Information Criterion (AIC). (See Burnham and Anderson (2002) and chapter 3 for more background on AIC.) Here we simply show the results (box 8.3) that led to our pruned final model (figure 8.7b).

8.3.4 *A graph-theoretic approach using local-estimation methods*

A graph-theoretic approach to SEM is non-parametric in the sense that the rules of causal modeling are compatible with any form of statistical specification (Pearl 2012; Grace et al. 2012). There is a great relaxation of restrictive assumptions that occurs if we can work with the original data instead of the derived covariance matrix as in the global-estimation approach. For our SE model (figure 8.7a) there are two aspects of specification we can now reconsider: (a) the distributions of variable responses, and (b) the form (linear or other) of relations between variables. Box 8.4 presents equations for our model that address these issues. Figure 8.8 shows the response distributions for the variables. For hypothesis testing, we are concerned about having residual variation that meets statistical assumptions. Often investigators using a global-estimation approach will use adjustment procedures to normalize residuals or will use resampling procedures to obtain bootstrapped parameter estimates. Local estimation permits us to do much more, as shown in box 8.4.

Box 8.3 MODEL SIMPLIFICATION: EXAMPLE R CODE AND SELECT RESULTS

```
# PART H: PRUNED MODEL ACCEPTED AS FINAL MODEL
mod_8b3 <- 'buffer ~ landuse
            hydro ~ landuse
            flooding ~ hydro
            soil ~ buffer
            cond ~ soil + landuse
            cattails ~ cond
            richness ~ flooding + landuse
            cattails ~ ~ 0 * richness'
fit_8b3 <- sem ( mod_8b3, data = semdat_8)
summary ( fit_8b3, rsq = T, fit.measures = TRUE)

# PART I: FINAL MODEL FIT
lavaan (0.5-11) converged normally after  66 iterations
  Number of observations                            37
  Estimator                                         ML
  Minimum Function Test Statistic               23.180
  Degrees of freedom                                19
  P-value (Chi-square)                           0.230
  RMSEA                                          0.077
  90 Percent Confidence Interval        0.000  0.171
  P-value RMSEA <= 0.05                           0.325
```

   PART J: PARAMETER ESTIMATES

	Estimate	Std.err	Z-value	P(>|z|)
buffer ~				
landuse	1.048	0.089	11.743	0.000
hydro ~				
landuse	1.413	0.218	6.469	0.000
flooding ~				
hydro	35.595	8.116	4.386	0.000
soil ~				
buffer	0.211	0.040	5.289	0.000
cond ~				
soil	0.222	0.097	2.289	0.022
landuse	0.227	0.037	6.145	0.000
cattails ~				
cond	1.038	0.145	7.133	0.000
richness ~				
flooding	-0.080	0.011	-7.498	0.000
landuse	-4.804	1.266	-3.795	0.000

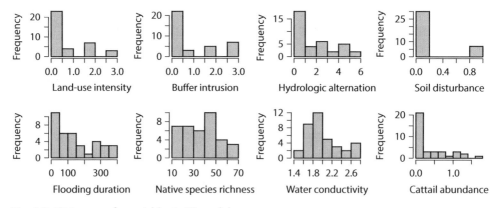

Fig. 8.8 Histograms for variables in SE model.

Box 8.4 INITIAL SPECIFICATIONS FOR LOCALLY ESTIMATED MODEL USING GLM AND LM FUNCTIONS IN R

```
# Buffer as a function of land-use intensity
glm_buffer <- glm ( buffer_prop ~ landuse, family = binomial)

# Hydrologic alteration as a function of land-use intensity
hydro_cat_pred <- ifelse ( landuse == 0, 0.739, 4.21)

# Soil disturbance as function of buffer intrusion
glm_soil <- glm ( soil ~ buffer, family = binomial)

# Flooding duration as function of hydrologic alteration and
    land-use

# intensity.
glm_flooding <- glm ( flooding_prop ~ landuse + hydro, family
    = binomial)

# Water conductivity as a function of soil disturbance and
    land-use intensity
lm_cond <- lm ( cond ~ soil + landuse)

# Native species richness as function of flooding duration and
    cattails
glm_richness <- glm ( richness ~ flooding + cattail,
    family = poisson)

# Cattail cover as change-point function of water conductivity
cattails_step1 <-lm ( cattail ~ cond)
cattails <- segmented ( cattails_step1, seg.Z = ~ cond, psi = 1.9)
```

One issue we address using local-estimation procedures relates to the fact that use of approximate methods for specifying response forms runs the risk of arriving at predicted scores that are not directly comparable to the data. For example, many of the variables in this model are bounded on one or both ends (e.g., species richness, percentage cover of cattails). To prevent predictions falling outside those bounds, particular distributions need to be specified for the responses (see chapters 3 and 6, and book appendix). Consider, for example, the relationships between land-use intensity and hydrologic alteration, and between water conductivity and cattails (figure 8.9). These data clearly do not fit linear models. When humans developed wetland areas, above some minimum value of land-use intensity they always put in structures (e.g., ditches, dams) to control the hydrology, resulting in a discrete (all or none) relationship (figure 8.9a). A more appropriate model representation is a two-level discrete response (figure 8.9c). For the case of cattails, they increase in abundance above some minimum threshold of water conductivity (figure 8.9b). This relationship can be specified with a change-point model (figure 8.9d) using local solution methods (box 8.4).

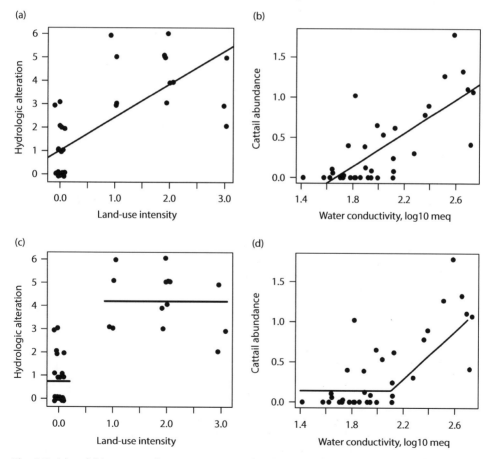

Fig. 8.9 (a) and (b) represent linear approximations of some key relationships, which are of the form typically supported by global-estimation methods (e.g., `lavaan`). (c) and (d) show more complex specifications of the same relationships as in (a) and (b), illustrating what can be included in locally estimated models.

Other response forms used in our model include a proportional odds specification (Agresti 2010) for the ordered categorical response of buffer intrusion (scored as one of four levels), and the proportion data of flooding duration (proportion of days flooded a year). For these variables, we used a logit link to represent the odds of observing maximum versus minimum values (box 8.4). Only two levels of soil disturbance were observed, so this was modeled as a Bernoulli outcome with a logit link (essentially a logistic regression). The degree of hydrologic alteration, although measured on a 6-point scale, behaved as a dichotomous response and was so modeled. Native species richness was modeled as a Poisson (count) variable while log water conductivity was treated as a Gaussian response. For details on the local-estimation procedures for these models, the reader should consult the appropriate R documentation. More justification of the forms used is in Grace et al. (2012). For more information on types of models, see chapter 6 and Bolker (2008).

Under a graph-theoretic approach for determining whether there are missing linkages, the key criterion is whether non-adjacent (unlinked) variables are conditionally independent. To determine whether a link between two variables should be added, we can use procedures that are illustrated in appendix 8.3. First, we obtained the residuals of the current model and then examined relationships among those residuals for unlinked variables. For variables with no predictors, the raw values were used in place of the residuals. Because we wished to consider all functional forms, we used both computational and graphical approaches. In our example, we detected residual associations (figure 8.10) that ultimately led us to a revised model (figure 8.11, box 8.5). This model is slightly different from that based on a global approach, specifically the inclusion of a link between buffer intrusion and native species richness, and the lack of a link between land-use intensity and richness (compare figures 8.7 and 8.10). The differences between the models arise from the use of linear Gaussian specifications in the global model, but more complex forms under local estimation (appendix 8.3).

8.3.5 *Making informed choices about model form and estimation method*

In practice, analyses are conducted by investigators with individual backgrounds, training, and scientific motivations. Analyses are also conducted with different software packages that have their own implementation of methods. Thus, one size does not fit all when it comes to SEM. There are two schools of thought as to which estimation method—global or local—is more appropriate. Pearl (2012) suggests that local estimation is more fundamental because it does not propagate errors caused by misspecification in one part of a model to other parts of a model. However, sometimes models are best seen as a single hypothesis. Consider psychology research where highly abstract concepts are represented exclusively using latent variables and the hypothesis is about how the latent machinery can cause the observed data patterns. In this case we can see an investigator preferring an estimation method that demands a simultaneous global solution. Biological problems with similar priorities might be expected for studies of behavioral ecology, life-history evolution, or organismal physiology. For example, Tonsor and Scheiner (2007) used SEM to relate physiological traits to traits involving morphology and life history, and ultimately to fitness. As the goal of that study was to investigate trait integration, a global-estimation procedure was necessary.

Global-estimation methods are inherently more constrained with regard to detailed statistical specification. With local-estimation methods we can implement highly specific and various response forms and complex non-linear linkages while avoiding propagating

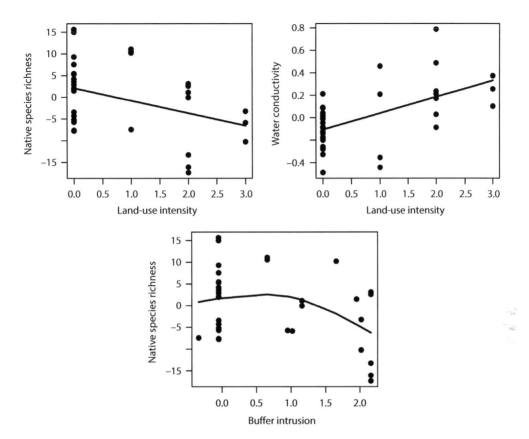

Fig. 8.10 Residual relationships between non-adjacent variables in initial model as revealed using local evaluation methods. The code for producing these graphs and associated results is presented in appendix 8.3.

misspecifications to the estimation of the whole model. In global estimation, declaring one response variable to be non-Gaussian causes the method of estimation of the entire model to change (e.g., from maximum likelihood to weighted least squares) and such changes can have undesirable features. Some software packages have the capacity to address statistical complexities within the global-estimation framework; however, these are always some form of approximate method, so local-estimation methods are generally more flexible in this regard.

When it comes to modeling with latent variables, global-estimation methods excel. By summarizing data and model implications as covariances, global-estimation methods permit estimation and evaluation of very complex latent hypotheses. To include latent variables with local estimations, one must use Bayesian MCMC methods. This approach permits greater flexibility (Lee 2007), but comes with a greater cost of time and expertise for setting up models, running them, and diagnosing problems.

Global estimation also facilitates modeling other types of relationships, including feedbacks, causal loops, and correlations among error terms. These types of relationships can be estimated for observed variable models using local-estimation methods, but with greater effort (appendix 8.3). Bayesian MCMC procedures do not readily permit estimation of models having causal loops. Thus, the choice between global-estimation and

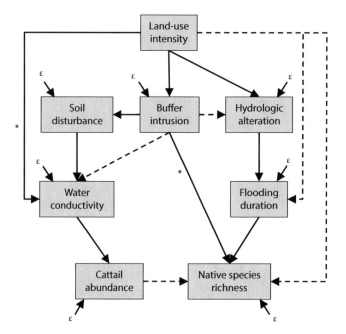

Fig. 8.11 Revised model based on local estimation and evaluation (see appendix 8.3). Paths with asterisks were added based on the discovery of non-independence. Paths represented by dashed lines were not supported by the data. Epsilons signify error variables representing the influences of unspecified factors on each endogenous variable in the model. The model differs slightly from that arrived at using global-estimation methods and linear relations (see figure 8.7b).

local-estimation approaches will depend on the structure of the model and its underlying theory.

Sample size requirements are a complex topic with a large and somewhat inconsistent literature. In general, careful consideration of the relationship between raw sample size (which is generally equal to the number of observations in SEM studies) and model complexity is very important in SEM. When power is low, one might consider reducing model complexity by including only the most important relationships (Anderson 2008). While general recommendations for standard sample size requirements are sometimes found in the older literature, it is inappropriate to require some fixed number of samples because sample adequacy depends on model complexity. Thus, the important issue is the number of samples per parameter (d). Our general advice is that a d of 20 is plenty, a d of 5 is on the low end, and a d of 2.5 is marginal. In our example, we had 37 samples and 9 estimated parameters in our final SE model, for a $d = 4.1$, a ratio of samples to parameters far less than ideal. Often the SEM analysis is intended to provide motivation for further studies that are stronger in numerous regards, including sample size. Because of the dependence of power on both sample size and model structure, it is important to consider the motivating theory and the possible model prior to data collection. Many things influence sample size, including the purpose of the study, the feasibility of large samples, and the need to use previously collected data, so one must be flexible while remaining

Box 8.5 PREDICTION EQUATIONS

```
# buffer response in logits
buffer_hat <- -2.9723 + 2.3232 * landuse
# transform from logits to proportions and rescale to (0:3)
buffer_hat_pr <- ( 1 / ( 1 + 1 / ( exp ( buffer_hat ) ) ) ) * 3

# hydro response: if landuse = 0, hydro = 0.739 else hydro = 4.21
hydro_cat_pred <- ifelse ( landuse == 0, 0.739, 4.21)

# soil response in logits
soilp_hat <- - 4.1617 + 1.6707 * buffer
# transform from logits to proportions and rescale to (0:1)
soilp_hat.pr <- ( 1 / ( 1 + 1 / ( exp ( soilp_hat ) ) ) )

# flooding duration response in logits
floodingp_hat <- -1.3146 + 0.4212 * hydro + 0.6845 * landuse
   - 0.55 * buffer
# transform from logits to proportions and rescale to (0:1)
floodingp_hat_pr <- ( 1 / ( 1 + 1 / ( exp ( floodingp_hat ) ) ) )
   * 365

# water conductivity response, in log-transformed units
cond_hat <- 1.8056 + 0.222 * soil + 0.2266 * landuse

# native species richness as poisson response; (log units)
richness_hat <- 3.998 - 0.00271 * flooding - 0.4556 * buffer3
# transform to linear units
richness_hat_tr <- exp ( richness_hat )

# cattail abundance as segmented/change-point response to
   conductivity
cattails_hat <- 0.14 + 1.4903 * ( ( cond - 2.104 ) > 0 )
   * ( cond - 2.104 )
```

skeptical about conclusions based on limited sample sizes. Not all estimation methods are equally defensible for small samples. Maximum likelihood global estimation is based on large sample theory and can lead to over-fitting when sample sizes are small (Bollen 1989). Local estimation is considered by some to be more tolerant of small sample sizes when a Bayesian MCMC approach is used (Lee and Song 2004).

8.3.6 *Computing queries and making interpretations*

Once final parameter estimates have been obtained for a SE model, it is time to use those estimates for interpretive purposes. A general technique for post-estimation analysis is the *query*. In the context of SEM, a query is a computation made using the prediction equations to yield some specific quantity or set of quantities. Such queries are extremely valuable in summarizing effects in models with non-linear linkages. They are also very useful for expressing the causal predictions implied by a model. Finally, queries allow us to use our hard-earned prediction equations to consider broad ecological implications that emanate from our scientific investigations.

There are four basic kinds of queries. Two are *retrospective*, looking backward in time. One of these is the query of *attribution*, which asks, "What prior conditions and processes led to the observed outcomes?" The second retrospective query is the *counterfactual*, which asks, "What would have happened if?" For example, what if wetland #12 had been exposed to a different set of conditions? There are also two *prospective* queries about future conditions. The *forecast* or prediction is a forward-looking query that extrapolates from current conditions and processes to future outcomes. Weather forecasts are a familiar example. *Interventions* are another kind of prospective query which ask, "What would happen if we changed (intervened upon) the conditions?"

When linear Gaussian models are used, information transfer through the network can be summarized by the multiplicative and additive properties of simple path coefficients (Wright 1921). We can calculate the strength of indirect and total effects, for example, by multiplying the coefficients along compound pathways. Using the `lavaan` model results (box 8.3, part J), the total effect of land-use intensity on native species richness can be computed as the sum of the direct (-4.804) and indirect ($1.413 \times 35.595 \times -0.08 = -4.024$) pathways (total effect = -8.828). When we use more complex model specifications, the computation of effects is more complex as well. In this situation, effects of varying land-use intensity from its minimum to maximum values on native richness through various pathways must be computed using the prediction equations given in box 8.5. Illustrations of such computations are given in the online supporting material for Grace et al. (2012).

One of the most common ways of expressing results is through the computation of standardized effects. An unstandardized effect represents the responsiveness of y to x in raw form. For the case of the total effect value of -8.828 computed above, this is a loss of nearly 9 species per unit increase in land-use intensity category. Such values have stand-alone interpretability and are the raw materials for our prediction equations. In contrast, standardized effects facilitate comparisons and can be used to talk about the relative importance of different variables and pathways. Classical standardization takes an unstandardized slope and multiplies it by the ratio of the standard deviations of x and y. Since slopes are in units of y/x, multiplying by $SD(x)/SD(y)$, puts the standardized coefficients into common units (standard deviations).

While standardized coefficients are widely used and advocated for (e.g., Schielzeth 2010), others have strongly criticized their use (e.g., Greenland et al. 1999), primarily because standard deviation estimates are strongly sample-dependent and using them in computations introduces multiple sources of error into the resulting estimates. In response to these criticisms, Grace and Bollen (2005) proposed an alternative method of standardization that uses the "relevant ranges" of the variables in place of their standard deviations. This changes the interpretation of a standardized coefficient into "the predicted change in y as a proportion of its range of expected values if we were to vary x across its range." Aside from providing a coefficient that has a more intuitive interpretation, this latter method lets the investigator decide the ranges over which slopes are relevant, giving further control to interpretation and comparison. This approach is especially relevant when comparing populations, where standard deviations will certainly not be constant. The use of relevant-range standardization is particularly useful in SEM when using complex model specifications because standard deviations are not good summary statistics for highly non-normal variables. Illustrations of the computation of both classical and relevant-range standardization are given in the online supplement to Grace et al. (2012).

The most basic forecast asks what would be the characteristics of a new sample from the same population. For example, we can ask, "If we took a random sample of wetlands from the same distribution and if cattails respond to conductivity as before, what would be the distributions of conductivity and cattail abundances?" Our interest here is extrapolating from a small sample. What we discover is that our simple extrapolation predicts a broader range of conductivity than seen in our sample but the current range of values for cattails covers the span we might expect from a much larger sample (figure 8.12).

Interventions are alterations of conditions, whether by humans or nature. An intervention-based query of interest is, "What would happen to the distributions of conductivity and cattail abundances if we could control influences on water conductivity?" We answer this question using a simplified model (figure 8.13). What we discover (figure 8.14) is that even when action is taken to prevent human influences, there is still an influence from other unspecified factors, which could be bedrock sources or effects or other past human activities.

The most difficult question to answer from a causal analysis is "What would have happened to a particular individual in our sample if in the past it had been subjected to different conditions?" Consider the scatter of points in figure 8.9b where wetlands have greater or lesser cattail cover than predicted. One wetland has 100% cover even though its conductivity is about 1.8, well below the estimated threshold. An informal examination of the evidence suggests that this wetland would have high levels of cattails even if it had never been influenced by human activities. Our evaluation here is informal; for formal analysis of such counterfactuals we refer the reader to Morgan and Winship (2007).

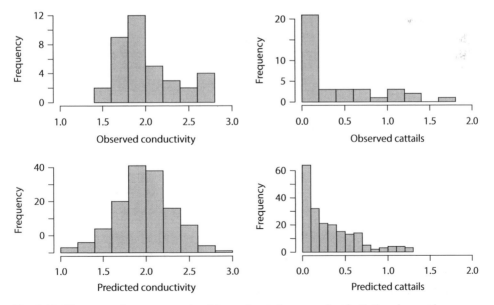

Fig. 8.12 Histograms for water conductivity and cattails, presenting both the observed distributions (upper) and the distributions that would be predicted for a new sample of 200 wetlands (lower).

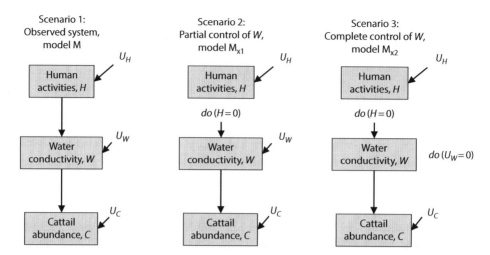

Fig. 8.13 Queries about predicted effects of interventions on water conductivities and cattail abundance. Scenario 1 is status quo; scenario 2 is the elimination of buffer intrusion and soil disturbance; scenario 3 is a reduction of water conductivity to reference conditions. The "U" variables refer to unspecified causes of variation. The operator "do($H = 0$)" refers to reducing the values of land use and soil disturbance on conductivity to 0. The operator "do($U_W = 0$)" refers to reducing the value of the unspecified influences on conductivity to 0. From Grace et al. (2012).

8.3.7 Reporting results

The guidelines in figure 8.5 (see also section 8.3.1) provide some advice for the general features of the study that should be reported. In particular, it is important to present the modeling rationale as explicitly and clearly as possible. Including a conceptual meta-model in your paper can help. Often, it is desirable to show both the initial and final SE models, unless the study was highly exploratory and presenting the initial model would just confuse the reader. In such cases it is important to declare that this was an exploratory, model-building exercise. When applications are model-comparing, however, SEM convention calls for describing each model examined and all modifications made (e.g., Laughlin et al. 2007).

A key aspect of any SEM analysis is to verify that the results are based on a model that is justified based on the data. Model-fit statistics, such as those presented in box 8.2 part F, are typically expected for cases of globally estimated models. For studies employing local methods, the number we are looking for is zero, i.e., no missing linkages that should be in the model. So, it is important to describe the examinations conducted and reassurances of model–data consistency.

Regarding model parameters and quantities, often final models and results are summarized graphically, though sometimes only the main findings (e.g., total effects) are summarized in the paper when many models are examined. Graphical presentations of final models typically include some representation of the parameters. If graphs are based on standardized parameters, then the primary unstandardized estimates and their properties (e.g., box 8.3) should be presented in a table or at least in online supporting materials. Readers typically find visual enhancement of SE model results, such as the inclusion of drawings or inset images, quite helpful (e.g., Alsterberg et al. 2013). Extrapolations such as those given in figure 8.14 have seldom been presented, but we feel this is an underutilized opportunity.

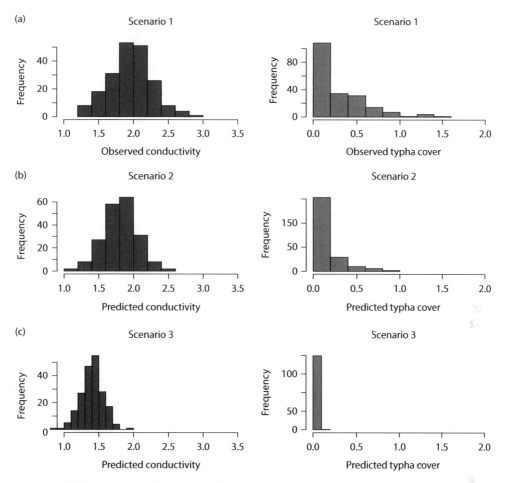

Fig. 8.14 (a) The observed distributions of water conductivity and cattail abundance (as cover). (b) The predicted distributions if effects of human activities are eliminated (Model M_{x1} in figure 8.13). (c) The predicted distributions if conductivity were completely controlled (Model M_{x2}). From Grace et al. (2012).

8.4 Discussion

The development of ideas and techniques that are emphasized in current treatments of SEM have been fueled by the needs of particular subject-matter disciplines primarily outside the natural sciences (specifically, in the human sciences). As a result of its multi-origin evolution, SEM has accumulated a large literature, much of which is not very accessible to biologists and ecologists. We believe a next-generation implementation is needed, both to incorporate transformative advances in methodology and to expand the practical potential of SEM (Grace et al. 2012).

The form of the presentation in this chapter differs from most other presentations of SEM currently available in that it is more focused on general capabilities than on technical details. This emphasis of ours is reflected in our consideration of both standard, matrix-based (global) and new, graph-theoretic (local) implementations. In this chapter we suggest fundamental principles for developing models and provide a formal exposition of causal diagrams as a useful step toward that goal, as causal diagrams support and encourage a rigorous consideration of causal assumptions. We contrast the classical SEM implementation using matrix techniques and global estimation with graph-theoretic methods using graphical analysis and local estimation. We further illustrate the use of both approaches, as each has its strengths and weaknesses. For example, global models are more constrained in the used of complex, non-linear, and non-Gaussian specifications, while local estimation permits a greater variety of specification forms. On the other hand, global estimation facilitates the inclusion of latent variables and non-recursive relationships. Another point of emphasis in our presentation is that obtaining parameter estimates is just the first step in exploring the implications of model results. Because of the tremendous flexibility and potential complexity of SEM, we only present core concepts and a limited illustration. In the following paragraphs, we advise the reader about additional issues and resources.

Bollen (1989) provides a historic treatment of the subject while Hoyle (2012a) summarizes recent advances. For those in the ecological and natural sciences, we recommend Mitchell (1992), Shipley (2000), and Grace (2006). For the measurement of theoretical entities see Raykov and Marcoulides (2010) for a psychometric perspective. Practical aspects of measurement in the social sciences are presented by DeVellis (2011) and Viswanathan (2005). For an ecologist's perspective, see Grace et al. (2010), especially with regard to the use of latent variables. Pearl (2009) deals with the theoretical development and use of causal diagrams. Practical presentations can be found in the field of epidemiology (Greenland et al. 1999). Recent summaries of specification complexities for global-estimation or local-estimation applications can be found in Gelman and Hill (2007), Zuur et al. (2007, 2009) , Edwards et al. (2012), and Hoyle (2012b), as well as in various chapters in this book. The relationship of data characteristics to specification is an important related topic (Graham and Coffman 2012; Malone and Lubansky 2012). For model evaluation and selection, West et al. (2012) have summarized choices under global estimation, while Shipley (2013) further illustrates the use of graph-theoretic methods for local estimation. The topic of interpreting the final model is one where a few technical aspects have received general treatment, though much of interpretation becomes discipline-specific. Grace and Bollen (2005) provide a general summary of coefficient types and their interpretations. More modern topics, such as queries and counterfactuals, are covered by Morgan and Winship (2007) and Pearl (2009).

There has been a notable expansion in the number and variety of applications of SEM in the natural sciences in recent years. SEM studies of trophic interactions (e.g., Lau et al. 2008; Riginos and Grace 2008; Laliberte and Tylianakis 2010; Beguin et al. 2011; Prugh and Brashares 2012), plant communities (e.g., Weiher 2003; Seabloom et al. 2006; Laughlin 2011; Reich et al. 2012), microbial communities (e.g., Bowker et al. 2010), animal populations (e.g., Janssen et al. 2011; Gimenez et al. 2012), animal communities, (e.g., Anderson et al. 2011; Belovsky et al. 2011; Forister et al. 2011), ecosystem processes (e.g., Keeley et al. 2008; Jonsson and Wardle 2010; Riseng et al. 2010), evolutionary processes (e.g., Scheiner et al. 2000; Vile et al. 2006), and macroecological relations (Carnicer et al. 2008) have been conducted. While SEM has been most commonly applied in observational studies, there have been numerous applications involving experimental manipulations (e.g., Gough and Grace 1999; Tonsor and Scheiner 2007; Lamb and Cahill 2008; Youngblood et al. 2009). To date, relatively few studies have applied Bayesian methods to ecological applications of SEM (e.g., Arhonditsis et al. 2006; Grace et al. 2011; Gimenez et al. 2012).

Finally, we reiterate that a confident, causal understanding requires pursuit via a series of studies that pose, test, and revise hypotheses. SEM, through both its philosophy and procedures, is designed to be used in that way. When you can turn your scientific understanding into a network hypothesis and evaluate it against data, a learning process is triggered that can change not only the way data are analyzed, but also how studies are designed and conducted. It is our experience that SEM can contribute significantly to this process, thus advancing our understanding of ecological systems. This chapter omits discussion of many possibilities and technical issues. We refer readers to our educational website http://www.nwrc.usgs.gov/SEM.html for more information.

Acknowledgments

We wish to thank Jessica Gurevitch, the editors of this volume, James Cronin, and E. William Schweiger, for helpful suggestions for improving the manuscript. JBG and DRS were supported by the USGS Ecosystems and Climate & Land Use Programs. The manuscript includes work done by SMS while serving at the U.S. National Science Foundation, though the views expressed in this paper do not necessarily reflect those of NSF. Any mention of software does not constitute endorsement by the U.S. Government.

## CHAPTER 9

Research synthesis methods in ecology

Jessica Gurevitch and Shinichi Nakagawa

9.1 Introduction to research synthesis

Meta-analysis was introduced to the fields of ecology and evolution in the early 1990s, using methods borrowed largely from the social sciences, and then adapted for the questions and nature of the data in ecology and related fields such as evolutionary and conservation biology (Gurevitch et al. 2001; Lau et al. 2013). Meta-analysis is the quantitative synthesis of the results of different studies addressing questions about the same thing in comparable ways, carried out in an unbiased and statistically and scientifically defensible manner.

9.1.1 *Generalizing from results*

Ecologists have debated the question of generalizing ecological data for generations: to what extent is each case study unique, and to what extent can we reach general conclusions from the results of specific studies? Ecology is also synthetic science in another sense. It rests on and combines many other sciences: geology and geomorphology, chemistry, genetics, physics, climate science, soil science, evolutionary biology and systematics, physiology, of course statistics and mathematics, and others. The issues of specificity, generalization, and synthesis are thus integral to the science of ecology, and thinking about them is an essential part of becoming and being an ecologist. Not surprisingly, ecologists and philosophers of science have a range of views on these issues. Research synthesis and meta-analysis bring these issues front and center, and we will discuss them further throughout this chapter.

When we carry out (or read) a scientific study in ecology, we hope that the results will reveal something valuable about a particular question or effect. This raises some surprisingly profound issues, though. What can we learn from a single study? Does it "reveal truth" about a scientific phenomenon by itself? How do the results of this study mesh with those of others that have considered this effect? Are the results unique or do they contribute to a general understanding of the problem by filling in missing parts of a broader pattern, corroborating other findings, or providing more conclusive evidence of something suggested in previous work? If a study is believed to stand alone, with results that are unique

Ecological Statistics: Contemporary Theory and Application. First Edition. Edited by Gordon A. Fox, Simoneta Negrete-Yankelevich, and Vinicio J. Sosa. © Oxford University Press 2015. Published in 2015 by Oxford University Press.

to the circumstances, treatments, and organisms particular to that study, then can this study add to our knowledge of anything beyond those specific circumstances and setting? To what extent are ecological studies replicable?

Research synthesis is based on the idea that the value of individual studies is enhanced by putting their results into a more general context. This implies a subtly different and unfamiliar way of thinking, in that a study is considered a member of a population of studies on similar questions, rather than a unique piece of information. Of course, this raises many practical questions about what studies can be sensibly combined, how similar is similar enough, and what scientific questions are being addressed. All of this requires that both the scientific and statistical issues are carefully considered in any attempt at carrying out a high quality, meaningful research synthesis. This perspective has been referred to as 'meta-analytic thinking' (Cumming and Finch 2001; Thompson 2002; see also Jennions et al. 2013a).

9.1.2 *What is research synthesis?*

Research synthesis is the application of the scientific method to reviewing and summarizing evidence about a scientific question. Research syntheses should ideally be as unbiased, repeatable, and transparent as anything else that we do in scientific research, with clearly and explicitly stated methods and results. In contrast, narrative, expert reviews do not use systematic or specified methods for searching or selecting literature to include; consequently, their results cannot be repeated or updated, and may be subject to various types of biases. Research synthesis includes two components: *systematic review* and *meta-analysis*. The goal of a systematic review is to carry out a complete, unbiased, reproducible literature search according to clearly specified criteria for searching for and selection of evidence.

Meta-analysis and systematic reviews are usually conducted together, although one can legitimately carry out one without the other under special circumstances. Sometimes a systematic review may reveal valuable information about the nature of the scientific evidence on a question, but the data are insufficient for a quantitative synthesis of the evidence, or the scope of the research goals do not lend themselves to a meta-analysis. Meta-analyses are sometimes conducted to directly synthesize the results of a group of researchers carrying out similar experiments without incorporating published literature or a systematic review. While thousands of systematic reviews have been carried out in medicine (and are also common, although not as extensive, in the social sciences), they have only recently been formally adopted in the ecological literature, primarily in conservation ecology (see examples in section 9.2).

9.1.3 *What have ecologists investigated using research syntheses?*

Meta-analysis has been used to address both basic and applied problems in ecology, medicine, and the social sciences, and to ask important and pressing questions in all of these fields. The answers to these questions may change our views, alter our conceptual understanding of our science or study systems, change medical practice, or influence policy (to the extent that data rather than politics are the basis for informing policy). A key aspect of the development of the statistical methodology for meta-analysis is weighting studies by their precision (technically, by the inverse of their sampling variances; see below). Large studies with less variation are weighted more heavily than smaller, more variable results when the results are combined. There are several statistical advantages to doing this, but, perhaps most importantly, it takes into account the fact that studies differ in how accurately they estimate the effects being synthesized.

People have addressed a wide range of questions at many different scales in ecology using meta-analysis. These include, for example: whether competition among animals in harsh environments is greater than in less stressful environments (Barrio et al. 2013); the relative effects of top-down and bottom-up control of species diversity in marine rocky shore communities (Worm et al. 2002); the effects of fungal root endophytes (Mayerhofer et al. 2013) and ectomycorrhizae (e.g., Correa et al. 2012; Hoeksema et al. 2010; Karst et al. 2008) on plants; how biodiversity affects ecosystem function and services (Balvanera et al. 2006); and the impact of intimidation by predators on prey populations (Preisser et al. 2005). The results of these meta-analyses also range greatly, from providing clear and conclusive answers, to complex results where the answer is "it all depends." Ecological meta-analyses have not always been conducted with the highest standards (e.g. Koricheva and Gurevitch 2014).

The responses of individuals (people, animals, plots, sites) vary for all sorts of reasons, both for reasons we think we can explain (age, environmental conditions, genetics, etc.) and reasons we can't or don't wish to explain, but the study of groups of individuals may reveal overarching patterns that help answer important questions. We use the statistical tools described in this book to answer these compelling questions about *populations of individuals*. The results of individual studies may also vary for many reasons, both those we think we can explain (variation in the nature and scope of the studies and in the way they were carried out) and for other reasons (both sampling variance and variation in the real outcomes of the studies, explained below, sections 9.4.3 and 9.4.4). We can use the statistical tools described in this chapter to answer compelling questions about *populations of studies*. Meta-analysis can help us gain much clearer answers to our questions. It can increase the precision and scope of our answers because it includes more information than is in any one study. It can also provide valuable information by combining studies that ask similar questions in similar ways, but differ in ways that we think might influence the answer (these study characteristics may be possible to model as moderators in a meta-analysis; see section 9.4).

9.1.4 *Introduction to worked examples*

We will illustrate the practice of meta-analysis in ecology with two published examples.

Plant responses to elevated CO_2

Peter Curtis and collaborators (Curtis 1996; Curtis and Wang 1998) carried out a number of important meta-analyses on the responses of plants to elevated CO_2. The example here was extracted as a heuristic subset of the data on elevated CO_2 responses of various grass species. We separate this into two data sets. The first data set is on photosynthetic responses (net photosynthetic carbon uptake, PN) in experiments in which CO_2 was elevated, while the second is on responses of leaf nitrogen concentration (N) to experimentally elevated CO_2. Note that these data are incomplete and arbitrarily selected outcomes chosen for demonstrating the methodology. We will use the log response ratio (lnR) as the effect size metric (section 9.4.1), as researchers who study elevated CO_2 responses typically express their results as the ratio of the experimental to control group responses. We treat the plants in the "ambient CO_2" treatment as the control group, against which the elevated CO_2 group is compared. Studies are weighted by their precision, determined here as the inverses of their calculated sampling variances (section 9.4).

Plant growth responses to ectomycorrhizal (ECM) interactions

Correa et al. (2012) carried out meta-regressions (section 9.4.4) of plant responses to ectomycorrhizal associations, reporting some surprising and counter-intuitive results. We will examine a small subset of their data to address the question of how plant dry weight responds to infection by ECM fungi. The data are the mean dry weights of mycorrhizal plants compared with those of non-mycorrhizal plants.

9.2 Systematic reviews: making reviewing a scientific process

When ecologists carry out primary research, we generally have a pretty clear idea of how to follow the scientific method. Literature reviews, though, are sometimes done in opaque, biased, and unstructured ways. While elements of systematic reviewing have been used in ecology, the whole process has been formally introduced to ecology only recently (e.g., Lowry et al. 2012) although it was introduced earlier in conservation practice (e.g., Centre for Evidence Based Conservation, http://www.cebc.bangor.ac.uk/). Unlike conventional narrative reviews, systematic reviews have Methods and Results sections, and aim to make the search and selection process explicit and repeatable.

One major advantage of systematic reviews is that they can be updated in a more rigorous manner than traditional reviews. Systematic reviews in medical science follow a detailed protocol, like that outlined in the *PRISMA* (Preferred Reporting Items for Systematic Reviews and Meta-Analyses) statement (Moher et al. 2009; http://www.prisma-statement.org), the MOOSE standards (Stroup et al. 2000) and the EQUATOR network guideline summaries (http://www.equator-network.org/home/). The PRISMA statement includes a checklist of 27 items, which are deemed as requirements for transparent and updatable systematic reviews. The statement also proposes a PRISMA flow diagram with which the four stages of paper collection processes (identification, screening, eligibility, and inclusion) are visualized; most systematic reviews in medical and social sciences are accompanied by a PRISMA flow diagram. While a detailed description of the methodology for carrying out and reporting the results of systematic reviews is beyond this chapter, these references can provide a good start into this literature; for an ecological example and discussion of methodology as applicable to ecology, see Lowry et al. (2012).

Conservation biology has played a leading role in the implementation of systematic reviewing in applied ecology (e.g., Stewart et al. 2005; Pullin and Stewart 2006). An excellent example is the systematic review of strategies for the reintroductions of wild dogs *(Lycaon pictus)* in South Africa (Gusset et al. 2010). Because this critically endangered species has been managed without considering methods that predict introduction effectiveness, Gusset et al. (2010) undertook a systematic review of the survival of wild dogs and their offspring. In contrast to a narrative review, their paper includes a *Methods* section, which had the subheadings *Search strategy*, *Study inclusion criteria*, and *Study quality assessment, data extraction strategy and data synthesis*. The *Results* and *Conclusions* confirm the importance of pre-release socialization in improving subsequent survival in the wild. Interestingly, they demonstrated that only two out of 40 proposed factors were statistically significant in explaining variation in mortality among packs (Gusset et al. 2010, see table 1). Their results emphasize that reliance on the best available evidence as identified by systematic reviews and meta-analysis, rather than reliance on "expert opinion," is a foundation for the most effective management and also for public outreach.

9.2.1 *Defining a research question*

This is the most important part of a research synthesis. If you do not have a clearly defined, interesting, focused question, nothing else will save the results. The scope of the study has to be clearly thought through. While valuable meta-analyses in ecology can be narrow or broad, you need an adequate number of studies to synthesize (that is, adequate to address the research questions), a compelling issue that is of general interest in your field, and the means to carry out the synthesis. You will get poor results if you attempt to carry out a synthesis without enough resources to do the work needed.

9.2.2 *Identifying and selecting papers*

This step should be defined by a systematic review protocol (such as the PRISMA statement). In most cases, papers will be selected from the published literature using a search of scientific databases. For ecological questions, the three most often used searching platforms are probably *ISI Web of Knowledge*, *Scopus* and *Google Scholar* although others such as *Agricola* may also be useful for certain topics. The use of more than two such search platforms is recommended in medical meta-analysis (Moher et al. 2009). Including papers in more than one language will result in a stronger and more complete meta-analysis. Unpublished reports (e.g., government reports and dissertations; generically called "gray literature") have been used more often in the social sciences than in ecology; there are efforts in medicine to register all trials so that they can be included in research syntheses regardless of whether the results are published or not. The search and selection of papers should be carried out using very specifically defined search and exclusion criteria. You should use a number of key word combinations and *wild cards* in your search (see, e.g., Côté et al. 2013). Of course, your question determines your key words. The key words and their combinations will need to be reported in your final publication. Search strategies using key words for meta-analysis are detailed by Côté et al. (2013).

A key word search may yield any number of articles, depending on the scope of your question. You will need to screen these articles based on your selection criteria. Typically one begins by searching key words and titles, then examining abstracts for relevance, then reading the papers (particularly the Methods and Results sections) to decide if the papers are appropriate and contain the necessary information to be included in the synthesis. Selection criteria can be based partly on the nature of the articles. For example, although narrative reviews would typically not be included in a systematic review or meta-analysis, reviews as well as other key articles (e.g., early empirical studies) play critical roles in what is termed as *forward search* and *backward search*. In a forward search one looks up papers citing a relevant article; in a backward search one examines papers cited in a relevant article. Literature cited sections from key papers may provide additional sources. Other issues related to searching for papers are covered in detail by Côté et al. (2013).

9.3 Initial steps for meta-analysis in ecology

Meta-analysis can be used to synthesize various types of response data, but has been typically used to compare the magnitude of a response to an experimental treatment (summarized by a metric called the *effect size*; section 9.4.1), where the mean of an experimental group is compared with that of a control group in each study. Ecological meta-analyses use statistical tests to ask questions like:

(1) What is the magnitude of the grand mean effect across studies, and is it significantly different from zero?

(2) Is this effect consistent across studies, given random sampling variation, or is there real heterogeneity among study outcomes?

(3) To what extent can heterogeneity among studies be explained by hypothesized covariates (categorical or continuous explanatory factors or moderators) that characterize different study outcomes?

(4) If outcomes differ among categories of studies, what are the means and their confidence intervals for the effects in those categories of studies? What is the magnitude of the slope for continuous covariates, and are their interactions among covariates in a meta-regression?

While these questions are answered statistically, one must always consider their biological meaning and biological significance as well.

In ecology and evolution, we are generally most concerned with addressing Question 4, given that the responses vary among studies after taking sampling variation into account (Questions 2 and 3). We wish to account for sources of variation in the outcome among studies. In contrast, medical meta-analyses tend to focus on the magnitude of the grand mean and its confidence interval (CI; Question 1), even if the meta-analyst wants to take into account factors modifying that grand mean in the statistical model (e.g., Lau et al. 2013).

9.3.1 *What not to do*

While we are big fans of the value and importance of meta-analysis, it is not always possible or advisable to carry out a quantitative synthesis. If a quantitative synthesis of a body of information does not make sense, or it is not possible to address a question with the literature available, or combining data is not justifiable scientifically (see section 9.6.1), it is probably not a good idea to attempt to do a meta-analysis. If you do not have time to do a meta-analysis well—and it almost always takes a considerable amount of time and effort—it is probably not a good idea to do a meta-analysis. If the data to do a meta-analysis are not available, and you are tempted to count up the number of significant outcomes and weigh those against the number of non-significant outcomes, then stop—it is definitely not a good idea to proceed with the project. This is called *vote-counting* and it is almost always a bad idea (e.g., Gurevitch and Hedges 1999; Koricheva and Gurevitch 2013; Vetter et al. 2013); people often justify its use by arguing that it was too difficult to obtain the data to calculate weighted effect sizes (or any effect sizes). If the data do not exist or cannot be obtained, move on and do something useful with your time. Do not count up significant and non-significant outcomes, which will give you biased and meaningless results.

While systematic reviews are new to ecology, carrying out a systematic search of the literature and using well-defined criteria for selecting papers to summarize is important. Documenting what you did clearly and explicitly is important. Casual searches and vague documentation of the methods you used to find the data to synthesize is no longer an acceptable way of doing research syntheses in ecology.

9.3.2 *Data: What do you need, and how do you get it?*

Once you have defined your research question and identified and selected the relevant papers (sections 9.2.1 and 9.2.2), you can acquire and extract data. Data required for meta-analyses are typically means, sample sizes, and some measure of variation from which standard deviations can be calculated. Data may be provided in a variety of formats, including tables, figures, and online supplements. If descriptive statistics are available, one can plug these values directly into formulas (table 9.1). If the data are unavailable in the published paper, you can try writing to the authors to obtain data; this is more likely to be successful with recent papers. It is very common to have to digitize data from figures. This can be done using a digitizing program such as ImageJ (http://imagej.nih.gov/ij/); other software includes GraphClick (http://www.arizona-software.ch) and GetData (http://www.getdata-graph-digitizer.com). There are some disadvantages to digitizing data from published figures (e.g., loss of data where some points are obscured by others) but it is a widely accepted and generally accurate approach.

Some papers only have inferential statistics such as t, F, and χ^2 values (section 9.2). Often we can still obtain effect sizes from these (Lipsey and Wilson 2001; Nakagawa and Cuthill 2007). The most useful and comprehensive resource for such conversions is the program "Effect Size Calculator," available online at www.campbellcollaboration.org/resources/effect_size_input.php, which is based on Lipsey and Wilson (2001). The R libraries `metafor` and `compute.es` (Del Re 2012) can also be used to carry out effect size calculations and conversions. While there are a

Table 9.1 Four common effect size metrics and the estimates of corresponding sampling variances ($\hat{\sigma}^2$)

Metric	Effect size	Sampling variance	Note
Hedges' d	$d = \dfrac{m_e - m_c}{s} J$ $s = \sqrt{\dfrac{(n_e - 1)s_e^2 - (n_c - 1)s_c^2}{n_e + n_c - 2}}$ $J = 1 - \dfrac{3}{4(n_e + n_c - 2) - 1}$	$\hat{\sigma}_d^2 = \dfrac{n_e + n_c}{n_e n_c} + \dfrac{d^2}{2(n_e + n_c)}$	m_e and m_c are the sample means for experimental and control groups, n_e and n_c the sample sizes and s_e^2 and s_c^2 are sample variances for the two groups; note that d is also known as Hedges' g
log response ratio $\ln R$	$\ln R = \ln\left(\dfrac{m_e}{m_c}\right)$	$\hat{\sigma}_{\ln R}^2 = \dfrac{s_e^2}{n_e m_e^2} + \dfrac{s_e^2}{n_c m_c^2}$	The symbols are the same as above
Fisher's z-transformed r Zr [back-transformation]	$Zr = 0.5 \ln\left(\dfrac{1 + r}{1 - r}\right)$ $\left[\dfrac{e^{2Zr-1}}{e^{2Zr+1}}\right]$	$\hat{\sigma}_{Zr}^2 = \dfrac{1}{n - 3}$	r is Pearson's correlation coefficient; n is the sample size for r
Log odds ratio $\ln OR$	$\ln OR = \ln\left(\dfrac{n_{c1} n_{e2}}{n_{c2} n_{e1}}\right)$	$\hat{\sigma}_{\ln OR}^2 = \dfrac{1}{n_{c1}} + \dfrac{1}{n_{c2}} + \dfrac{1}{n_{e1}} + \dfrac{1}{n_{e2}}$	n_{c1}, n_{c2}, n_{e1} and n_{e2} are count values in a 2 by 2 contingency table where the subscript c and e represent experimental and control groups and 1 and 2 stand for two different outcomes.

number of assumptions and limitations to converting effect sizes, this can be a very useful approach under some circumstances. Means and standard deviations can sometimes be estimated even when we only have information on medians or ranges (see Hozo et al. 2005). More comprehensive coverage of effect size extraction can be found in Borenstein 2009, Borenstein et al. 2009, Fleiss and Berlin 2009, Lajeunesse 2013, Rothstein et al. 2013, and Rosenberg et al. 2013.

It is also important to have a description of how the data were obtained and what they represent. For example, is the reported mean the mean of the individual organisms, or the mean of plot means? How was the experiment set up, and how were the reported values calculated? If the data are taken from complex experimental designs, are they comparable to data in other studies? It is a good idea to collect any data that might be relevant later (e.g., additional covariates categorizing the data); otherwise you may need to go back to the papers multiple times.

9.3.3 *Software for meta-analysis*

In this chapter, we use the `metafor` libraries in R (Viechtbauer 2010) to carry out meta-analysis and meta-regression. A recent review of computer software for meta-analysis was published by Schmid et al. (2013). Many ecological meta-analyses have been carried out using MetaWin (Rosenberg et al. 2000), which has limited capabilities and is now out of print. A powerful open-access program for meta-analysis and meta-regression in ecology with a flexible and accessible graphical user interface (GUI), OpenMEE (Open-access Meta-analysis for Ecology and Evolution, Dietz et al. 2014), has very recently become available (http://www.cebm.brown.edu/open_mee) and fully functional.

9.3.4 *Exploratory data analysis*

As with any statistical data analysis, it is recommended that you first carry out exploratory data analysis to understand the nature and structure of the data. Many of the graphical methods used for exploratory data analysis of primary data are useful for meta-analysis (such as quantile–quantile (QQ) plots; appendix 9A). Normal QQ plots are among the most frequently used tools in exploratory meta-analysis (Wang and Bushman 1998; Rosenberg et al. 2000). Gaps in a QQ plot may indicate publication bias (section 9.5.3); outliers or strongly curvilinear patterns may also indicate issues to be investigated with the structure of the data, such as distinctly non-normal distributions. Often outliers may be detected at the two ends of the graph. The meta-analyst should check to see if they are the result of an error in data extraction or recording. Outliers, particularly if they are orders of magnitude greater than the other data points, can drive the results of a regression (including a meta-regression); whether to eliminate outliers is a substantive topic in statistics that is beyond the scope of this chapter.

Funnel plots (which are specific to meta-analysis) are scatter plots where the effect sizes are plotted against either sample size or the inverse of sampling variance (known as the "precision" in the meta-analytic literature; see Sterne and Egger 2001). The idea is that one might expect a lot of scatter due to low precision in estimating the true effect size when study sizes are low, but that as study sizes increase, the sample estimates of the effect size should "funnel down" to values close to the true effect size in a symmetrical manner (Egger et al. 1997). Deviations from a symmetrical funnel, "holes" in the plot where all data are conspicuously missing (particularly where effect sizes are close to zero and sample sizes or precision is low), or if many data points are far from the funnel, may indicate problems with data structure or publication bias (see Egger et al. 1997).

However, ecological meta-analyses typically have one or more hypothesized explanatory variables, small sample sizes, and uneven data distribution, making funnel plots often less useful. One alternative is to calculate the residuals from the chosen model, and plot them against the sample sizes or inverse of the weights in the funnel plot (Viechtbauer 2010; Nakagawa and Santos 2012). Such residual funnel plots are a

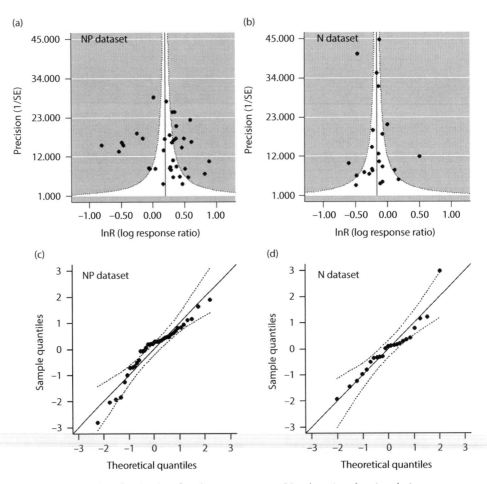

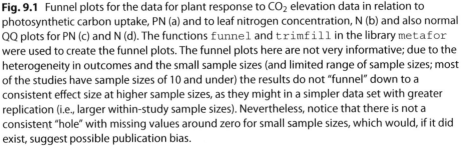

Fig. 9.1 Funnel plots for the data for plant response to CO_2 elevation data in relation to photosynthetic carbon uptake, PN (a) and to leaf nitrogen concentration, N (b) and also normal QQ plots for PN (c) and N (d). The functions `funnel` and `trimfill` in the library `metafor` were used to create the funnel plots. The funnel plots here are not very informative; due to the heterogeneity in outcomes and the small sample sizes (and limited range of sample sizes; most of the studies have sample sizes of 10 and under) the results do not "funnel" down to a consistent effect size at higher sample sizes, as they might in a simpler data set with greater replication (i.e., larger within-study sample sizes). Nevertheless, notice that there is not a consistent "hole" with missing values around zero for small sample sizes, which would, if it did exist, suggest possible publication bias.

better option when one expects heterogeneity (variation in true effect sizes among studies; section 9.4.3) because heterogeneity will often generate asymmetry in a funnel plot (Egger et al. 1997). The potential and limits to the usage of funnel plots are described in detail elsewhere (Sterne et al. 2005; Sutton 2009; Jennions et al. 2013b). Although these visual methods are conceptually intuitive, they do not provide statistical tests to evaluate whether publication bias exists in a data set; some tests are given in section 9.5.3.

The R library `metafor` has dedicated functions `funnel` and `qqnorm` for these plots. Both kinds of plots are shown for the elevated CO_2 (figure 9.1, section 9.4.2) and mycorrhizal plant (figure 9.2, section 9.4.3) data sets.

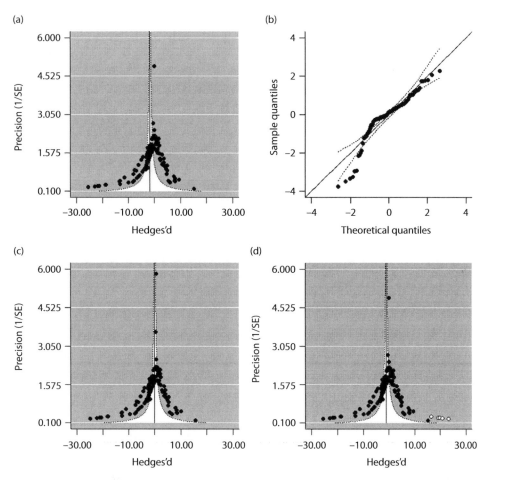

Fig. 9.2 A funnel plot for the data for the data on plant response to ectomycorrhizae, (a); a normal quantile–quantile (QQ) plot (b); a funnel plot with residuals after fitting a moderator (light level; PPFD) (c); and a funnel plot with added data from the trim-and-fill method (d). The funnel plots (a and c) show signs of potential publication bias with noticeable asymmetry (the left side contains more data points) and a hole in the non-significant region. The trim-and-fill method (section 9.5.3) detected potentially missing data, which restores symmetry (d). The functions `funnel`, `qqnorm`, and `trimfill` in the library `metafor` were used to draw these.

9.4 Conceptual and computational tools for meta-analysis

9.4.1 *Effect size metrics*

The choice of an *effect size* metric (i.e., an index that expresses the magnitude of the outcome of a study) depends on several issues: what makes the most sense ecologically, what metrics are commonly used and understood in a particular research area, the data available, and whether the statistical properties of the metric are well-behaved (e.g., normally distributed, unbiased) and well-understood, so that common meta-analysis procedures can be carried out reliably (e.g., Rosenberg et al. 2013). In particular, the distribution and sampling variance of the effect size metric should be known.

Effect size metrics which are commonly used in ecology (table 9.1) include bias-corrected standardized mean differences (Hedges' *d*, also known as Hedges' *g*), log response ratios (*lnR*), and Fisher's *z*-transform of Pearson's correlation coefficient (particularly common in evolutionary studies). Odds ratios are more common in medicine but are occasionally used in ecology (table 9.1). Other metrics also may be used (Mengersen and Gurevitch 2013). The choice of the effect size metric can influence the outcome of the meta-analysis and the interpretation of the results; this has led to some debate about which are most appropriate, accurate, and straightforward to interpret (Osenberg et al. 1997; Osenberg and St. Mary 1998; Houle et al. 2011; Mengersen and Gurevitch 2013; and see the empirical comparison of results expressed as *lnR* and *d* by Hillebrand and Gurevitch 2014). While there is no universally agreed-upon resolution as to which metric of effect size is "best," both *lnR* and *d* are commonly used in ecological meta-analysis; interested readers should consider the arguments made in the referenced publication above for more detail.

9.4.2 *Fixed, random and mixed models*

The original meta-analyses used fixed-effect (fixed) models, in ecology as in other disciplines. Random-effects models were introduced somewhat later (e.g., Hedges and Vevea 1998). Random- and mixed-effects models have become the most widely used in recent years; in ecology, we are generally interested in mixed models. These models are analogous to, but not identical with, the concepts of fixed, random, and mixed models in the family of linear models such as least squares and method of moments (as used in familiar ANOVA and regression models), generalized linear models (chapter 6), and linear mixed-effects models (chapter 13). Fixed, random, and mixed models in meta-analysis are described in detail elsewhere (e.g., Borenstein et al. 2009, chapters 11–13; Mengersen et al. 2013b; Rosenberg 2013). The choice of model affects the analysis and the interpretation of the results.

A *fixed-effect model* in meta-analysis makes the assumption that the differences in outcome among studies are due only to random sampling variation (table 9.2). Random sampling variation is the variation in outcome one would expect if the identical experiment was conducted many times. Another way of thinking about this is that a fixed-effect model assumes that all of the studies have the same "true" outcome (effect size), but the outcomes we see (sample outcomes) estimate this true effect with some error. If all studies had infinitely large numbers of replicates, the outcomes would be identical (but see Viechtbauer (2010) and Borenstein et al. (2009)). *Random-effects models* (table 9.2) include an extra source of variation, variation in the "true" effects among studies, due to unspecified causes. In other words, in a random-effects model in meta-analysis,

Table 9.2 Three types of statistical meta-analytic models

Model	Formulation	Note
Fixed-effect model	$Y_i = \theta + e_i$ $e_i \sim N(0, \sigma_i^2)$	Y_i is the ith effect size, θ is the grand mean of the fixed-effect model (or the intercept, β_0), e_i is the ith within-study effect (sampling error) and distributed normally with the mean of 0 and the variance of σ_i^2 (sampling variance).
Random-effects model	$Y_i = \mu + \varepsilon_i + e_i$ $e_i \sim N(0, \sigma_i^2)$ $\varepsilon_i \sim N(0, \tau^2)$	μ is the grand mean of the random-effects model (or the intercept, β_0), ε_i is the ith between-study effect and distributed normally with the mean of 0 and the variance of τ^2 (the between-study variance estimated as T^2; see section 9.3.3). The other symbols are the same as above.
Mixed-effects model (Meta-regression)	$Y_i = \eta_i + \varepsilon_i + e_i$ $\eta_i = \beta_0 + \beta_1 X_{1i} + \beta_2 X_{2i} + \beta_3 X_{3i} \cdots$ $e_i \sim N(0, \sigma_i^2)$ $\varepsilon_i \sim N(0, \tau^2)$	β_0 is the intercept for meta-regression (see section 9.3.4), β_1, β_2 and β_3 are regression coefficients for respective explanatory variables, X_1, X_2, and X_3 (moderators, estimated as b). When X is a continuous variable, the interpretation of β is straightforward while when X is a categorical variable, the interpretation of β may be complicated. For example, with a categorical variable with two levels, β_0 is the mean of the firstlevel and β_1 is the contrast between the first and the second level with X_1 as a *dummy variable* with value either 0 or 1 (0 for the first level and 1 for the second level; see the Introduction chapter). The other symbols are the same as above.

one accounts for both real differences among studies and for random sampling variation among the studies' outcomes.

A *mixed-effects model* (table 9.2) adds another source of variation: variation that can be accounted for by one or more known explanatory factors (called covariates or moderators in the meta-analytic literature) that categorize the studies. These can be categorical (e.g., trophic level) or continuous factors (e.g., temperature at which an experiment was conducted). These explanatory factors are typically modeled as fixed effects. Because such models are essentially weighted multiple regression models, mixed-effects models are also called *meta-regressions* (section 9.4.4). The goal of most meta-regressions is to account for and explain among-study variation, which is known as heterogeneity (section 9.4.3).

9.4.3 *Heterogeneity*

The total amount of variation across all studies includes both within-study (sampling) variance and *heterogeneity* in the true effect sizes among the studies. Heterogeneity is modeled only in random- or mixed-effects models. The existence of heterogeneity can be tested by the Q statistic, the sum of weighted squared deviations of each study's effect size from the mean effect across studies:

$$Q = \sum_{i}^{k} W_i (Y_i - M)^2,$$

$$W_i = \frac{1}{V_i},$$

where k is the number of effect sizes (i.e., the number of outcomes being synthesized), Y_i is the effect size in the ith study, V_i is the ith effect size's sampling variance, W_i (the inverse of V_i), is its weight in the meta-analysis, and M is the mean effect size of the meta-analysis. Q is distributed as chi-squared, with degrees of freedom given by the number of studies, k, minus 1.

But the Q statistic is useful for much more than tests of heterogeneity. If we assume that all of variation among studies is due to within-study sampling variance (as in a fixed-effect model), the expected value of Q would be the df. We can estimate the variance of the true effect sizes, τ^2, based on the sample estimate T^2

$$T^2 = (Q - k - 1)\left(\sum_{i=1}^{k} W_i - \frac{\sum_{i=1}^{k} W_i^2}{\sum_{i=1}^{k} W_i}\right)^{-1}. \tag{9.1}$$

T^2 is an estimate of the true between-studies variance in random-effects models. T^2 is added to each study's sampling variance to calculate inverse variance weights (Der-Simonian and Laird 1986). In addition to this method, inverse variance weights can also be calculated by restricted maximum likelihood (REML; Hedges and Vevea 1998; section 9.5.4).

Another useful index for describing the heterogeneity among studies is I^2, the ratio of the between-studies variance to the total variance (Higgins and Thompson 2002; Higgins et al. 2003):

$$I^2 = \frac{T^2}{T^2 + S^2}$$

$$S^2 = \frac{\sum_{i=1}^{k} W_i(k-1)}{\left(\sum_{i=1}^{k} W_i\right)^2 - \sum_{i=1}^{k} W_i^2},$$

where S^2 is the sample estimate of the within-study variance. I^2 is a measure of the proportion of the total variance that is due to the true variance (that is, the heterogeneity) between studies. Higgins et al. (2003) suggested that 25% could be considered low, 50% medium, and 75% a high value for I^2.

Q can be partitioned additively into its components. We can use this property to evaluate whether the effect sizes of different categories of studies differ from one. This has been expressed as:

$$Q_{\text{total}} = Q_{\text{between}} + Q_{\text{within}},$$

where Q_{between} is the heterogeneity between the mean effect for categories of studies (e.g., carnivores and herbivores; Q_{between} can be considered as a sub-class of Q_M, the model heterogeneity, which is the heterogeneity accounted by moderators; see section 9.5), Q_{total} is the total observed dispersion (weighted sums of squares around the grand mean), and Q_{within} is the sum of the within-group variances (Q_{within} is a sub-class of Q_E, the error heterogeneity, which is the heterogeneity unaccounted by the moderators; see section 9.5). Either fixed- or mixed-effects models can be used in these analyses. We generally test Q against a χ^2 distribution; an alternative is to use randomization tests (Adams et al. 1997;

Higgins and Thompson 2004). The Q statistic tends to have low power, however, so finding a non-significant Q may be due to lack of heterogeneity, or to the inability to detect true heterogeneity.

9.4.4 *Meta-regression*

In most ecological meta-analyses, some of the heterogeneity among studies may be attributable to covariates (section 9.4.2; table 9.2). Ecologists carrying out meta-analyses have commonly evaluated many hypothesized covariates one by one, in separate models. But this can amount to a "fishing expedition" in which one searches for and only reports statistically significant outcomes, inflating Type I errors (Forstmeier and Schielzeth 2011). Terminology has been somewhat different in ecology than in other disciplines. Ecologists have generally used the terms "meta-analysis" to include calculation of weighted mean effect sizes and heterogeneity tests with or without a covariate, and "meta-regression" to refer to meta-analysis models with more than one covariate. Researchers in other disciplines have typically used "meta-regression" where there are any covariates included in the model, restricting "meta-analysis" to calculating grand means, their CIs, and heterogeneity; sometimes "subgroup analysis" is used to refer to the calculation of means and CIs for different groups (or categories) of studies.

Ideally, one should examine hypothesized covariate effects in a well thought-out meta-regression model (often called a full model; cf. Forstmeier and Schielzeth 2011). On the other hand, the results of full meta-regression models where there are several covariates, with more than two levels per covariate, can be difficult and complicated to interpret (section 9.5.1). Ecological meta-regression is often complicated by confounded and correlated covariates (multiple collinearity), and by the problem that different covariates are measured and reported in different studies, resulting in "sparse" matrices with many zeros (such covariates may be non-informative, at least with a given data set). Although careful choice of covariates can sometimes avoid these issues, such problems remain to be fully resolved.

9.4.5 *Statistical inference*

Once we have a statistical model, we can use various approaches to estimate the parameters (Mengersen et al. 2013b). Least squares and method of moments approaches using parametric methods are conceptually and computationally simplest (Rosenberg 2013). Randomization tests and bootstrapping were first introduced as tools for inference in meta-analysis by ecologists (Adams et al. 1997) and were used extensively in our field for many years before being adopted in social sciences and medical research (Higgins and Thompson 2004). They have a number of advantages (e.g., normality of residuals is not assumed) and limitations (e.g., they are ill-advised when the number of studies is small—as a rule of thumb, fewer than 5 studies, and still assume independence of observations). Maximum likelihood and restricted maximum likelihood (REML) approaches have become very common for solving random-effects models (Mengersen and Gurevitch 2013), and have some advantages over moment-based approaches such as the DerSimonian and Laird method (DerSimonian and Laird 1986). Bayesian methods have also been used in meta-analysis, especially to deal with complex meta-analyses with hierarchical structures (Schmid and Mengersen 2013) because accurate estimation for such complex models is difficult or impossible using moment-based or maximum likelihood approaches. The treatment of these issues and approaches is beyond the scope of this chapter; interested readers can consult Cooper et al. (2009) and Koricheva et al. (2013a).

9.5 Applying our tools: statistical analysis of data

We divide the analytical process into four steps, each of which is contingent on the preceding step: 1) fitting a fixed-effect or random-effects model to find a meta-analytic mean;
2) evaluating statistical heterogeneity; 3) fitting a mixed-effects model with potential explanatory variables or moderators (both categorical and continuous); and 4) quantifying
and evaluating effects of different categories or magnitudes of slopes. These four steps correspond with the four key questions that we typically address in conducting meta-analysis
(section 9.3).

Below, we go through the analysis of the two worked examples by combining snippets
of R code and explaining how to interpret statistical outcomes to give biologically meaningful results. The complete R code, data sets, and more explanations of the two examples
are all found in the online appendix.

9.5.1 *Plant responses to elevated CO_2*

*Question 1: What is the magnitude of the grand mean effect across studies, and is it
significantly different from zero?*

We will use a random-effects model, because we believe it is likely that there are real
differences among the study outcomes due to study conditions and other unidentified
characteristics of the studies; also, we may wish to extend the results beyond this group
of outcomes to make generalizations about the responses of plants to elevated CO_2. Here,
we use the function `rma` in the `metafor` library to carry out the meta-analysis and meta-
regression. (All of this can also be done using OpenMEE without the need to program
this in R.)

```
# random-effects model using REML
> Model1 <- rma(yi = lnR_PN, vi = VarlnR_PN, method = "REML",
    data = Data1)
```

The grand mean effect (intercept) of response of log(PN) (net photosynthetic carbon uptake) to elevated CO_2 is $M_{lnR} = 0.198$ with a 95% CI = (0.080, 0.316). The back-transformed
values are grand mean effect size = 1.219, 95% CI = (1.083, 1.372); meaning that there is
about a 20% average increase in net photosynthesis across studies. The 95% CI does not
overlap 0. Alternatively, the z value is 3.28 with $p = 0.001$, confirming that the grand mean
effect is significantly different from zero. The result will be slightly different depending
on the method used to calculate τ, the "true" variance among studies. Under a fixed-effect
model, the results are somewhat different, because the additional variation among studies
due to real variation in the responses is not taken into account; the grand mean then is
$M_{lnR} = 0.178$, 95% CI = (0.157, 0.199). The z value for the fixed-effect model is 16.63 with
$p < 0.0001$. There is an important caveat here, though: we eliminated studies from this
data set if we could not find the sample size or calculate the standard deviation. In real
meta-analysis, you would be certain to encounter such missing data (see section 9.6.3).

For the response of leaf N to elevated CO_2, we have $M_{lnR} = -0.167$ (95% CI = -0.265 to
-0.070), $z = -0.336$, and $p < 0.0008$ for the random-effects model using REML. That is,
elevated CO_2 has a negative effect on leaf N concentration that is significantly less than
zero. This makes sense biologically: elevated CO_2 increases carbon uptake and "dilutes"
leaf N. The meta-analytic results from the random-effects models for the two data sets
along with each data point are visualized in what are called *forest plots* in figure 9.3. Forest
plots are a common way of presenting meta-analysis results graphically, and consist of

means with CIs, typically graphed horizontally, and stacked vertically. Forest plots may be used to graph the effect sizes of each study with its CI (based on the sampling variances), or they may be used to graph means for different groups of studies with their CIs. They also typically show a line of "no effect" and sometimes a line for the grand mean; the size of the symbol for the mean may be used to indicate study weight or sample size. The derivation of the name "forest plot" is obscure.

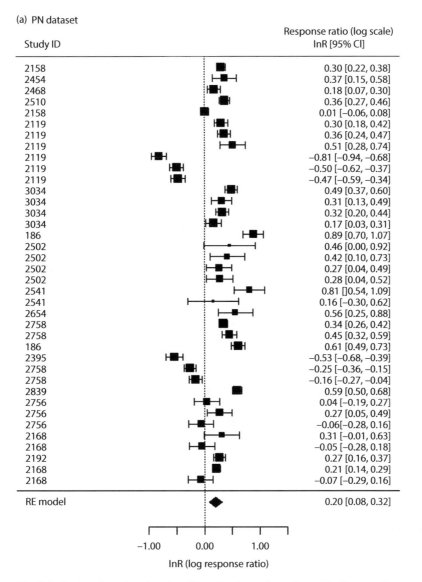

Fig. 9.3 Forest plots showing the effect size for each study and its CI, as well as the grand means and their CIs (diamonds) from random-effects (RE) models for (a) plant response to CO_2 elevation in photosynthetic carbon uptake, PN, and (b) leaf nitrogen concentration, N. The function `forest` in the library `metafor` was used to draw the plots. The size of each square is proportional to the precision of the effect size estimate.

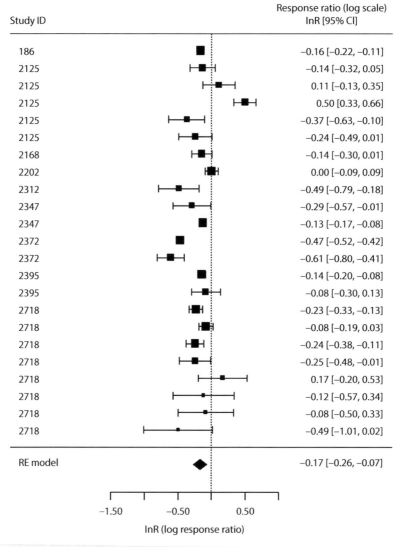

(b) N dataset

Fig. 9.3 *(Continued)*

Question 2: Is this effect consistent across studies, given random sampling variation, or is there real heterogeneity among study outcomes?

The heterogeneity test for PN indicates that there is substantial additional "true" variation among studies ($Q = 999.81$, $df = 37$, $p < 0.0001$). The percent variability due to true heterogeneity among studies, I^2, is 96.64%, accounting for most of the observed dispersion among outcomes. For leaf N, the Q test reveals significant heterogeneity ($Q = 264.04$, $df = 22$, $p < 0.0001$). $I^2 = 93.52\%$, indicating that most of the total dispersion is among (rather than within) studies.

Question 3: To what extent can heterogeneity among studies be explained by hypothesized covariates (categorical or continuous explanatory factors or moderators) that characterize different study outcomes?

Question 4: If outcomes differ among categories of studies, what are the means and their confidence intervals for the effects in those categories of studies?

There are a number of potential explanatory variables that we might expect to influence the PN responses: plant species, grasses vs. sedges, C_3 vs. C_4 species, the amount of CO_2 augmentation, and the experimental method. Unfortunately, we cannot include all of these in a full meta-regression model. First, they are seriously *confounded*—they can't be separated in a statistical model because values for some of these variables always correspond to particular values of other variables. For example, there is only one sedge species, and it lives in wetland habitats, it is a C_3 species (although there are also many C_3 grass species), and the sedge experiments were conducted in open-top chambers. Another problem is that there is a very uneven distribution of studies estimating the effects of the different explanatory variables; we have much more information on some systems and species and very little information on others. This is always a challenge in research synthesis, because you have to deal with the literature that exists, rather than designing an experiment in which you control the factors hypothesized to affect the outcome. Another problem is that most of the sample sizes are very small. That means that the estimates of effect sizes have low precision.

We will compare two of the covariates of greatest interest, and for which we have more information: photosynthetic type (C_3 vs. C_4 species) and experimental method (growth chambers vs. open-top chambers). We will first fit photosynthetic type (`PSType`) and experimental method (`EMethod`) in separate meta-regression models for illustration.

```
# meta-regression fitting photosynthetic type
> Model2 <- rma (yi = lnR_PN, vi = VarlnR_PN, mods = ~ PSType,
    method = "REML", data = Data1)
> summary (Model2)
```

```
Model Results:
            estimate       se      zval      pval     ci.lb     ci.ub
intrcpt       0.2025   0.0738    2.7430    0.0061    0.0578    0.3472  **
PSTypeC4     -0.0148   0.1311   -0.1126    0.9103   -0.2718    0.2423
# meta-regression fitting experimental method
> Model3 <- rma (yi = lnR_PN, vi = VarlnR_PN, mods = ~ EMethod,
    method = "REML", data = Data1)
> summary (Model3)
```

Interpreting the estimate outputs from `metafor` is not completely straightforward because they are regression coefficients, so that in `Model2`, `intrcpt` (b_0) represents the mean value for C_3 plants, whereas `PSTypeC4` (b_1) represents the *difference* between the means of C_3 and C_4 plants. Thus, the difference between C_3 and C_4 plants is not significant ($p = 0.910$). This p value of the z statistic matches the value for the Q test for the model ($Q_M = 0.012$, $p = 0.910$, $df = 1$), which tests the significance of the covariate, `PSType`. On the other hand, the Q test for residuals is significant ($Q_E = 987.12$, $p < 0.0001$, $df = 36$), meaning much of the heterogeneity in the data is still unaccounted for. For `Model3`, the results are similar in that there is no significant difference between the two different experimental methods ($z = 1.000$, $p = 0.317$; $Q_M = 1.000$, $p = 0.910$, $df = 1$) with much of heterogeneity unaccounted for ($Q_E = 968.62$, $df = 36$, $p < 0.0001$).

Next, we will obtain the means and confidence intervals for the different groups for photosynthetic type and for experimental method. To do so, one needs to use a trick if one is using a regression program like metafor. That is, we need to remove the intercept to reparameterize the model as demonstrated below.

```
# Means for PS type without the regression intercept
> Model4 <- rma (yi = lnR_PN, vi = VarlnR_PN, mods = ~ PSType - 1,
    method="REML",data=Data1)
# Means for Experimental Method without the regression intercept
> Model5 <- rma (yi = lnR_PN, vi = VarlnR_PN, mods = ~ ExperMethod - 1,
    method = "REML", data = Data1)
```

The CIs of the two photosynthetic types largely overlap (Model4: C3, $M_{C3} = 0.203$, 95% CI = (0.058, 0.347); C4, $M_{C4} = 0.188$, 95% CI = ($-0.025, 0.400$)). The CI for C_3 plants does not overlap 0, while that for C_4 plants does overlap zero. We would expect that C_3 plants on average would respond positively to elevated CO_2, but C_4 plants on average would not. The CIs for the two experimental methods also largely overlap one another (Model5: Growth chambers, $M_{GC} = 0.125$, CI = ($-0.059, 0.310$); Open-top chambers, $M_{OTC} = 0.248$, CI = (0.0942, 0.402)). It is not clear why the response to elevated CO_2 is greater than zero in open-top chambers but not in growth chambers.

We can model these two different factors (photosynthetic type and experimental method) in a single, more complete model. However, the results of meta-regression models with more than one categorical covariate can be difficult to interpret; if the covariates have more than two levels and if they have interactions the output becomes challenging (see Model6 in appendix 9A). The software package OpenMEE makes the analysis of this sort of data much easier to understand and interpret than the output from metafor. One solution is creating a variable that has four levels [the two combinations of C_3 and C_4 with growth chambers (GC) and open-top containers (OTC), giving C3GC, C3OTC, C4GC, C4OTC].

```
# create a categorical variable with four levels
> Mix1 <- paste(Data1 $ PSType, Data1 $ EMethod, sep = "")
> Model7 <- rma (yi = lnR_PN, vi = VarlnR_PN, mods = ~ Mix1,
method="REML",data=Data1)
> summary (Model7)
```

```
Model Results:
            estimate       se     zval     pval    ci.lb    ci.ub
intrcpt       0.0538   0.1082   0.4976   0.6188  -0.1582   0.2658
Mix1C3OTC     0.2648   0.1446   1.8314   0.0670  -0.0186   0.5483   .
Mix1C4GC      0.2686   0.2097   1.2812   0.2001  -0.1423   0.6796
Mix1C4OTC     0.0634   0.1692   0.3746   0.7080  -0.2683   0.3951
```

The intercept is the mean for C3GC; the other terms are the differences in mean values between C3GC and corresponding groups. The output from Model7 provides Q statistics moderators ($Q_M = 4.222$, $df = 3$, $p = 0.239$). These Q tests for covariates examine the statistical hypothesis that the regression coefficients are all 0 (see table 9.2). This is not exactly the same as the hypothesis $\mu_{C3GC} = \mu_{C3OTC} = \mu_{C4GC} = \mu_{C4OTC}$, which you may really want to test in this case but which is awkward to carry out in `metafor`; see more in appendix 9A in the online supplemental material.

The confidence limits of all the groups can be easily obtained using the no-intercept model:

```
> Model8 <- rma (yi = lnR_PN, vi = VarlnR_PN, mods = ~ Mix1 - 1,
    method = "REML", data = Data1)
> summary(Model8)
```

```
Model Results:

            estimate      se     zval     pval     ci.lb    ci.ub
Mix1C3GC      0.0538  0.1082   0.4976   0.6188   -0.1582   0.2658
Mix1C3OTC     0.3187  0.0960   3.3201   0.0009    0.1305   0.5068  ***
Mix1C4GC      0.3224  0.1796   1.7953   0.0726   -0.0296   0.6745  .
tMix1C4OTC    0.1172  0.1302   0.9006   0.3678   -0.1379   0.3724
```

This no-intercept model (`Model8`) will also provide a Q test for the model ($Q_M = 15.305$, $p = 0.004$, $df = 4$), but this is a test of the hypothesis $\mu_{C3GC} = \mu_{C3OTC} = \mu_{C4GC} = \mu_{C4OTC} = 0$, which is not so useful for us because we are more interested in whether these four groups are different from each other rather than whether these four groups are different from zero.

For leaf N, the number of studies (k) available is smaller, the issues are somewhat similar to those for PN, and the factors we might wish to test are the same (photosynthetic type and experimental method). The difference is that experimental method has an extra level (Free-Air Carbon Enrichment, FACE) in addition to GC and OTC. Thus, we will combine these two variables to make five categories in a similar manner to `Model7` and their confidence limits can be obtained by taking out the intercept:

```
> Mix2 <- paste (Data2 $ PSType, Data2$EMethod, sep = "")
> Model9 <- rma (yi = lnR_N, vi = VarlnR_N, mods = ~ Mix2,
method = "REML", data = Data2)
> summary (Model9)
```

```
Model Results:

            estimate      se     zval     pval     ci.lb    ci.ub
intrcpt      -0.5273  0.1249  -4.2205   <.0001  -0.7722  -0.2824  ***
Mix2C3GC      0.2261  0.1545   1.4629   0.1435  -0.0768   0.5289
Mix2C3OTC     0.4414  0.1383   3.1915   0.0014   0.1703   0.7124  **
Mix2C4GC      0.5054  0.1569   3.2213   0.0013   0.1979   0.8129  **
Mix2C4OTC     0.3825  0.2202   1.7371   0.0824  -0.0491   0.8141  .
> Model10 <- rma (yi = lnR_N, vi = VarlnR_N, mods = ~ Mix2 - 1,
method = "REML", data = Data2)
```

In `Model9`, the intercept represents the mean for C_3 plants in FACE studies. You would interpret the other terms in the same way as in `Model7`, that is, each term is the difference between the mean value for C_3 plants in the FACE studies and each other group (e.g., C_3 plants in growth chambers). The difference between the C_3 plants in the FACE treatment and at least one of the groups is significantly different from zero, based on the Q test ($Q_M = 14.717$, $df = 4$, $p = 0.005$), which here is testing whether the statistical hypothesis, $\beta_1 = \beta_2 = \beta_3 = \beta_4 = 0$, is false. In `Model10`, we obtain mean and 95% CI estimates for five groups (or levels; C3FACE, C3GC, C3OTC, C4GC, C4OTC) and only two groups (C3FACE and C3GC) are significantly different from zero. The metafor output also includes R^2 = 40.41%, which is an estimate of the heterogeneity explained by the model.

9.5.2 *Plant growth responses to ectomycorrhizal (ECM) interactions*

What is the magnitude of the grand mean effect across studies, and is it significantly different from zero?

As with the CO_2 data, we will use a random-effects model, but in this case we will employ the other common metric of effect size, Hedges' *d*. As in most ecological research syntheses, there are problems with missing data and confounded explanatory variables. In this case, many of the authors reported ranges of sample sizes rather than exact sample sizes (e.g., 9 to 12 replicates), and where this was the case, Correa and colleagues used the median of the reported sample sizes as an approximation. But there are many studies without standard deviations, SD (in the data set, CS and ES stands for SD for control groups and SD for experimental groups, respectively) although mean weights for both groups (CM and EM) are always available. Also, several studies failed to report sample sizes (N in the data set for both groups), so we cannot calculate Hedges' *d* for the studies with this missing information unless we somehow impute the missing data (for the issue of missing data imputation, see section 9.6.3). We will start with only those studies with complete information, calculate *d*, and fit a model with the function rma.

```
> ModelA <-rma (yi = Hd, vi = vHd, method = "REML", data = Data3)
```

The grand mean effect in response to mycorrhizal infection is M_d = −1.730, 95% CI = (−2.658, −0.801). That is, the mycorrhizal plants had reduced dry weight compared with the non-mycorrhizal plants, and the effect was large and significant (the 95% CI does not overlap zero). Unlike *lnR*, there is a standard interpretation for the magnitude of standardized mean differences, first proposed by Cohen (1988). A *d* value of 0.2 standard deviation units is considered to be a small effect, 0.5 medium, and 0.8 a large effect (whether negative or positive). This is useful in making sense of the mean effect sizes, which are necessarily expressed on a common scale across studies, rather than in units of, say, dry weight (which may differ by orders of magnitude among studies). Thus, the responses of plant dry weights to mycorrhizal infection can be considered a very large negative effect. Although *d* is closely related to t-tests and ANOVAs, if it does not seem intuitive to you, you can choose another metric of effect size when carrying out your own meta-analyses.

Is this effect consistent across studies, given random sampling variation, or is there real heterogeneity among study outcomes?

The heterogeneity test indicates that there is a large and statistically significant amount of variation among studies in the random-effects model (Q = 1568.82, df = 119, $p <$ 0.0001). This large amount of heterogeneity among studies might be accounted for by covariates.

Are there categorical or continuous covariates that account for heterogeneity among studies?

While this data set contains fewer potential explanatory variables than the CO_2 data set, there are still several that might influence the outcomes. Light level (photosynthetic photon flux density, PPFD) is a continuous covariate that might affect the amount of photosynthesis that the plants can carry out. PPFD levels are surprisingly low in many of these studies (possibly because plant growth chambers, particularly older ones, often have low maximum light levels). We can treat PPFD as a continuous covariate in a mixed-effects model.

```
> ModelB <- rma(yi = Hd, vi = vHd, mods = ~ PPFD, method = "REML",
    data = Data3)
```

The slope of the effect of PPFD on the response to mycorrhizae is –0.013 [SE = 0.003; 95% CI = (–0.0019, –0.007)], and statistically significant, indicating that the higher the light level, the less the negative effect of mycorrhizae on plant dry weight. Presumably if plants are able to capture more carbon, this reduces the negative effect of mycorrhizal infection on dry weight. This is also indicated by the heterogeneity test, with $Q_M = 17.27$, $df = 1$, and $p < 0.0001$, showing that PPFD accounts for a significant part of the variation among outcomes.

The intercept tells us that this effect size [$b_0 = 2.702$, 95% CI = (–0.445, 4.959)] is positive, which is opposite the sign from the grand mean in `ModelA`! However, this intercept value does not make much biological sense because it is the effect size when PPFD is zero (that is, in darkness)! To make the intercept meaningful, we center the variable; that is, we reset the mean value of PPFD ($M_{PPFD} = 350.5$) to zero (see Schielzeth 2010).

```
> cPPFD <- scale (PPFD, scale = FALSE) # centering
> ModelC <- rma (yi = Hd, vi = vHd, mods = ~ cPPFD, method = "REML",
    data = Data3)
```

Now, the intercept value [$b_0 = –1.768$, 95% CI = (–2.647, –0.889)] is the effect size at the mean PPFD value, and is similar to the grand mean from `ModelA` ($M_d = –1.730$).

The full data set includes 120 effect sizes across 26 plant and 38 fungus species. Although their effects are confounded to some extent (in these studies and in nature), variation in both fungus and plant species could potentially explain a large amount of heterogeneity in response. Unfortunately, we cannot fit these covariates (as categorical effects) because, for example, we would be asking a model to estimate 25 additional parameters from a data set with 120 outcomes. In other words, estimating mean values for each plant species would use 25 degrees of freedom. One alternative is to use meta-analytic software packages which can model such categorical variables with a large number of levels as random factors (i.e., estimating variance components rather than regression coefficients; e.g., `MetaWin`, Rosenberg et al. 2000; `MCMCglmm`, Hadfield 2010, Hadfield and Nakagawa 2010). Another important issue is that our meta-analysis does not control for correlated structure among studies using the same species or for different degrees of phylogenetic relatedness across species (Lajeunesse et al. 2013; Nakagawa and Santos 2012; Chamberlain et al. 2012). Ignoring such correlated structure in data sets might increase the Type I error rate; see section 9.6.3.

9.5.3 *Is there publication bias, and how much does it affect the results?*

Publication bias is the hypothesized tendency for papers showing no significant effect to remain unpublished, due to either reviewer/publisher bias or to author self-censorship (section 9.6.1). Sensitivity analyses are statistical tests in which we check robustness and generalizability of our results in relation to potential publication biases. There is a large literature on methods for the detection and correction of publication bias in meta-analysis (e.g., Møller and Jennions 2001; Jennions et al. 2013b; Rothstein et al. 2005). Sutton (2009) categorizes the analytical tools for addressing publication bias into methods that detect publication bias, and methods that assess the impact of publication bias.

The methods for detecting publication bias include exploratory methods (section 9.3.5), and statistical methods based on regression/correlation. Statistical methods are the best-known sensitivity analyses; they test the relationship between effect sizes and their weights. The two most widely used methods are Egger tests (Egger et al. 1997) and the rank correlation test (Begg and Mazumdar 1994). Both test for statistically significant funnel plot asymmetry (section 9.3.4). These tests can be done in `metafor` using the functions `regtest` for Egger tests and `ranktest` for the rank correlation tests. For the plant/mycorrhizae data, both tests indicate statistical evidence for publication bias or funnel asymmetry (Egger test: $t = -3.895$, $df = 118$, $p = 0.0002$; rank correlation: Kendall's $\tau = -0.261$, $p < 0.0001$). However, funnel asymmetry can be caused not only by publication bias but also by heterogeneity (e.g., Egger et al. 1997). Therefore, for data with heterogeneity (most data), it is more appropriate to test for funnel asymmetry after fitting the meta-regression model (the same can be said for visualization methods; see figure 9.2c). In our example, the test for funnel asymmetry is still significant with residuals from `ModelB` or `ModelC` (Egger test: $t = -4.105$, $df = 117$, $p < .0001$).

If publication bias is identified by the above methods, we need to use methods that assess the impact of such publication bias. Sutton (2009) classifies these as: 1) the fail-safe number; 2) the trim-and-fill method; and 3) selection model approaches. *The fail-safe number* is supposed to indicate how many studies would be required to nullify a significant meta-analytic mean. The fail-safe number attempts to resolve the "file drawer problem" (Rosenthal 1979) where non-significant results remain unpublished in researchers' file drawers. It was once common to provide the fail-safe number in meta-analytic studies, but due to recent criticism of this method, fewer current publications (or software) provide this quantity (Becker 2005; Jennions et al. 2013b; and see Rosenberg 2005).

The trim-and-fill method proposed by Duval and Tweedie (2000a, b) uses the funnel symmetry to impute potentially missing data points and provide a corrected meta-analytic mean (see Duval 2005; Jennions et al. 2013b). This method is best viewed as a sensitivity analysis that asks whether our original result is robust against potential publication bias. For ecological meta-analysis, Jennions and Møller (2002) used this method to check around 40 ecological meta-analyses and found that the conclusions of over 15% of studies would change if the trim-and-fill method were applied. This method can be implemented using the functions `trimfill` and `funnel` in `metafor`. Applying a trim-and-fill procedure to the plant/mycorrhizae data set changes the main conclusion of the meta-analysis (section 9.5.2). The meta-analytic mean (plant response to mycorrhizal infection) is no longer significant; the corrected result is $M_d = -1.099$ [95% CI = (-2.225, 0.026)]. In figure 9.2d, the five imputed data points from the trim-and-fill method are plotted with the original data points in the funnel plot.

Selection model approaches are the best method for modeling missing data points, assuming the reasons such missing data are missing. If we use the terminology from missing data theory, missingness in the data is modeled as missing not at random (MNAR; chapter 4). This method is beyond the scope of this chapter, but we note there are an increasing number of R libraries, which are dedicated to selection model approaches for meta-analysis, such as `copas` (Schwarzer et al. 2010) and `selectMeta` (Rufibach 2011).

9.5.4 *Other sensitivity analyses*

Apart from significance-based publication bias, the best-known sensitivity analysis is a change in effect size over time. This phenomenon is called the time-lag bias, in which we see a declining trend of effect size over time, and often initially found strong effects disappear in newer studies (Trikalinos and Ioannidis 2005). We have many examples

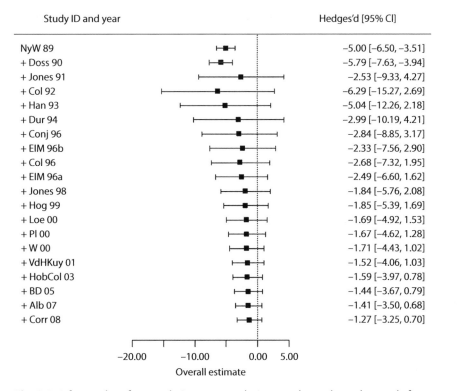

Fig. 9.4 A forest plot of a cumulative meta-analysis; note that only a sub-sample from the data set from the plant/mycorrhizae data was used, for illustrative purposes. For this example, we do not see a temporal trend. The functions cumul and funnel in the library metafor were used to draw the plot.

of the time-lag bias in ecology. Koricheva and colleagues compiled a list of 18 ecological meta-analytic studies, which identified the time-lag bias (see table 15.1 in Koricheva et al. 2013b). While we cannot discuss methods to detect the time-lag effects in any detail, readers are referred to the literature on cumulative meta-analysis (Koricheva et al. 2013b; see also Barto and Rillig 2012) for an evaluation of other biases. Meta-regression can also be used to evaluate other potential biases, including the effects of particular research groups or journal impact factors (e.g., Murtaugh 2002) on result magnitudes.

Finally, we may have outliers which influence the meta-analysis. These can be visualized by funnel plots or normal QQ plots (figures 9.1 and 9.2). While we could just conduct meta-analysis with these data points we could also use the metafor function leavelout to carry out meta-analyses repeatedly with one data point removed automatically each time. Additionally, we can use the metafor function influence to statistically quantify the effect of these data points.

9.5.5 *Reporting results of a meta-analysis*

The best practices for reporting the results of ecological meta-analyses are covered in detail by Rothstein et al. (2013). Results should include:

- the methods by which the literature was searched and papers selected (section 9.2),
- forest plots of the means and CIs of individual studies (section 9.3.4, figures 9.3, 9.4),

- forest plots or other graphical or tabular representations of the outcomes showing means and CIs for covariates,
- clear explanation of the models used to carry out statistical tests and the results of these tests (sections 9.4, 9.5),
- heterogeneity analyses including statistical tests (section 9.3.4), and
- the results of exploratory and sensitivity analyses (sections 9.3.4, 9.3.5, including evaluation of potential publication bias and its influence on the results.

In reporting results of meta-regressions, particularly those with two or more categorical covariates and where covariates have more than two levels, the outputs from `metafor` can be confusing and incomplete, as we have seen above. In addition to the outcomes detailed above, we suggest reporting the calculated model covariates (that is, the outcome of the calculated statistical model), the sums of squares (that is, Q values) for the covariates where there are several levels and their P–values, and categorical means and confidence intervals. These outcomes are available in the package OpenMEE.

Some ecologists (e.g., Michael McCarthy, personal communication) have suggested that weighted means and confidence limits resulting from meta-analyses should be back-transformed in reporting results so that they are expressed in natural units rather than mean effect sizes. Such natural units might be the units on which the effect sizes were originally calculated (e.g., biomass). Doing this might seem to make the outcomes easier to interpret biologically, but may be problematic. Since meta-analysis is based on expressing study outcomes originally expressed on different scales into a common scale by means of calculating effect sizes, back-transformation of combined effects is often going to be impractical in most cases. Where all studies express outcomes on the same scale, this would be straightforward, but that is rarely the case in meta-analyses. Perhaps future developments will reveal ways to make the outcomes of meta-analysis easier to interpret biologically as well as remaining statistically meaningful.

9.6 Discussion

9.6.1 *Objections to meta-analysis*

In nearly every field in which meta-analysis has been introduced, similar objections to its implementation have been debated (often strenuously); these discussions eventually decline, as the issues are successfully addressed and meta-analytic methods are broadly accepted and incorporated into contemporary practice. This has happened in ecology as well.

The major objections are summarized as follows. First, meta-analysis risks combining "apples and oranges." The answer, famously made by Gene V. Glass, who is responsible for coining the term meta-analysis, is that it is a trivial thing to combine apples and oranges if we want to know about fruit (in Rosenthal and Rubin 1978); if we want to reach generalizations, limiting our study to a single variety of apples would be poor practice (R. Rosenthal, stated verbally in a talk at the annual meeting of the Society for Research Synthesis Methodology, Providence, RI June 2013). One way of thinking about this is to realize that what you combine defines the scope of your answers: you may choose to do a narrow or broad meta-analysis, and your conclusions may apply narrowly or broadly as a result. But of course you should not combine anything that is not scientifically defensible to combine! To improve the comparability of data one can carry out heterogeneity analysis or meta-regression to assess the sources of variation in outcomes among studies with different characteristics (section 9.4.3). No other method of summarizing data across studies provides this kind of robust quantitative methodology for assessing differences in outcomes.

A second major objection is that the collection of information being summarized is biased, so that any summary will also be biased. One hypothesized major source of such bias is *publication bias* (Section 9.5.3). Publication bias exists when only those studies showing large and statistically significant effects are accepted for publication, while those reporting no significant effect are rejected. Any summary of such biased results would give a biased overestimate of the effect. Various meta-analysis methods have been developed to detect publication bias, estimate its magnitude, and attempt to correct for it (section 9.5.3). It is important to recognize that any attempt to make sense of the literature will be subject to these same biases, if they exist, including not only traditional formal narrative reviews but any informal impression of the literature. Meta-analyses are uniquely able to address publication biases quantitatively. Another possible source of bias in the literature has been called *research bias* (Gurevitch and Hedges 1999), which is the case when certain organisms, systems, etc. are well studied while others are not (e.g., Lowry et al. 2012); others forms of bias also exist Møller and Jennions 2001).

Other objections have been raised to the specific details of the way particular research syntheses have been conducted, and to specific effect size metrics (i.e., indices used to put the results of all studies on the same scale; see section 9.4.1), sometimes rejecting all attempts at meta-analysis as a result (e.g., Hurlbert 1994; standardized mean differences have been considered to be especially objectionable by some authors, but see Hillebrand and Gurevitch 2014). We would not suggest rejecting all attempts at ANOVA, for example, or for that matter molecular biology, because they are sometimes done sloppily or in an automatic and thoughtless fashion. If you hypothesize that a particular effect size metric might bias the outcome of a meta-analysis, that hypothesis can be examined in various ways, such as by comparing the results using a different effect size metric with different properties.

A final common objection is that the literature is poor in quality, so any attempt to synthesize it will be as well. Weighting results by study quality is a possibility that has been examined extensively in the meta-analysis literature (e.g., Higgins and Green 2011 section 8.8.4.1; Greenland and O'Rourke 2001) but again, if all of the literature is of poor quality that will affect any attempt to make sense of it or to generalize results from it. On the other hand, we might argue that some information may be better than no information: for example, we may have very low replication for studies on pangolins or honey badgers but no amount of well-replicated studies on mice is going to tell us much about pangolins or honey badgers, and so such studies may be invaluable for the evidence they provide despite possible limitations in experimental design.

The responses to the above objections in the largest sense are the same: as the available information in a field grows beyond a certain point, and the need to make sense of it becomes ever more compelling, robust means for synthesizing the information to reach general conclusions become essential to progress in science. There is no alternative. Objections to the use of meta-analysis are based on concerns that are relevant to all approaches to synthesis (formal or otherwise), yet meta-analysis provides the only formal methodological framework to address these concerns. Critiques of individual meta-analyses may or may not be valid. There are new methodologies being developed over time for quantitative research synthesis, and concerns that some can involve the kinds of flaws discussed here do not invalidate the major benefits of meta-analysis for conducting rigorous synthesis of scientific results.

9.6.2 *Limitations to current practice in ecological meta-analysis*

Perhaps the most profound contribution of meta-analysis to science is in challenging our thinking about the nature of generalization. The ramifications of this have begun to be explored, but much more remains to be done. The issue of generalizing is a longstanding one in ecology, and the tools of meta-analysis open up new doors for thinking about this problem.

In a more practical vein, one of the valuable contributions of experiments is the ability to disentangle confounded factors. Meta-analysis itself is non-experimental, and while it is used to summarize the results of experiments, it cannot separate confounded information, as we saw above (sections 9.5.1, 9.5.2). Ecologists often want to consider the effects of large numbers of potential explanatory factors on the outcomes of experiments, and either the data may be insufficient to address all of these effects, or the results of complex meta-regressions with many covariates at many levels are incomprehensible. As these methods become more widely used, better approaches may become available for interpreting such results.

Meta-analysis is a branch of statistics, not a branch of magic. It cannot magically create data where data do not exist. It is not a reasonable criticism of meta-analysis to expect it to be possible to summarize non-existent data (it sounds silly, but meta-analyses have been criticized for not including summaries of organisms, questions, systems, etc. where there is insufficient primary literature to summarize). However, research syntheses may be highly effective in pinpointing areas where more primary research is needed. They may also be highly effective, if controversial, in pinpointing areas where no more primary research is needed, because sufficient information already exists in abundance to satisfy particular research questions (e.g., Hyatt et al. 2003).

A very real and common limitation to carrying out meta-analyses is poor data reporting, although this may be improving. Common problems include failure to report some measure of variance or sample sizes, and such limited information on experimental design that it is difficult to determine what was done or what the reported results mean. Clearly that makes collecting data for a meta-analysis difficult. In addition, abstracts of papers may not accurately reflect the contents of the paper, and key words may sometimes be misleading, making searching for appropriate papers more difficult. Papers may call themselves meta-analyses that are not meta-analyses at all, but vote-counts or something else entirely (Vetter et al. 2013; Gurevitch and Koricheva 2013).

9.6.3 *More advanced issues and approaches*

Finally, we turn to four advanced topics: 1) correlated data, 2) missing data, 3) multivariate (multi-response) meta-analysis, and 4) network meta-analysis. First, although we have so far treated our data as if they are independent of each other, correlated data are common, and they arise in a number of ways in ecological data sets for meta-analysis (reviewed by Mengersen et al. 2013a). For meta-analysis encompassing multiple species, such correlation may arise from shared ancestry, so that effect sizes of closely related species may be more similar than those of distantly related species (chapter 11; Lajeunesse et al. 2013). There is a software package for implementing phylogenetic meta-analysis (phyloMeta), which can incorporate a phylogenetic tree as a correlation matrix (Lajeunesse 2011). The R library MCMCglmm not only incorporates phylogenetic correlations but also models other correlated structures as random effects (Hadfield 2010; Hadfield and Nakagawa 2010), and MCMCglmm can incorporate two or more phylogenetic trees for one data set (Hadfield et al., 2014). Nakagawa and Santos (2012), Mengersen et al. (2013a), and Lajeunesse et al. (2013) can be consulted for more information on models accounting for various types of correlated data.

Second, missing data in meta-analysis is a ubiquitous problem. This is of two types, missing effect sizes and missing values for covariates (Lajeunesse 2013). For example, the initial data set for the plant/mycorrhizae analysis had substantial numbers of missing data in sample size and standard deviations so that we could not obtain effect sizes and sampling variances (section 9.5.2). Many studies did not have information on light levels (PPFD); that is, data for a covariate were missing. We could potentially treat both types of missing data effectively by using either multiple imputation or data augmentation (explained in detail in chapter 4). In appendix 9A, we provide an example using the R library `mice` to conduct multiple imputation on the plant/mycorrhizae data set. As discussed in section 9.5.5, there is another kind of missing data in meta-analysis because of publication bias. Multiple imputation and data augmentation methods are not useful for this kind of missing data, and we need to use selection model approaches (reviewed in Hedges and Vevea 2005; Jennions et al. 2013b).

Third, multivariate (multi-response) meta-analysis is rarely used in ecological meta-analyses, but we see a number of potential applications (Nakagawa and Santos 2012; Mengersen et al. 2013a). In fact, our photosynthesis data sets could be run as a multivariate meta-analysis model because experiments on plant response to CO_2 elevation often report two types of effect sizes: one for photosynthetic carbon uptake (PN) and one for leaf nitrogen concentration (N). If each study has a pair of these two effect sizes (PN and N), we can run a bivariate meta-analysis where we can estimate a correlation (covariance) between the two effect sizes. For example, Cleasby and Nakagawa (2012) looked at the relationship between male age and paternity inside and outside social pair bonds in socially monogamous passerine species using Bayesian hierarchical modeling with correlated data and missing data (i.e., in many studies, only one type of effect sizes was available). Interested readers should see Congdon (2010) and Schmid and Mengersen (2013) for more information on implementing complex meta-analytic models.

Other methodological developments include network meta-analysis, used in medicine but not yet used in ecology, and structural equation modeling (chapter 8) within a meta-analysis framework, so far used mostly in the social sciences. Network meta-analysis is used where all of the studies do not compare the same treatments but each includes a subset of the treatments. In some cases one can model a network of studies where all studies share at least one treatment in common; the goal is to allow for indirect comparison of treatments that were not compared directly (Caldwell et al. 2005; Janson and Naci 2013).

Acknowledgements

Support (to JG) from the U.S. National Science Foundation program in Advances in Biological Informatics is gratefully acknowledged. SN gratefully acknowledges support by the Rutherford Discovery Fellowship (New Zealand). Insightful comments by Michael McCarthy, Marc Lajeunesse and the editors of this book improved the text of this chapter.

CHAPTER 10

Spatial variation and linear modeling of ecological data

Simoneta Negrete-Yankelevich and Gordon A. Fox

10.1 Introduction to spatial variation in ecological data

An ecologist is interested in studying hummingbird feeding strategies. She has observed that species A and B dominate two neighboring forest areas. Both areas are dominated by the same flowering plant. The ecologist wants to know what determines the success of each species in the different areas. She samples the two areas and uses a series of *t*-tests to compare them, but finds no differences between areas in flower density, nectar availability or any other characteristic. On conducting some spatial analyses, she finds that the feeding plant is *aggregated* in small patches in the area dominated by A, and in large patches in the area dominated by B. This makes biological sense: A is a strongly territorial species, and is only able to defend (with aggressive behavior) small flower patches as territories. In contrast, B is more successful in areas where there are large flower patches and many birds of *trapliner* species can feed simultaneously. The initial studies of flower density in these two areas (which did not account for the difference in the scale of aggregation) missed this biological point. Moreover, they were problematic statistically because observations were not independent at the same distances in the two areas; this violates the assumptions of linear models (like the *t*-test) and leads to increase in type I errors or biased parameter estimation. Perhaps more important, in this story (simplified from Jiménez et al. 2012), spatial variation plays a central role in explaining the ecological questions of interest.

Spatial variation has been often considered as undesirable noise in ecological studies because many statistical methods we use assume a *random spatial distribution* (spatial independence). This assumption is partly the result of the low computational power that limited the complexity of analyses available to most scientists until the 1990s. We now have powerful computers and software that facilitate storing, displaying, and analyzing spatial data sets, rapidly and cheaply. Today most ecologists still design their studies to avoid considering spatial structure. It is time to begin changing this. Empirical observations and simulations show that spatial structure is the rule and not the exception in natural systems (Tilman and Kareiva 1997). Organisms, sites or cells that are close together are similar because they are driven by the same processes or because they influence each other. Plants tend to grow in patches of good soil conditions, hummingbirds aggregate where flowering resources are clumped (and away from territories of dominant species),

Ecological Statistics: Contemporary Theory and Application. First Edition. Edited by Gordon A. Fox,
Simoneta Negrete-Yankelevich, and Vinicio J. Sosa. © Oxford University Press 2015.
Published in 2015 by Oxford University Press.

and diseases radiate from a focal geographical point. Therefore, spatial dependence among observations is difficult to avoid and a random spatial distribution is, in most cases, an unreasonable assumption.

But there is a deeper reason to begin incorporating spatial structure in our analyses: numerous results underline the conclusion that spatial structure plays an important role in sustaining biodiversity, driving evolution, and determining the outcome of many interspecific interactions like biological invasions (Levin 1992). Indeed, there are many ecological processes that are driven, in part, by spatial structure. Moreover, many ecological processes generate spatial patterns—for example, patterns in flux of matter and energy.

Thus modeling spatial structure is a key objective in ecological research. Ecological models that consider spatial processes are generally more realistic and predictive than those that do not (Legendre and Fortin 1989; Skidmore et al. 2011). Many ecological theories incorporate spatial terms, and their predictions include spatial distributions of organisms, interactions, genes, propagules, etc. (Tilman and Kareiva 1997). Confronting these theories with evidence often requires spatially explicit study designs and spatial statistical tools. Consequently, in recent years there has been rapid growth in the use of spatial statistics in ecological studies. Consider the long-studied problem of why communities are composed of a few common and many rare species. The metacommunity approach now proposes four paradigms to explain community composition and dynamics: patch dynamics, species sorting, mass effects, and the neutral model (Leibold and McPeek 2006). These models differ in the degree to which metacommunity structure is determined by spatial processes such as dispersal limitation and by species' responses to environmental factors (Siqueira et al. 2012). To apply these models in a given situation, it is necessary to determine whether dispersal rates are large enough to affect local population abundances, and whether the environmental conditions in patches are homogeneous (Holyoak et al. 2005). Doing so requires using spatially explicit research designs and analyses. For a start, *scale*, *patch*, and *homogeneity* all need to be operationally defined and statistically tested.

Regardless of whether spatial patterns are part of the research hypothesis or a source of noise to be reduced, ecologists need to be familiar with a conceptual framework (and associated jargon) that will allow them to understand and analyze spatial structures. In this chapter we have included a brief introduction to the basic concepts. Of these, probably the most transcendent is *scale*. One source of confusion is that researchers in different disciplines (geography, geophysics, or biogeography) operationalize it differently, which has led to different meanings of the same term in the literature (box 10.1). The spatial scale that we refer to in this chapter is not the familiar cartographic scale in maps (a ratio between the distance on the map and the real-world distance) but the concept as used in ecological studies, which works in the opposite direction from the cartographic scale. A small cartographic scale means there is less detail available about small objects (e.g., a map of a continent as compared with a city), while a small ecological scale means that the focus is on detail occurring in a small area.

The choice of scales for an ecological study is of utmost importance. A mismatch between the scale of analysis and the scale of the phenomenon being studied (e.g., a species' response to its environment) leads to a decrease in (a) the portion of variation explained by competing models; (b) the rigor of statistical analyses; and ultimately (c) the ability to understand the processes underlying the spatial patterning. To see this, consider the study of plant density-dependence by Gunton and Kunin (2009). They showed that for their plant population, there was a positive relationship between plant aggregation

---

Box 10.1 WARNING: THE TERMINOLOGY OF SPATIAL STATISTICS CAN BE CONFUSING!

Spatial statistics have been developed independently and simultaneously, in the context of diverse disciplines, including economics, epidemiology, mining, geology and geophysics, geography, and (more recently) ecology. This parallel development has led to terminology that can be confusing: there are many terms, and some terms are defined somewhat differently by authors, even within a discipline. This problem is particularly complex in ecology, because several terms have also been defined ecologically in ways that differ from the statistical definitions inherited from other disciplines. A simple example is *scale*; as noted in the main text, ecologists typically use this term to mean the inverse of what geographers mean.

But the problem is much more widespread than this; some technical terms (that lend themselves to explicit mathematical definitions) turn out to be defined differently by different authors. Here we provide a couple of examples of central terms that have multiple definitions which are alike, but not quite the same. Our goal is to call your attention to this issue; we hope that this awareness will help prevent confusion while you familiarize yourself with spatial statistics. We expect (and hope) that eventually researchers will converge to unified—and clear—terminology.

- *Autocorrelation*. Fortin and Dale (2005) define spatial autocorrelation simply as the correlation of a variable with itself, at a given spatial distance. Under this definition, spatial gradients can be considered cases of autocorrelation. By contrast, Legendre and Legendre (1998) consider the non-stationary part of correlation due to environmentally induced gradients to be "spatial dependence."
- *Heterogeneity*, *homogeneity*, *uniformity*, and *stationarity*. Legendre and Legendre (1998) say that "heterogeneity is the opposite of homogeneity which means the absence of variation . . . In spatial pattern analysis heterogeneous refers to variation in the measurements, in some general sense that applies to quantitative, semiquantitative, or qualitative variables." From this definition, one would conclude that if there is a within-plot spatial pattern, there is spatial variation in measurements and thus heterogeneity. Pfeiffer et al. (2008), accordingly, call local dependence of spatial processes meso-scale heterogeneity. By contrast, several authors (Burrough 1987; Fortin and Dale 2005) use homogeneity or uniformity as synonymous with stationarity, even when spatial pattern exists.

We could provide further such examples. The point is not that any particular definition given by particular researchers in this field is wrong—and certainly not that our terminology is better—but that some key terms in spatial statistics are not yet consistently defined.

The bottom line here is that, no matter what you do, you should report what you mean by your terminology, and, in reading the literature on spatial statistics, you should be aware that some terms are used in varying ways. We think the ideas and tools are important, even if the terminology is not yet standard.

---

and reproductive output (through pollinator attraction to more concentrated flower resources), but it operates on a different spatial scale than the negative effect of aggregation on plant survival through competition. Ecological researchers often choose a spatial scale for their studies not based on biological considerations like these, but rather according to data availability, budget capability, the measuring equipment they have on hand, or human perception of the system. One goal of this chapter is to highlight the importance of examining the appropriateness of the spatial scales used in study design and data analysis. A long-term goal for every ecologist should be developing an understanding of spatial problems that is sufficient to provide you with some intuition about how to approach these problems. As daunting as spatial statistics can seem, this is just a matter of experience; this chapter will introduce you to basic tools that will aid this understanding.

There are several valuable sources of technical information on spatial statistics for ecologists in general (e.g., Legendre and Legendre 1998; Fortin and Dale 2005; Ives and Zhu 2006) and in specific areas (Plant 2012; Diggle 2013), as well as treatments oriented to other disciplines (e.g., Cressie 1993; Griffith 2012). The challenge for us is to integrate these tools into our thinking, so that we aren't afraid of spatial issues and have some intuition as to when they are important. Spatial statistics may initially seem daunting, but they often open the possibility of analyses for more types of data and experimental designs. In this chapter, we will focus on drawing the link between the type of spatial hypothesis and the selection of statistical modeling tools. We hope it will help readers incorporate spatial processes in their ecological thinking, establish the basic conceptual basis, demystify some of the tools, and start incorporating spatially explicit statistical tools in their work.

Many ecological papers that use spatial statistics concern observational (and often massive) data sets, as they are obtained from remote sensing. In this context, a common objective is to recognize matching spatial patterns in maps. The use of spatial statistics in this task is important, because, as Plant (2012) puts it, "the human eye has a famous ability of seeing patterns that are not really there"; therefore statistical confirmation is necessary. We suggest that there are two dangers in such practice. The first is a result of what Plant (2012) describes as high spatial resolution but low ecological resolution: in remotely sensed data the indicator variables are often related to the phenomena of interest only indirectly. For example, the data consist of reflectance over a range of wavelengths, but a researcher may be interested in asking how photosynthetic rates vary across a landscape. Because of the sophistication of the technological tools, ecologists may lose track of the low explanatory power that this entails. Interesting experimental studies have tackled spatial questions in ecology, but these often involve small data sets (Gunton and Kunin 2009). The second danger is related to the fact that spatial statistics (especially geostatistical analyses) are built into much geographical information systems (GIS) software; if the user is not careful, default analyses (with corresponding untested assumptions) are performed and map results presented in a surprisingly beautiful, but sometimes badly incorrect, manner (see Fortin et al. 2012 for a good review of how spatial statistics are correctly used with GIS).

The data sets used in this chapter are small, univariate, and not managed in GIS. This is because the focus is in principles and basic modeling. Most ecologists work with relatively small data sets and consider that spatial analysis is for large data sets, often with dramatic spatial variation. One of our goals is to convince you of the value of spatial analysis even for much smaller data sets and more subtle spatial variation. Therefore this chapter will be closer to their needs. Once you have the basic ideas, it is much easier to learn the techniques needed to use spatial statistics correctly in the context of GIS or multivariate data.

This chapter introduces a basic set of tools for exploratory data analyses and graphs that will make spatial patterns apparent and help you make analytical decisions. If spatial structure in the data is an important variance component, techniques discussed in chapters 12 and 13 may prove useful in helping account for the influence of position in space. Spatial regressions and autoregressions are widely used by ecologists; we have included two examples of autoregressive models. There are many other useful tools for spatial analyses, including Bayesian hierarchical models (Gelfand 2012). Your knowledge of spatial statistics should not end with this chapter; rather, we hope that you can use it as a springboard to find, understand, and use methods from the wide range of possibilities now available.

10.2 Background

10.2.1 *Spatially explicit data*

In this chapter we deal with statistical techniques that use information on the spatial distribution of data (geographic location, distance matrices, or frequencies in grid cells) to answer *spatially explicit questions*—questions about the shape and scale of spatial patterns. For example, one might ask whether trees (and the leaves they shed) create an aggregated pattern of species-specific litter and detritivore species under their canopies; answering this requires estimating the scale and shape of this patchiness. It is not enough to do a significance test and show that that detritivores do not occur at random in space. The latter is a spatially implicit question, because it does not involve modeling the actual information on spatial distribution.

Spatially explicit studies involve information on location of observations. Usually, we are interested in spatial patterns in *random variables* (chapter 1). Random variables that are associated with location information are called *random fields*. Location variables themselves—for example, location in a study grid—are assumed to be measured without error; they are fixed, rather than random. Spatial data has been classified in different ways by different authors (Cressie 1993; Fortin and Dale 2005; Plant 2013). Here we recognize three types of data sets, according to the role that location variables play:

- *Geostatistical data* involve precise coordinates of data points, with a value of the response variables for each location. An example might be a map of individual trees, along with their sizes.
- *Point pattern data* consist solely of locations; this is useful in studying spatial patterns of occurrence (e.g., a map of dung locations; chapter 12).
- *Areal data* (or lattice) use spatial units like grids or mosaics, where each cell has a single value of the response variable, and there is information about which cells are neighbors. One example is the mean plant cover in pixels in a satellite image.

We will concentrate mostly on geostatistical data, with a few references to the other types; see Perry et al. (2002) for a good review of these classifications and the statistical methods that can be used with them.

10.2.2 *Spatial structure*

Imagine that we have data on the nitrate content of soil samples on a grid. The measured concentrations will certainly vary spatially; is the variation consistent with random variation, or is there a spatial pattern to the data? Because we are studying a single realization of a process (the distribution of a random field in a plot), it will take some work to answer this question formally.

Spatial structure occurs in patches or clusters (which we use as synonyms). A *patch* is an area where values of a variable are more similar to each other (or individuals more aggregated in the case of point pattern or count data) than expected at random. *Spatial autocorrelation* or *spatial covariance* is, formally, the degree of correlation of a variable with itself, which is also a function of the spatial position. Informally, this means that if it is possible to predict our measurements based on their location, there is some spatial autocorrelation. A little thought about word meanings may be helpful here: a variable (like soil nitrate concentration) can be correlated with itself because nearby values are similar (since they have similar positions). In section 10.5 we present two measures of autocorrelation. Ecologists mostly encounter *positive autocorrelation* or *aggregation* (in the case of count data); because high or low values (or counts) often tend to occur near one another *Negatively autocorrelated* spatial patterns also occur in ecology, but they are less common. This kind of pattern means that the data are more uniform spatially than expected at random. For count data this is known as *disaggregation* of individuals, and occurs when counts are evenly spaced (Perry et al. 2002), as is the case in populations of some highly competitive plants like some desert shrubs. Because neighboring locations typically have contrasting shrub occupancy, local plant density is negatively autocorrelated. We have avoided using the term *overdispersion* in this context, although it is sometimes used in the spatial ecology literature, because it can easily be confused with the statistical term referring to random variables that have greater variance than expected (chapters 1 and 6). The terms *trend* or *gradient* describe patterns that are just what they sound like: a measured variable changes monotonically over the entire area being studied (box 10.2). These patterns fall within our definition of autocorrelation (but see box 10.1 and Legendre and Legendre 1998, for an alternative view). The process generating the trend might occur on a scale much larger, or only slightly larger, than the sample plots. For example, a study of plant species richness in the forested area of a nature preserve might find a gradient—but the underlying causes of this pattern might be operating on a scale of hundreds of meters (e.g., a hillside larger than the sampled area) or many kilometers (e.g., distance from the seashore) beyond the sampled area. To say anything about the scale of this gradient would require information. The distinction between trend and patch is only one of study scale. If spatial autocorrelation is not stronger than expected at random, we can tentatively conclude that observations are *spatially independent*.

Having identified spatial structures, we are often interested in determining their characteristics (such as shape, number, and pattern); there is, of course, much terminology for these characteristics. Patches may have a *contagious* character, i.e., autocorrelation increases and decreases smoothly through space. By contrast, patches may have abrupt edges called *steps*. Patches may consist of particularly high or of particularly low values; the latter are also called *gaps* (box 10.2).

It is important to know whether spatial structures change with direction; for example, whether patches are elongated in one direction (box 10.2). The answer often dictates the kinds of models and statistical tests that are appropriate. Systems in which structure does not change with direction are *isotropic*, while those that do change with direction are *anisotropic*. To see why this is important, realize that anisotropy implies a different degree and distance of autocorrelation or aggregation in different directions. Plant (2012) compares anisotropy to the texture of corduroy fabric. When you rub it the degree of friction perceived in one direction is different than in the other.

We need to consider a final issue about pattern: is the spatial process that generates the pattern the same everywhere in the area studied? First, the terminology:

Box 10.2 COMPONENTS OF SCALE: A HYPOTHETICAL EXAMPLE

Here we describe a hypothetical example (a descriptive study of the biomass of leaf-cutting ants, *Atta mexicana*), to illustrate the components of scale and how they are used. These ants occur in both natural and agroecosystems, and the researchers aim to understand how their distribution is affected by a mosaic of crop fields and forest. They use two nested sampling schemes (*a* and *b*; see the figure). Both schemes have the same grain size, a 5 m radius sampling point (78.5 m²), but *a* has much larger intervals than *b* (1 and 0.3 km, respectively), and *a* has a much greater extent than *b* (20 and 1.89 km², respectively). The scale of scheme *a* was suggested by the hypothesis that at a larger scale, ant biomass would aggregate in or near the most disturbed areas, since most cultivated plants lack the chemical defenses against *A. mexicana* found in forest plants. On the other hand, the scale of scheme *b* was chosen because the researchers hypothesized that the distribution of ant biomass within the productive plot was determined by the planted citrus trees forming a live fence to the right of the plot, suggesting an aggregation of sampled biomass on that side.

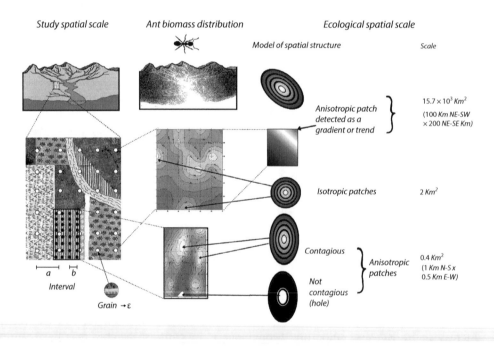

Relationship between the study and ecological scales in a leaf-cutting ant spatial study

Identifying the *ecological scale(s)* is often a goal of spatial studies. If spacing is possible between patches with high values (say, spaces between colonies of a species), the size of these "empty structures" or low-value patches also counts as a part of the ecological scale and can be reported separately. In this ant study, the biomass of ants is modeled as a continuous variable with no empty spaces. Three relevant ecological scales were found for the spatial structure of ant biomass. The largest was defined by a single patch of 15.7×10^3 km² covering the whole catchment and related to the influence of the branching river. This structure was only partially detected in the form of a gradient departing from the creek that crosses the sampling area *b*. The second ecological scale was defined by patches with a mean area of 2 km^2, and is probably related to the location of large ant nests. The third ecological scale was defined by patches with mean area 0.4 km^2 within the productive plot, which may locate the position of the most-consumed crops by ants.

In this example, *A. mexicana* biomass turns out not to match the predictions of the hypotheses described above; the spatial analyses suggest the river, the position of nests, and the positions of food plants as drivers of variance at the three nested ecological spatial scales detected as relevant.

This example illustrates how different processes may operate at different scales to determine spatial structure of a single ecological variable. Moreover, different spatial sampling schemes in the same area can reveal different spatial structures. Conclusions derived from one study spatial scale most often cannot be extrapolated to another.

Spatial data sets are usually very dense in within-site information and sparse or null in between-site replication; they are often unreplicated. While this violates basic principles of traditional statistics (Plant 2012), it does not mean that these data do not provide important information. Phenomena (or interactions between them) that are observed at one spatial scale may not be observed at other scales. Thus, many small plots, each containing sparse data, will not reveal these patterns.

- *Second-order stationary* (or *stationary*) processes have constant mean and variance within a study area. The autocorrelation depends on the distance and direction between locations (and not on their absolute locations).
- In *non-stationary* spatial processes, both the mean and the autocovariance change with the location within the region studied.

As you might guess, models for stationary systems are simpler to build than models for non-stationary systems (because one can use the same autocorrelation functions everywhere), and we will not discuss non-stationary systems further.

Whether a system is stationary depends on the spatial process, but what we observe are the data (Fortin and Dale 2005). This may seem to create a truly annoying problem: stochasticity can mean that we may observe an irregular spatial pattern even when the spatial process is stationary. Since we usually have data from only one realization of the process, we must typically *assume* stationarity. Though this may sound like a reckless way to proceed, Haining (1990) suggests it is analogous to using observations in different spatial locations the same way replicates are used in a controlled experiment. Exploratory graphs or maps should definitely be used to detect very strong non-stationary patterns (such as distinct regions or gradients in patch size or orientation across the study area) which are an unlikely result of chance alone and will clearly violate the assumption. Many spatial statistical methods are based, to different degrees, on stationarity assumptions. This is particularly true for statistical tools that estimate model parameters globally (from all data points) instead of locally (for each observation and its surroundings).

A more general form of stationarity is sometimes assumed when spatial patterns are being studied with descriptive purposes and the generating process is known to have a distribution with undefined variance. This relaxed assumption is called the *intrinsic hypothesis* and states that the variance of the difference between the values at pairs of locations depends only on the distance d between them and the direction, but not on their specific location. It also implies that (a) the mean of these differences is 0 and (b) stationarity holds only within the range of a relatively small neighborhood (within a moving window for example; Fortin and Dale 2005).

10.2.3 *Scales of ecological processes and scales of studies*

We have used the term *spatial scale* a bit loosely; it is used so widely in ecology that it sometimes *sounds* like a simpler concept than it is. There are multiple definitions in the literature, and these involve varying degrees of formality (see Dungan et al. 2002 for a review). Indeed, there are some subtle ideas involved in thinking about spatial scale; we need to define it operationally (see box 10.2 for an illustrated example).

It is useful to begin by distinguishing between the scale(s) at which a study is performed (*study spatial scales*) and the scales (usually more than one) at which the structure of the ecological phenomenon occurs (here called *ecological spatial scales*). Although these two concepts both involve spatial dimensions, they are operationally defined in different ways. Identifying the *ecological scale(s)* is often a goal of spatial studies and it is expressed by the sizes of the *spatial structures* (also called patterns and patches) in a single variable or in the co-occurrence of two or more variables. If spacing is possible between patches with high values (say, spaces between colonies of a species), the size of these "empty structures" or low-value patches also counts as a part of the ecological scale and can be reported separately. The study spatial scale is entirely defined by the researcher; the choice is informed by previous knowledge about the study subjects or driven by hypotheses, but constrained by budget and instrumental capabilities. In geostatistical data, three dimensions need to be identified to define the study scale:

- The *grain* (sometimes called *resolution*) refers to the size of a single observation: for example a 2 m^2 sampling quadrat in a study of herbaceous vegetation, the 50 m radius of a hummingbird observation station, or, in the hypothetical example presented in box 10.2, a 5 m radius ant sampling point.
- The *interval* is the minimum distance between the grains.
- Finally, the *extent* or domain comprises the total area (or distance in the case of transects) covered by the study. The extent is often expressed as the length of the larger axis of the sampling area, but in irregularly shaped sites all dimensions or surface must be reported.

In point pattern data, the study scale is solely defined by the extent; the grain is generally not meaningful because the coordinates locate individuals, and the interval is not defined by the researcher. In areal data the study scale is defined by the grain (e.g., a 10 m^2 pixel of an aerial photograph) and the extent, as the interval is zero.

The sizes, number, and types of patches detectable by a study depend on the sizes and shapes of grains, and on the interval and extent of sampling (Dungan et al. 2002). Therefore, matching your study scale to the targeted ecological scales is a crucial issue for the inferential capability of the study. Failure to detect spatial patterns does not imply they don't exist; it may only mean that the study spatial scale did not match any of the relevant scales for the phenomenon under study.

Choosing study scales has many of the same challenges as choosing other sampling and experimental design elements, but designing spatial studies requires considerations additional to those required for non-spatial ecological designs. A carefully posed hypothesis reduces the difficulty. For an introduction to the theory of spatial sampling design see Wang et al. (2012).

The following guidelines are a summary of those from more extensive works on spatial ecology (Legendre and Legendre 1998; Dungan et al. 2002; Fortin and Dale 2005). These guidelines are only a starting point; there is no substitute for knowledge of the study system.

(1) Sample size. Both empirical and simulation studies (summaries in the sources cited in the preceding paragraph) have established that 20–30 samples is the minimum required to detect strong spatial autocorrelation; 100 samples may be needed to reliably estimate parameters on the spatial patterns. In the ant example (box 10.2) both sampling schemes are near the minimum necessary for detecting spatial patterns, and such a study would not yield enough information to reliably model the spatial pattern.

(2) Grain. For frequency data, grains should be large enough to contain at least a few elements (e.g., individuals, dung piles, nests). For continuous variables, grains should be two to five times smaller than the patch size to be detected. The shapes of grains should be as symmetrical as possible in all directions, unless there is substantial reason to expect a strong anisotropic pattern. In the ant example both sampling schemes, *a* and *b*, meet these conditions; box 10.2).

(3) Interval. To detect spatial patterns, intervals should be smaller than the patch size. If gaps are expected, intervals should be smaller than the interval between patches. When the ecological spatial scales are completely unknown, nested sampling designs with two or more interval sizes are useful to help identify pattern.

(4) Extent. The extent should be two to five times the size of the largest patch to be detected. In the case of the catchment size patch in the ant example (box 10.2), this rule was not followed, and therefore this spatial structure was detected only as a gradient. For a constant sample size (often dictated by limited resources!), the extent can be maximized by using a linear transect rather than a two-dimensional area. If anisotropy is expected, one can use two or more intersecting transects to form a star.

What should be reported in papers about spatial scale? Dungan et al. (2002) put it well when they urged "ecologists to be explicit on their usage of the term scale and to report the decisions that were implemented in the sampling design and in the statistical analysis." This also means reporting all three components of the study spatial scale (grain, interval, and extent), and the basic information used to match the study and ecological scales. See box 10.2 for an example of how to do this.

10.3 Case study: spatial structure of soil properties in a *milpa* plot

We are ready to consider an empirical example. We will use a data set of soil properties sampled in a plot of traditional maize polyculture (*milpa*) in Los Tuxtlas, Mexico (see Negrete-Yankelevich et al. 2013 for details). An influential hypothesis in soil conservation science is that there is a relationship between the diversity of plants above ground and the spatial heterogeneity of soil organisms and environmental conditions below ground (Wardle 2002). Plant species generate different environments in their rhizospheres, resulting in distinct environments and soil communities. A diverse plant assemblage is also thought to make soil ecosystem functioning more resilient to disturbance and spread of soil-borne plant pathogens (Naeem 1998). Under this hypothesis, we would expect that the high agrodiversity in traditional *milpas* will increase spatial heterogeneity in the soil, as compared with the less diverse *milpas* planted in recent decades. The data we use here are from a highly diverse *milpa* where maize is planted with common bean, pumpkin, leek, chayote

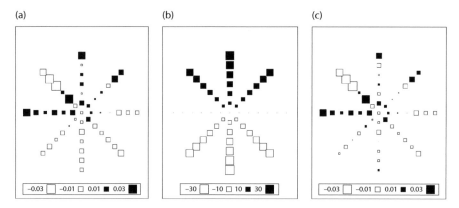

Fig. 10.1 Bubble maps can be used to explore spatial patterns. These are representations of the values of a measured variable (symbol size) located in their spatial coordinates. To aid visual exploration, bubble maps are often constructed with the centered variables ($y_{ci} = y_i - \bar{y}$), then the symbol size becomes proportional to the observation's deviation from the mean. Black and white points indicate values above and below the mean, respectively. Here the bubble maps are for (a) the centered and transformed C:N data, (b) a model of a pure gradient where the values of the observed variable depend solely on the position along the y-axis, and (c) the bubble map of the detrended, centered, and transformed C:N data (removing the trend along the y-axis).

squash, radish, sweet potato, banana, mango, and guava. The plot was sampled with a star design (four transects crossing each other at 45° angles; Fortin and Dale 2005). Thirteen samples (one every 5 m) were taken along each transect, yielding a total of 49 samples (figure 10.1a). The study scale thus has a 0.002 m^2 grain, 5 m interval, and 2,827.4 m^2 circular extent (60 m radius).

10.4 Spatial exploratory data analysis

As with all data, you should begin with exploratory plotting. Bubble maps are one of the simplest and most useful graphs for exploring spatial patterns. These display the data with symbols proportional to the value of the measurement at each spatial coordinate. To aid insight, bubble maps are often built with centered variables ($y_{ci} = y_i - \bar{y}$): the symbol size is proportional to the observation's deviation from the mean. Values below the mean are shown in white, and those that are higher, in black. Figure 10.1a shows the bubble map of the centered C:N ratio (one of the most commonly used soil fertility indicators) in the *milpa* plot. Bubble maps can easily reveal spatial gradients. One reason to identify gradients at the start of data analysis is that many statistical analyses (including Moran's test of autocorrelation, variogram models, and spatial autocorrelated models) assume stationarity (i.e., there are no gradients; the only spatial structures in the data occur at scales within the study extent). Ecological processes are often structured in spatial patterns at different scales, and therefore one can detect patches within a sampling site that are nested within an overall trend (e.g., downhill). It is possible that patches in C:N ratio within the *milpa* plot are driven by the presence of crops, particularly legumes, that increase the concentration of N in their rhizospheres, decreasing this quotient. But it is also well known that C:N changes along mountain slopes and between large patches of different soil types.

Because such trends may be strong, it is often useful to detrend the data before analyzing within-site structures. Detrending consists of removing the variation explained by a linear combination of the spatial coordinates. For example, in a study of soil moisture along a hillside, one might detrend by regressing moisture against the distance from the hill top and use the residuals to model the patchy structures. Unfortunately, when a trend is removed, some of the spatial variation accountable by the within-site spatial structure may be removed with it, making within-site pattern less detectable (Diggle and Ribeiro 2007). In short, this is a matter of the balance between the strength of the spatial pattern being modeled, the amount of information available and the power of the statistical tools being used. If one has information on the drivers of the trend (e.g., temperature, altitude, humidity) it is better to detrend the data by modeling the variable with this driver, and using the residuals as the detrended variable (a linear model will often suffice, but see other modeling tools in chapters 6, 13).

If no information is available to formally explain the trend (or evidence of a trend is still present after candidate covariates have been included in the model) then detrending needs to be performed using only the position in space. We illustrate this with the C:N data using trend-surface analysis. This technique can also be used to estimate a trend's parameters, but here we use it only as a detrending tool. In appendix 10B you will find the details of the procedure in R. It is useful to start by using polynomial equations to create visual models of how trends in different directions would look on your data design, and compare these with the bubble maps of your centered data (figure 10.1a). Figure 10.1b shows how a very simple trend would look on the sampling star, if the values of the transformed (arcsin-transforming the proportion data to make its distribution more symmetrical; Legendre and Legendre 1998) and centered C:N variable (*cntc*) depended entirely and linearly on their location along the y-axis ($cntc = a + by$). It is apparent that large positive deviations from the mean in C:N data (large black symbols in figure 10.1a) occur at the top of the star and negative deviations (white symbols) are most often at the bottom. One could even suspect that the trend is slightly tilted to the left, also involving the position along the x-axis as in $cntc = a + b_i y + b_j x + b_k x^2$. Although both models could be reasonable since their AICs (chapter 2) differ by only two, we decided to keep the simplest (the *y*-only model), since we may risk extracting too much of the within-site variation by trying to remove all traces of gradients (figure 10.1c shows the detrended bubble map; see appendix 10B for details).

Finally, when you begin a spatial analysis you should know whether your response variable deviates from the Normal distribution (if continuous) or other known density distributions such as Poisson or binomial. Contrary to other modeling tools (e.g., chapters 6, 13), many spatial analyses do have distributional assumptions on the *original variable*, not the residuals. Data transformations may be required (see appendix 10B).

10.5 Measures and models of spatial autocorrelation

In this section we present two measures of autocorrelation, Moran's *I* and the semi-variance. Each has been used historically in the context of different families of linear models. For example Moran's *I* often accompanies the modeling process in spatial autoregressive models (section 10.6.2) and semi-variance model parameters are incorporated in generalized least square models to account for spatial dependence (section 10.6.1). In the past, semi-variance has been preferred over the Moran's *I* because it is a less biased estimator of the true autocorrelation. However, if you are using *parametric statistics*, this has less importance (Cressie 1993).

10.5.1 *Moran's* I *and correlograms*

Moran's *I* —one of the most commonly used spatial statistics—measures the degree of spatial autocorrelation of a variable. It is based on the notion that a portion of the variance around the mean in a random field *y* (denominator in equation 10.1) is explained by the covariance between nearby observations (numerator in equation 10.1), because observations near each other will have more similar values than more distant pairs. In this sense it is similar to Pearson's correlation coefficient, which measures the proportion of the total variance that is shared by two variables (covariance); *I* instead estimates the proportion of variance shared by observations of the same variable at a given distance. *I* is also called an autocorrelation function, structure function, or spatial correlation coefficient. The trick is that *I* can be computed separately for different distance classes *d* (e.g., observations separated by 1–2 m or 2–3 m; usually between 12 and 15 evenly spaced distance classes are arbitrarily selected, see Diggle and Ribeiro 2007 for a discussion); *I* is a function, not a scalar like Pearson's coefficient, therefore it includes a binary indicator variable w_{hi} whose value is set to one for pairs of sites belonging to distance class *d*, and $w_{hi} = 0$ otherwise. *W* is the number of pairs of points used to compute *I* for the distance class in question, the sum of the w_{hi}:

$$I(d) = \frac{\frac{1}{W}\sum_{h=1}^{n}\sum_{i=1}^{n} w_{hi}(y_h - \bar{y})(y_i - \bar{y})}{\frac{1}{n-1}\sum_{i=1}^{n}(y_i - \bar{y})^2} \quad \text{for } h \neq i. \tag{10.1}$$

Having computed the values of *I*, we can plot *I* against distance classes (this is called a *correlogram*) to determine at what distance observations of *y* are most correlated, and examine other patterns that may appear in the autocorrelation function (figure 10.2). We can compute hypothesis tests for each distance class: *t*-tests can tell us how likely the observed level of autocorrelation is if observations are randomly drawn from a normally distributed population, while randomization tests can tell us how likely the autocorrelation is if observations are randomly redistributed over the sampling locations. The *t*-tests for correlograms assume stationarity and that *I* is normally distributed. The randomization approach assumes stationarity but does not require normality; therefore it should be preferred when sample size is small (say, $W < 100$ pairs in the distance class), as *I* converges to a Normal distribution as *W* becomes large (Legendre and Legendre 1998).

Most correlograms have descending values of *I* at short distances, followed by negative values and then oscillations that level off to non-significant fluctuations around zero. If oscillations occur at regular distances, the patchy pattern is regular in terms of patch size and distance between patches; otherwise correlogram oscillations are fairly irregular (Fortin and Dale 2005). Patch size (diameter) is often interpreted as the distance at which the correlogram crosses the *x*-axis. In the C:N data, the bubble map suggests that the pattern is probably irregular (as is often the case with real ecological data, because it is structured simultaneously at different spatial scales) and the correlogram is only able to detect small patches of around 16 m in extent (figure 10.2). The latter are the patches reflected by the ups and downs (about two patches per arm) within the star in the bubble map (figure 10.1c). There are two important pieces of evidence in the C:N correlogram: (1) autocorrelation is low ($-0.18 < I < 0.18$); (2) positive *I*'s at short distances are not significant at the 0.05 level (they are nearly so, but the correction for sequential hypothesis tests performed by the R function *correlog* renders them non-significant); and (3) the only significant value is a negative autocorrelation at a distance of 23 m. These three points may result from the small variation observed in the raw data, the fact that we are analyzing

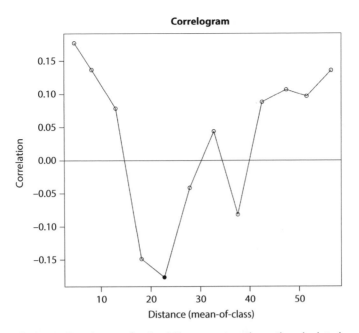

Fig. 10.2 Correlogram for the C:N content in *milpa* soils, calculated by the function *correlog* from the *ncf* library. If the permutation test is requested (*resamp > 1*), values significant at a nominal (two-sided) 5% level are represented by filled circles and non-significant values by open circles (incorporating a correction to alpha for multiple testing). Use of a test based on resampling makes the analysis robust to deviations from the normal distribution, but not against heteroscedasticity: data must be detrended if there is a significant trend.

a very small data set (spatial analyses in general are very information hungry), and the possibility that the selected interval for the study scale might not be small enough and, therefore, we observe abrupt rather than contagious fluctuations.

We can conclude that there is some spatial pattern in the distribution of C:N in a *milpa* plot, but it is weak. It is characterized by variation at two scales: (1) a gradient along the *y*-axis that reflects patches with diameter larger than the study extent and (2) within that gradient there is some heterogeneity with a weak patchiness (*c*. 16 m in diameter). However, if we wanted to map and model this distribution in detail, we would probably need a more intensive sampling scheme (smaller intervals).

When variables are strongly influenced by trends, correlograms show a single descending pattern from very high to very low values (see appendix 10B for an example). This implies that the second-order stationarity assumption is not being met, but the intrinsic assumption is. Therefore, we are only able to compute and explore the correlogram, but parameter estimation and significance tests should not be used, as they will not be sound. Also, because *I* is the average correlation of all pairs of observations at a distance class (covariance is divided by *W* in equation 10.1), it is highly sensitive to outliers and small sample sizes. Thus the values for large distance classes are unreliable, as is obviously the case for the *c*.56 m point in the C:N correlogram (figure 10.2), which represents only 16 pairs of observations.

Ecologists often use correlograms to estimate the distance at which all of their samples are significantly independent from each other and therefore contribute a full degree of freedom to their analysis. In the case of C:N data, points become statistically independent when they are more than 23 m apart, because the last significant value of I occurs at that distance. Recalling that we are working with the *detrended* C:N data, ask yourself whether we might get a different result with the raw data. What would it tell us?

Although we have concentrated here on describing autocorrelation on continuous variables, Moran's I and correlograms are also useful in modeling and testing for autocorrelation in areal (contiguous polygons) and binary data (Plant 2012; Bivand et al. 2013).

10.5.2 *Semi-variance and the variogram*

In the geostatistical tradition the most commonly used function to describe autocorrelation is the semi-variance $\hat{\gamma}(h)$ (equation 10.2), which has several characteristics in common with Moran's I. It is also a function of distance (h) and it compares the values of observations separated by a particular *lag* (a fixed distance or distance class, which is chosen arbitrarily as with correlograms). To study changes in autocorrelation as the distance between samples increases, we can plot the variogram (technically, the experimental semi-variogram) $\hat{\gamma}(h)$ as a function of lag. However, semi-variance is based on the mean square *difference* between pairs of observations separated by lag h. A plot of the differences (without averaging) against lags is called a variogram cloud (figure 10.3a). If observations close to each other are more similar, then on average the squared differences should be larger for observations further apart. Therefore, $\hat{\gamma}(h)$ increases as spatial autocorrelation decreases (figure 10.3b). The semi-variance measures square difference between measured values and the mean; there is no distinction between values that are x units smaller or larger than the mean. Finally, $\hat{\gamma}(h)$ is called semi-variance because it is the spatial variance divided by 2. This correction was introduced because every observation (y_i) is compared with every other observation separated by $h(y_{i+h})$, and therefore each pair of observations is actually compared twice:

$$\hat{\gamma}(h) = \frac{1}{2W(h)} \sum_{i=1}^{W(h)} [y_i - y_{i+h}]^2. \tag{10.2}$$

Variograms assume stationarity. Constructing any variogram involves using no more than about 2/3 of the maximum distance available from the data (some authors suggest 1/3 as a more appropriate limit; Dungan et al. 2002). Moreover, a rule of thumb of at least 50–100 observations (Webster and Oliver 1992; Fortin and Dale 2005) has been proposed as an advisable minimum to model variograms, or otherwise too few pairs are involved in the semi-variance estimation and the pattern becomes unreliable. Omnidirectional (in all directions) variograms assume isotropy, but directional variograms can also be built, in which case h becomes **h** in equation 10.2, a vector that contains the distances and directions represented by the semi-variance $\hat{\gamma}(\mathbf{h})$. Studying and modeling anisotropic phenomena, with multiple directional variograms, is a data-hungry enterprise, if you consider that the data for each direction should comply with the minimum data rules above.

When a variable is spatially structured on a scale within the study extent, semi-variance increases with distance (autocorrelation decreases) until it reaches (or slowly approaches) a plateau (figure 10.3b). This indicates the range, the distance at which observations are no longer spatially correlated. Semi-variograms that decline with distance or oscillate can be indicative of a patchy pattern that is not contagious or is very irregular in terms of

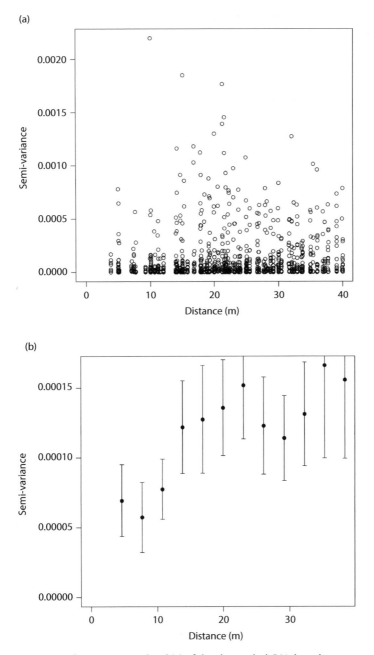

Fig. 10.3 The variogram cloud (a) of the detrended C:N data shows the unaveraged squared differences between all points at a given lag; the semi-variogram (b) shows the mean-square differences at each lag, with their 95% confidence intervals (appendix 10B). The semi-variogram has a shape often seen when a variable is spatially structured on a scale within the study extent (box 10.3).

patch size (Fortin and Dale 2005). When variograms do not level off, it indicates that the variable is structured at a scale larger than the study extent and stationarity is not met, yet the variogram can still be defined for descriptive purposes under the intrinsic assumption.

One of the advantages of variography for studying changes in autocorrelation with distance is that several models can be fit to this relationship, and the estimated parameters provide useful information (e.g., it can be incorporated in linear models such as the generalized least square; section 10.6) . Models fitted to the experimental variogram are called theoretical variograms. Box 10.3 gives a brief summary of the models commonly used in ecological studies and the interpretation of their estimated parameters. Once changes in autocorrelation with distance have been modeled with a variogram (or several, in the case of anisotropy) then one can use the model to predict values between observations (interpolate) to produce surface or contour maps of the variable on the sampling area. The most common family of techniques used to interpolate are called kriging (Isaaks and Srivastava 1989; Bivand et al. 2013). Kriging functions are available in R library geoR. A minimum of 100 data points is recommended for reliable interpolation (Webster and Oliver 1992); kriging is very sensitive to changes in parameters and shapes of variogram models. This leads us to a warning. Ecologists and environmental managers often find themselves able to interpolate maps with a mouse click within GIS software. These interpolations do not use data to model variograms, and they assume an omnidirectional linear model by default, which is a very unreasonable assumption for most ecological phenomena. The picture you get may be pretty, but worthless! There is an extensive coverage of kriging by Stein (1999), and the spatial statistics texts by Bivand et al. (2013), Fortin and Dale (2005), and Plant (2012) introduce the topic.

The sample size (number of pairs) typically declines with the lag h. The confidence intervals in figure 10.3b vary substantially, but are generally wider as distance increases; for some sampling designs like the star we are analyzing, there are also few pairs for the first lag. Most modelers choose the theoretical variogram that minimizes the weighted residual sum of squares, using the number of pairs in each lag as weights (this is called *weighted least squares* or WLS; Webster and Oliver 1990). For the transformed and detrended C:N data, the best model was spherical with a range of 24.6 m (nugget of 1.6×10^{-5} and sill of 1.4×10^{-4}, figure 10.4), suggesting a patchy distribution with patches of $\sim$24 m in diameter. This is rather different from the $\sim$16 m calculated from the correlogram, because the variogram model was constructed giving more importance to those points in which we have more confidence.

More recently, maximum *likelihood* (see chapters 2 and 3) has also become a popular modeling approach to fit theoretical variograms. Variables being modeled need to have (or be transformed to have) a Gaussian distribution (Rossi et al. 1992).The parameters of a random field are estimated by fitting a model of the form

$$Y(x, y) = \beta(x, y) + S(x, y) + \varepsilon, \tag{10.3}$$

where Y is the variable being modeled, and x and y are spatial coordinates. $\beta(x, y)$ describes the mean component of the model, and may be taken to be a polynomial representing a trend. The within-plot spatial structure is modeled by $S(x, y)$, a stationary Gaussian process characterized by the parameters of the particular theoretical variogram model being used (box 10.3). Finally, ε is the error (nugget variance).

Using maximum likelihood for the detrended C:N data in the *milpa* plot, the best-fitting model with a sill was the spherical, and the worst was the non-spatial model (pure nugget). However, the difference in AIC between these two models was only 5.3, suggesting (chapter 3) that spatial structure within the study site is not strong. This is consistent

Box 10.3 COMMON VARIOGRAM MODELS USED IN ECOLOGICAL STUDIES

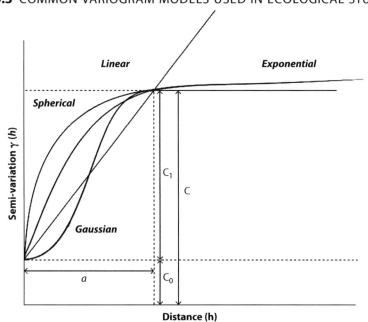

There are five models commonly used for semi-variance.

There are five models commonly used for semi-variance. They are defined as follows:

Nugget effect: $\gamma(h) = C_0$

Exponential: $\gamma(h) = C_0 + C_1\left[1 - \exp\left(-3\frac{h}{a}\right)\right]$

Spherical: $\gamma(h) = C_0 + C_1\left[1.5\frac{h}{a} - 0.5\left(\frac{h}{a}\right)^3\right]$ if $h \leq a$; $\gamma(h) = C$ if $h > a$

Linear: $\gamma(h) = C_0 + bh$

Gaussian: $\gamma(h) = C_0 + C_1\left[1 - \exp\left(-3\frac{h^2}{a^2}\right)\right]$

Here $\gamma(h)$ is the predicted semi-variance at distance lag (h). The nugget model is often a null model; it represents the variation due to sampling error or autocorrelation at smaller distances than the minimum lag, which cannot be explained by the model. In the exponential and spherical models the sill ($C = C_0 + C_1$) represents the maximum semi-variance (inverse of the maximum autocorrelation) estimated. The sill provides an estimate of the range, the distance at which observations become spatially independent. For spherical models a is the range; for exponential models (which approach the sill asymptotically) the range is often considered to be where semi-variance reaches 95% of the sill. Different models can be combined. Often the nugget effect (C_0) is used in combination with another model to provide the variogram with a y-intercept greater than 0, and account for sampling error and autocorrelation at smaller scales. When data are structured with nested patchy distributions at different scales, two models may be added to model both structures; however, since many more parameters must be estimated, this should only be done with fairly large data sets.

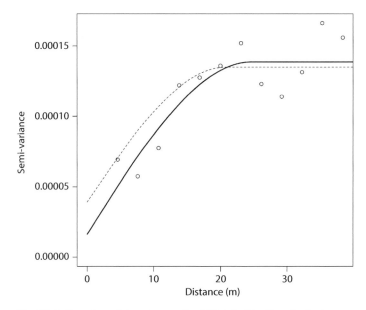

Fig. 10.4 Experimental variogram for C:N data fit with spherical models. Dashed line is fit with maximum likelihood, and continuous line with weighted least squares.

with the correlogram. The spherical model fit with maximum likelihood suggests a range (patch diameter) of 20.7 m, while the same model fit with weighted least squares suggests a patch diameter of 24.6 m (see appendix 10B). This difference is probably due to the fact that WLS gives more importance to points supported by more pairs. With the small amount of data available, reporting a spatial structure on the scale of approximately 20–25 m is the best we can do. It might be desirable to calculate confidence intervals on these parameter estimates to compare how different they might be. However, given the spatial dependence between observations and the inherited uncertainty from the experimental variogram, the best strategy to calculate confidence intervals is still an unresolved problem (Clark and Allingham 2011). In this case, we suggest that the WLS model is more reliable because we have points in the variogram that are represented by very few pairs, and the data set is in general quite small. On the other hand, an advantage of using maximum likelihood for variograms is that trends can be modeled simultaneously to stationary (within-site) spatial patterns, eliminating the need to detrend, and permitting comparison of AIC values of models with and without the trend. Commented R code on all of these steps is provided in appendix 10B.

10.6 Adding spatial structures to linear models

Ecologists would like to model the drivers of major ecological patterns, like those strongly affecting organism distribution, population parameters, or indicators of ecosystem function. But only a limited number of explanatory variables are available, and sometimes important endogenous or environmental drivers of the response are missed. If these drivers are spatially correlated on the scale of the study, errors will also be correlated and each observation will not contribute a full degree of freedom. In this section we will examine how to account for spatial autocorrelation in linear models, when the drivers of this correlation are not known and there is residual spatial autocorrelation. For normally distributed data two main approaches exist. The first is *generalized least squares* (GLS, also used in chapter 11), which directly models the spatial covariance structure with a variance–covariance matrix incorporating the range and sill of a variogram function fitted to the residuals. The other approach is *spatial autoregressive models* (SAM); we will discuss only conditional autoregressive models (CAR) and simultaneous autoregressive models (SAR). CAR and SAR operate with weight matrices that specify which observations are considered neighbors, as well as the strength of interaction between them. These weighted matrices are used to regress observed values on the response values of the specified neighbors (Dormann et al. 2007).

10.6.1 *Generalized least squares models*

Suppose we hypothesize that C:N variability within our study *milpa* can be explained by species-specific conditions that plants create under their rhizospheres. We have information about what plant species was planted closest to each soil sample (figure 10.5a), but now we know that our C:N observations are not independent in space. There is a trend across the site and a pattern of patches of approximately 20–24 m in diameter. Suppose that we are not interested in the spatial patterns, but only focus in modeling the mean effect of plant species on C:N (note that this is not an experiment and we are assuming a cause-effect relationship that may not be justifiable). A common approach would be first to detrend the data and then to sub-sample the site until you have the maximum number of observations possible, which are at least 20–24 m apart, all of this in order to conform to the "independence of observations" assumption of most statistical analyses. This approach is not our best option because it throws out the baby (or at least a substantial part of it) with the bath water. It is a better idea to introduce terms in your model that will account for the spatial dependence of observations and therefore avoid inflating type I errors. This is what generalized least squares (GLS) models can do (R-library *nlme*, function *gls;* see the appendix for computational details). These models are also known as kriging regressions. They use the linear mixed-models framework (detailed in chapter 13, section 13.3.1) to account for the correlation structure in the data within the error term (an *R-side effect*). This structure is incorporated by introducing the parameters of a variogram model (sill and range) in the covariance matrix (see Pinheiro and Bates 2000 for details).

The bubble map and boxplot in figure 10.5 a and b suggests that samples nearest sweet potatoes have lower C:N ratios than other samples. We might fit a linear model to the transformed C:N data to test whether crop is a significant predictor of variation in C:N. But the observations are not independent in space, and do not contribute a full degree of freedom; type I errors are inflated. This is evident from the variogram of the linear model residuals (figure 10.6). Semi-variance increases with distance and there is a "hump" within that increase. This pattern suggests that there may be a gradient and some stationary patchy structure. Nor are the residuals normally distributed (appendix 10B). We can handle these issues by turning the model into a GLS: we can introduce a correlation structure in the error term, introduce the positions in space (*x* and *y*) as co-variables to account for a gradient, or both. To decide on the most appropriate model, we tested several alternatives and compared the results in table 10.1 using AIC. The first

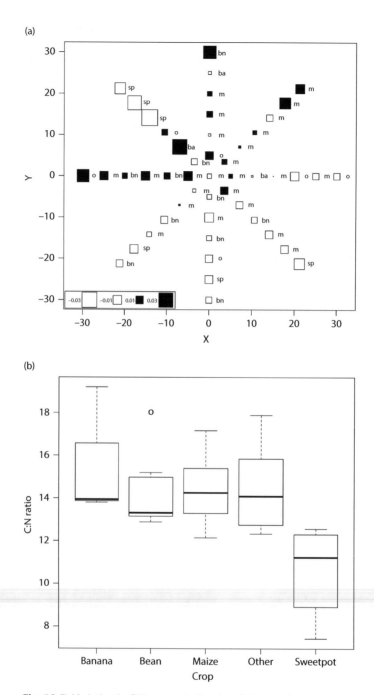

Fig. 10.5 Variation in C:N concentration in relation to the crop species planted in a *milpa* plot. (a) Distribution of the C:N centered data in a bubble map with the species of the nearest crop (ba = banana, bn = bean, m = maize, o = others, and sp = sweet potatoes). Black and white symbols show positive and negative deviations from the overall mean, respectively. (b) Box plot of the C:N concentration according to the species of the nearest crop.

thing to note is that the `gls (nlme)` function in R works with a correlation structure and not with semi-variance; this means that the nugget parameter is standardized (it is 10^4 times larger than what you would expect from the variogram in figure 10.6). Second, the model that includes the gradient and the correlation structure together produces quite an unreasonable nugget. In fact, the parameters are unstable, because they change depending on the initial parameters given to the algorithm (appendix 10B). Finally, the smallest AIC is for the linear model with a gradient and plant species, but is it really the best model? Figure 10.6 shows the variograms of the residuals of the best of these models (the gradient and no correlation structure model) compared to the simple linear model (with crop only). The variogram for the gradient model is flatter than that of the linear model. The change in AIC between these models is modest (table 10.1), but the variogram provides persuasive evidence that the gradient model is better: it accounts for the spatial correlation of observations, even if it does not greatly improve the explanatory power (Legendre and Legendre 1998). If you are not sure whether the variogram is flat enough to consider residuals independent, fit a likelihood model to the variogram itself and see if the AIC of the spatial model is smaller than the non-spatial model (appendix 10B explains how).

The approach used here can be extended to include random grouping variables; for a linear mixed model this is done with the `lme` function of the `nlme` library (Pinheiro and Bates 2000; Plant 2012); there is also a spatial version of GLMM (Bivand et al. 2013). The modeling is performed by first constructing your *G-side* effect mixed model (chapter

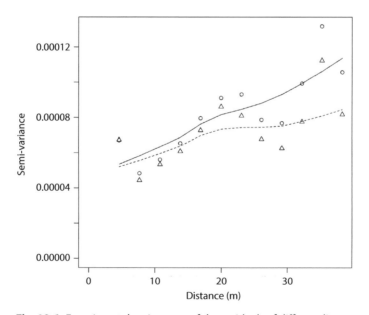

Fig. 10.6 Experimental variograms of the residuals of different linear models where soil C:N ratio (arcsin-transformed) is explained by the species of the nearest crop. The lines are a lowess fit (used here only informally to explore the shape of the variogram). Solid line and open circles correspond to the residuals of a simple linear model, triangles and dashed line correspond to the linear model with a gradient. See appendix 10B for computational details and table 10.1 for model fits.

Table 10.1 △AIC (max AIC–model AIC) and correlation structures of different models explaining C:N ratio in *milpa* soils with the species of the nearest plant crop. GLS models include a correlation structure that accounts for stationary spatial dependence (within-site patchiness) of observations; models which include x and y positions in space of the samples account for spatial dependence in the form of a gradient. The dependent variable (cnt) is the arcsin-transformed C:N ratio

Model	△-AIC	Residual SE	Correlation structure	
			range	nugget
Linear model (cnt~crop)	13.3	0.0092	—	—
GLS (cnt~ crop and corr: Spherical)	12.0	0.0093	2.92	0.47
GLS (cnt~ crop and corr: Exponential)	11.3	0.0092	2.27	0.52
GLS (cnt~ crop and corr: Gaussian)	12.0	0.0093	2.60	0.84
Linear model (cnt~crop + x + y)	17.0	0.0086	—	—
Linear model (cnt~ x + y)	0.0	0.0111	—	—
GLS (cnt~crop + x + y and corr: Spherical)	13.0	0.0085	2.02	17.82

13) and then updating it with a variogram correlation structure like the one exemplified above (for details see Pinheiro and Bates 2000; Plant 2012). An advantage of GLS is that the inclusion of a model of the spatially correlated error enables the possibility of predicting values of the response variable within the study range where no observations are available, based on the values of observed sites (e.g., kriging). The same holds true for the spatial version of GLMM (see Bivand et al. 2013 for details).

A brief ecological summary

So far, we have learned that C:N content in *milpa* soils is structured on at least two different ecological scales. It is structured in patches within the site with a mean diameter of 20–24 m, but this pattern is not very pronounced. The other structuring factor occurs at a scale beyond those covered by the study. We also found out that changes in the mean C:N can be explained by the species of crop planted nearest the soil sample, mainly because samples near sweet potato have lower C:N ratios. Because soil C is usually fairly homogeneous within plots, this suggests that N may be increased around sweet potato. Since this species forms root associations with the nitrogen-fixing soil bacteria *Bradyrhizobium* (the dominant fixing bacteria in nearby tropical forests) the symbiotic association of native soil organisms with particular crops may be generating the spatial structure we observe. If we explain some of the variation in the data by the nearest crop, the patchy structure fades from the residuals but the gradient is still clearly there; we need to incorporate it into the model. This leads us to the hypothesis that the within-plot patchiness may be imposed on the C:N ratio by the influence of the crop species (particularly sweet potato).

10.6.2 *Spatial autoregressive models*

We turn now to spatial autoregressive models (SAM), which can model spatial patterns in ecological data as imposed by unknown factors that are either external or endogenous to the variable. Warning! We use "endogenous" here (and it is widely used in the literature on spatial statistics) in its common sense, not in the statistical sense used in chapter 7. We often want to know why a variable (say, population abundance) has a spatially structured pattern. It makes sense to first try to account for structure explainable by known environmental variables (say, moisture availability). Then we can ask if there is remaining structure that could be caused by (1) unknown factors, (2) distance-related endogenous sources like dispersal patterns, competition, or social interactions, or (3) non-linear relationships with the environmental variables already in the model. To explore this topic, we will turn from the C:N *milpa* soil data to a different empirical example.

Spatial autoregressive models can help address these questions. Besides improving estimation of regression coefficients, SAMs are able to provide a picture of the residual spatial pattern, providing insight into what omitted drivers may be operating (Wall 2004). Thus these models are useful tools to formalize hypotheses about unaccounted drivers of spatial pattern, which can point the way for further research.

SAMs consider that the response at each location is a function not only of the explanatory variables but the values of the response at neighboring locations (Cressie 1993). While GLS estimate spatial structure and effects of covariates simultaneously, SAM assume that at least the scale of spatial structure is known *a priori*. Which observations are considered neighbors, and the strength of influence neighbors have on one another, is specified by the analyst in a matrix of spatial weights **W**. Three main approaches exist: the conditional autoregressive (CAR), simultaneous autoregressive (SAR), and moving average (MA) models. The MA approach (including SARMA, mentioned in the Lagrange multiplier test output, `lm.LMtest` in R library `spdep`, see appendix 10C) is not widespread in the ecological literature, and we will not discuss it further.

CAR and SAR models have different technical requirements. In particular, CAR models need symmetrical **W** matrices and so are unsuitable when directional processes such as stream flow, wind direction, or slope aspect are at work. SAR models do not have this requirement. CAR models should be preferred in terms of estimation and interpretation, but they lack the flexibility that SAR models currently have in terms of accepting asymmetric covariance matrices; they also lack the possibility of modeling autocorrelation associated with different components of the model (Cressie 1993). Both models are used extensively in ecology to address spatial correlation as a nuisance. But they also have considerable potential to provide a clear picture of the residual spatial pattern, thus yielding insight into omitted variables and spillover effects (see below) driving spatial patterns. There is active research on using these models (Wall 2004).

The models involve some new notation, and can be a bit confusing at first. Formally, the CAR model is:

$$Y = \beta X + \lambda W(Y - \beta X) + \varepsilon$$
$$\varepsilon \sim N(0, \sigma^2(I - \lambda W)^{-1}), \tag{10.4}$$

where the first term on the right-hand side describes the mean, the second term describes the spatial pattern, and the third term is the error. β is the usual regression coefficient for explanatory variables, the λ is a spatial autoregression coefficient and **W** is the neighborhood matrix. **I** is the identity matrix.

SAR models can be formulated in three different ways: including autoregressive terms for the error (SARerr), the response variable (SARlag) or both the response and explanatory variables (SARmix or Durbin model). These possibilities have attracted ecologists because they can be used to represent different hypothesized sources of spatial patterns in distributions of organisms. The three model formulations are

$$\text{SARerr}: \boldsymbol{Y} = \boldsymbol{\beta X} + \lambda \boldsymbol{Wu} + \boldsymbol{\varepsilon}$$

$$\text{SARlag}: \boldsymbol{Y} = \rho \boldsymbol{WY} + \boldsymbol{\beta X} + \boldsymbol{\varepsilon}$$

$$\text{SARmix}: \boldsymbol{Y} = \rho \boldsymbol{WY} + \boldsymbol{\beta X} + \theta \boldsymbol{WX} + \boldsymbol{\varepsilon}$$

$$\boldsymbol{\varepsilon} \sim N(0, \sigma^2). \tag{10.5}$$

Here u is a spatially dependent error term, and λ, θ, and ρ are autoregression coefficients. θ is often denoted γ, but we have changed the notation to avoid overlap with semi-variance notation above.

Case study: spatial structure in a pine population

In this section, we will explore spatial structure in a sample of longleaf pine (*Pinus palustris*) in Florida (Ford et al. 2010; Fox unpublished data), using CAR and SAR. Researchers have observed that young trees grow spatially aggregated and away from adults (figure 10.7a; Platt et al. 1988; Grace and Platt 1995; Brockway and Outcalt 1998). They hypothesize that there are endogenous and exogenous factors determining this spatial distribution of sizes. First, there are few seedlings under adults because wind takes seeds away from the parent tree, and because the bed of accumulated pine needles prevents germination. In addition, competition for resources makes young trees survive better away from large trees, and eventually natural thinning leaves adult individuals fairly isolated. On the other hand, the site slopes from south to north, and soils are drier in the higher areas. Therefore, juveniles are more aggregated in low areas (figure 10.7a). Additionally, experimental fires have been conducted on different plots on average every 1, 2, 5, or 7 years, or never burned in 30 years (coded as 0 in the data set in appendix 10C). There are two plots for each burn frequency, between 6,500–19,000 m^2 (figure 10.7b). Very frequent fires (every 1–2 yrs) may reduce the survival of seedlings and young trees (Ford et al. 2010). But less frequent fires increase their survival and growth, mainly by reducing competition, and perhaps through the availability of nutrients from ash.

Building a model for the distribution of longleaf pine size (diameter at breast height, DBH) should thus include altitude and burn treatment as explanatory variables. We also need to include plot nested within burn treatment, because we have many observations within each plot (chapter 13). We Box-Cox transformed ($y^{-0.78}$) DBH because it had an unusual distribution of residuals. This is because the effects of burn regime include an effect on tree growth, as well as an effect on recruitment. This complexity is beyond the demonstrative scope of the models in this chapter (Ford et al. 2010). Because this transformation reverses the signs of the estimated coefficients, we back-transformed before estimated parameters are interpreted (appendix 10D). We suspect that even after accounting for the influence of elevation and burn regime on the spatial distribution of DBH, there will still be small-scale autocorrelation related to factors like dispersal and competition.

Defining neighbors: distance and weight matrices

The matrix **W** in autoregressive models defines which observations are considered neighbors of each data point. Usually researchers work with the most proximate neighbors. For geostatistical data like these, first a matrix of distances (usually Euclidian) of all pairs of observations needs to be constructed. Because all tests of autocorrelation, and all autoregressive models assume that all observations have neighbors and all observations are

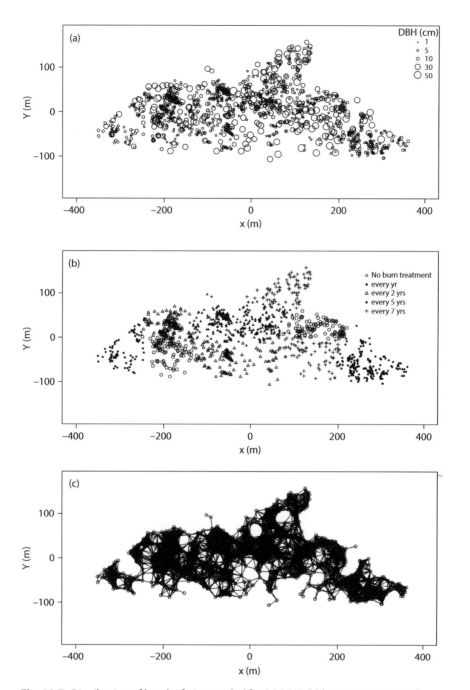

Fig. 10.7 Distribution of longleaf pines coded for (a) DBH, (b) burn treatment, and (c) neighborhood connections within a minimum distance neighbors of 26.2 m (the smallest distance at which all observations have at least one neighbor and are connected).

connected (there are no "islands"), researchers often find the smallest distance at which all observations have at least one neighbor; this is called the minimum distance neighbors (MDN; see figure 10.7c). Bivand et al. (2013) provides a thorough explanation of different neighborhood matrices.

Having found the MDN, we create the weight matrix **W**, usually by assigning a 0 to any pair further apart than the MDN, and assigning a 1 to every other pair. It is also possible to give different numbers to more distant neighbors (Bivand et al. 2013). Most researchers then standardize the weights so that each row sums to 1 and the estimated autoregressive coefficients conveniently range from 0 to 1. Row standardization often facilitates interpretation; however, it also implies competition among neighbors, because fewer neighbors have individually greater influence on the observation. In the case of longleaf pines, it is reasonable to think that if a tree is near a single neighbor, that neighbor will have the strongest influence, and as more neighbors appear, the influence of any one of them may be diluted because competition will occur among more individuals (not only for the focal tree, but also for the neighbors). Row standardization (`style="W"` in function `nb2listw`, R library `spdep`) thus gives more importance to observations with few neighbors, while using **W** with only 0 and 1 (`style="B"`) gives the least importance to such observations. If there are irregular distributions of observations, or many observations on edges (say, trees next to roads), this trade-off needs to be considered carefully (see Bivand et al. 2013). Library `spdep` includes an option for creating weights that stabilize variances by summing over all links (`style="S"`), intended to partially compensate for these biases (Bivand et al. 2013). For the pine data we chose row standardization (see appendix 10D for further discussion). **W** then consists of zeroes on the diagonal, and weights for the neighboring locations in the off-diagonal positions.

Where, when, and how should autocorrelation be modeled?

These are difficult questions. The only sure answer is that the presence of residual autocorrelation in models should not be ignored; it indicates a lack of understanding of the spatial distribution of a phenomenon. For the longleaf pine data set, the correlograms (figure 10.8) show that some of the spatial autocorrelation in DBH is accounted for by explanatory variables in the OLS model, but there is significant patchiness in the residuals occurring at distances below the second lag (52.4 m, two times the MDN). This is confirmed by a significant Moran's I global test (appendix 10D). Autocorrelated residuals may be caused by endogenous phenomena (such as dispersal or competition) that generate a spatial structure in the response. In that case a spatially lagged term of the dependent variable needs to be introduced as an explanatory term. Residual autocorrelation may be caused by *misspecification* of the model—that is, an important (spatially structured) driver has not been included. If the unmeasured driver is independent of other explanatory variables, autocorrelation should be modeled in the error term (as in the SARerr or CAR models; LeSage and Pace 2009). If the unmeasured driver is known or suspected to be correlated with the explanatory variables in the model, a SARmix model should be used. These include a spatially lagged form of explanatory variables and a spatially lagged form of the dependent variable. Ecologically, this adds an additional local aggregation component to the spatial effect in the lag model above. For this reason, SARmix models are useful when it is clear that the explanatory variables are strongly structured in space. This is not so for the longleaf pine data; if we did not know the burn histories of the plots, but knew that burn histories would be correlated with altitude (due to moisture differences), a SARmix model would be sensible.

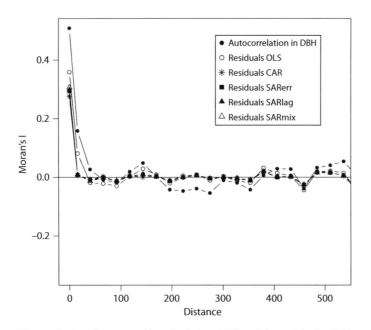

Fig. 10.8 Correlograms of longleaf pine DBH and the residuals of all the models that explain it. Note that spatial models reduce autocorrelation left in residuals that occurs at distances beyond the 26.2 m used as neighborhood definition. Accounting for this autocorrelation is important because the response at each location is modeled as a function not only of the explanatory variables but the values of the response at neighboring locations. However, these models are unable to incorporate spatial structures that occur at smaller distances (reflected in the drop between the first and second point).

It might be tempting to fit all possible models, use AIC to choose the most parsimonious, and then conclude that the model is telling you where the autocorrelation comes from in nature. Don't do it! Describing a pattern does not tell us about its causes. Residual spatial pattern may be the product of a combination of processes creating patterns that may be observationally equivalent (de Knegt et al. 2010). There is no statistical solution to this problem. In descriptive studies, the choice of model should come from interaction between theory, hypotheses, and statistics. The final model should be considered a working hypothesis that can be tested by subsequent research.

The analysis of these different models for our longleaf pine DBH data should help you see why this is a problem. From the competing spatial models (table 10.2) the SARerr, SARlag, and SARmix all have AIC values that qualify as feasible (chapter 3). Our explanatory variables did not account for all of the spatial autocorrelation, and suggested that this pattern may be the result of dispersal patterns or density-dependent mortality caused by competition (suggesting a SARlag model). This is consistent with reports of younger trees having a clumped pattern on a smaller scale than older trees (McDonald et al. 2003). But the same pattern can occur if seedlings are more likely to be discovered by herbivores or fungal pathogens when found near adult trees of the same species, corresponding then to a SARerr model (Skidmore et al. 2011). Like any descriptive tool, spatial analyses are generally unable to conclusively confirm hypothesized ecological processes when used in

Table 10.2 Spatial autocorrelated model comparison for longleaf pine DBH in Florida. Coefficients correspond to the model explaining the transformed DBH ($y^{-0.78}$). Note that this transformation reverses the signs of the coefficients. See appendix 10D for detailed procedures. Explanatory variables in all cases were burn treatment, elevation, and plot (nested in burn treatment). All SAR model parameters estimated with a row standardize matrix. The CAR model was calculated with a binary matrix because row standardization makes matrices asymmetric. Note that all of the parameters associated with the plot factor are not shown, thereby the large value for degrees of freedom in each case

Model	Intercept (α)	Burn (β_{B1-5})	Elevation (β_E)	Spatial coefficients	d.f.	AIC
OLS	0.17***	$\beta_{B1} = 0.032$* $\beta_{B2} = -0.004$ $\beta_{B5} = -0.020°$ $\beta_{B7} = -0.071$***	$-0.006°$	—	12	$-2,352$
CAR	0.12**	$\beta_{B1} = 0.028°$ $\beta_{B2} = -0.020$ $\beta_{B5} = -0.019$ $\beta_{B7} = -0.076$***	-0.009**	$\lambda = 0.01$***	13	$-2,419$
SARerr	0.13°	$\beta_{B1} = 0.024$ $\beta_{B2} = -0.027$ $\beta_{B5} = 0.005$ $\beta_{B7} = -0.074$**	$-0.009°$	$\lambda = 0.59$***	13	$-2,438$
SARlag	0.04	$\beta_{B1} = 0.011$ $\beta_{B2} = -0.013$ $\beta_{B5} = -0.008$ $\beta_{B7} = -0.049$**	$0.005°$	$\rho = 0.55$***	13	$-2,434$
SARmix or Durbin	0.08	$\beta_{B1} = -0.033$ $\beta_{B2} = -0.040$ $\beta_{B5} = 0.069°$ $\beta_{B7} = -0.084$**	0.007	$\rho = 0.55$*** $\theta_{B1} = 0.05$ $\theta_{B2} = 0.04$ $\theta_{B5} = -0.08°$ $\theta_{B7} = 0.05$ $\theta_E = 0.01$	23	$-2,435$

Probability code for Likelihood ratio tests: $P \leq 0.001$:***; $P \leq 0.01$:**; $P \leq 0.05$:*; $P \leq 0.1$:°

isolation. Dormann et al. (2007) discuss in detail the uses and misuses of autoregressive models in ecology.

Having recognized that it is difficult to use goodness of fit to make inferences about the nature of the structure-causing factors, it makes sense to explore different model structures. One tool that is very widely used is the Lagrange multiplier test; this is discussed in more detail in appendix 10D. An important limitation of this test is that it is restricted to comparisons of only the SARerr, SARlag, and SARMA models. Comparing other models requires using some other method, such as AIC (as in table 10.2). LeSage and Pace (2009) also suggest a Bayesian model averaging approach to obtain posterior probabilities for competing models.

Parameter interpretation and impacts

Because the CAR and SARerr models do not involve spatial lags of the dependent variable, estimated β parameters can be interpreted to represent, as in OLS, the total change in Y in response to a unit change in the explanatory variable. On the other hand, a very useful feature of SARlag and SARmix models is that the total effect of an explanatory variable can be partitioned between its direct effect and its *indirect impacts* (also called spillovers) caused by spatial autocorrelation in Y. For example, Folmer et al. (2012) and Folmer and Piersma (2012) studied shorebird densities. An increase in food resources in a sampling cell (a change in X) attracts birds; through social aggregation these densities increase further in the sampling cell and its neighbors. The coefficient for the regression of density on food represents only the direct effect, ignoring social aggregation. Using a SARlag model, the authors estimated the total effect of X (attraction to food plus social aggregation). This can be done by calculating an *overall spatial multiplier* $[(1-\rho)^{-1}]$, which is the average level by which the direct effect (β) is multiplied to account for the spreading of indirect effects in the system. Logically, indirect impacts can be calculated as the total effect minus β (appendix 10D). In SARmix models, the presence of autocorrelation in X means that the indirect impacts are heterogeneous across space and therefore the spatial multiplier also changes with location. The calculations for an average indirect impact become more complex (LeSage and Pace 2009), but it can be done automatically in the R `spdep` library (appendix 10D). Note that indirect impacts, both in SARlag and SARmix, will also account for the feedback effects that a region will receive from neighbors that it has affected. In the shorebird example, an increase in food resources in a sampling cell will attract birds, which in turn will attract birds to neighboring cells. This increase in bird density in neighboring cells will in return affect bird density in the cell where food resources originally increased. Indirect impacts represent what occurs once the system has reached an equilibrium. It is also possible to examine the rate of decay over space of the indirect impact of a factor (see LeSage and Pace 2009 for details).

Significance tests and standard deviations for total and indirect impacts can be obtained by drawing a large number of simulated parameters from a multivariate normal distribution to create an empirical distribution of all parameters required to calculate impacts. Alternatively, a Markov chain Monte Carlo (MCMC) simulation can be used (LeSage and Pace 2009; Bivand et al. 2013).

Table 10.2 shows spatial and non-spatial coefficients and AIC values for all models fit to the longleaf pine DBH data. It is clear from their AICs that the OLS and CAR models are inferior to the others (chapter 3). Is the SARmix model justifiable? Some care in answering this question is warranted, because it implies the fitting of a large number of coefficients. But clearly, this model is an over fit. None of the lagged explanatory coefficients (θ) are significant at $p < 0.05$, and so the impact analysis does not add anything to the SARlag model either (table 10.2, appendix 10D). This leaves us with the SARerr and the SARlag models. In our opinion, there are no statistical criteria that will help us decide between these two models. They are two competing ways to explain spatial structure in longleaf pine DBH. Researchers will have to investigate further whether other exogenous sources of spatial structure could be at work, or whether competition and dispersal are the explanations. Both models are consistent in several respects. There is very little effect of altitude (β_E is nearly 0) and the main effect of burning regime occurs when it is the least frequent (every 7 years). When altitude is removed from both models (appendix 10D), the SARerr model predicts an intercept for the unburned treatment of 0.24, which corresponds to a mean DBH of 6.23 cm ($[0.24]^{1/-0.78}$), while the intercept of the 7-year frequency treatment corresponds to a mean of 9.01 cm ($[0.24-0.06]^{1/-0.78}$). Therefore, there is a predicted *increase* in size for the 7-year burn treatment of 2.78 cm. The SARlag model suggests that the total impact of burning every seven years (total impact = -0.082) is about half direct ($\beta_{B7} = -0.037$) and the other half thorough the indirect spread of impact through space (indirect impact = -0.045), corresponding to endogenous interactions like competition or dispersal (table 10.3). Both models find a highly significant spatial autocorrelation of about 0.6.

There is an important, but often missed, limitation to what we can infer from these models. They are based on MDN and therefore cannot be used for inferences at smaller scales. For example even if many trees in the longleaf data are at smaller distances, none of the fitted models have the ability of explaining spatial patterns occurring below the 26.2 m. Yet, in this case, from variography there is evidence that these small spatial structures occur (figure 10.8). One conclusion here is worth highlighting: not considering the spatial structure of the DBH observations (OLS model) would have misled the researchers to believe that annual burning has a negative effect on DBH compared to not burning (Ford et al. 2010). Once spatial structure is accounted for, this effect is no longer substantial, since the impacts in SARlag or its regression coefficient (β_{B1}) in SARerr are not large or significant.

Model simplification and prediction

By now you have found that spatially explicit models can become complicated! Also, model simplification in general should be considered carefully (chapters 1 and 3). Unfortunately there has been little research on simplification of linear models in the presence of spatial autocorrelation. It is often a good idea to statistically select explanatory variables for the model before fitting autocorrelated structures, because the explanatory variables may remove or reduce the correlation structure of the residuals from the model (Hoeting et al. 2006). On the other hand, ignoring the autocorrelation structure of the error process during variable selection may mask explanatory variables with important effects on the mean. This may lead to a downward-biased contribution of environmental predictors and an upward-biased contribution of small-scale influences such as endogenous factors. The additional noise can overwhelm the information in the data, resulting in the identification of fewer important explanatory variables.

In other words, spatial regression methods are not a quick fix for spatially correlated data; treating them as such can lead to serious misinterpretation of results (de Knegt et al. 2010). Thus, during model simplification, autocorrelated parameters should be introduced and excluded at different stages of the process, and the results of non-spatial OLS should not be discounted in favor of those of spatial models when non-spatial coefficients differ substantially. One tool that can aid this process is calculation of a pseudo-R^2 (Nagelkerke 1991; implemented in the spdep library), which assesses the relative explanatory importance of each spatial model compared with an intercept-only spatial model. One can also calculate the pseudo-R^2 for the non-spatial models. There has been little research on simplifying SARmix models that include multiple factors and interactions.

There is an important complexity to SARmix models. All explanatory variables in the model must contribute to the spatial variation (lagged term). This is so, even if their contribution is estimated to be nearly zero; those terms cannot be removed from the model. However, it is possible to remove the non-spatial component of these terms. The technical reasons for this limitation are beyond the scope of this chapter (see Lesage and Page 2009). SARmix models are difficult to interpret if they include many terms and interactions, as is the case of the longleaf pine example (table 10.2).

Finally the descriptive nature of SAM models imposes an important limitation to the inferences one can make from them. One cannot use modeled spatial correlations to infer the spatial correlation structure for entirely unobserved regions. For example, our explanations of the longleaf pines spatial distribution cannot be considered a rule in unstudied regions, because it is unreplicated at that scale. The R library spdep includes a predict

Table 10.3 Analysis of impacts for the simplified SARlag model of longleaf pine DBH in Florida (see appendix 10A for detailed procedures). Elevation is no longer included due to its small explanatory power. Note that mean total impact is the sum (besides rounding) of direct (or β-coefficient) and indirect impacts because the spatial multiplier is uniform across space (see text). The s.e. of model coefficients and direct impacts are different because the first are calculated with a standard t-test while the second with a multivariate normal randomization test (100 draws)

	Model coefficients	Direct impact	Indirect impact	Total impact
No burn treatment	$\alpha = 0.109$*** s.e. $= 0.016$	—	—	—
Burning every year	$\beta_{B1} = 0.020$° s.e. $= 0.012$	0.020 s.e. $= 0.001$	0.024 s.e. $= 0.001$	0.044 s.e. $= 0.002$
Burning every 2 years	$\beta_{B2} = -0.002$ s.e. $= 0.012$	-0.002 s.e. $= 0.001$	-0.002 s.e. $= 0.002$	-0.005 s.e. $= 0.003$
Burning every 5 years	$\beta_{B5} = -0.006$ s.e. $= 0.011$	-0.006 s.e. $= 0.001$	-0.007 s.e. $= 0.001$	-0.0134 s.e. $= 0.002$
Burning every 7 years	$\beta_{B7} = -0.037$** s.e. $= 0.014$	-0.038** s.e. $= 0.001$	-0.045** s.e. $= 0.003$	-0.082** s.e. $= 0.004$

Probability code: $P \leq 0.001$:***; $P \leq 0.01$:**; $P \leq 0.05$:*; $P \leq 0.1$:°

function that will generate predictions of the dependent variable (for details use `?predict.sarlm` in R). These predictions can be useful within the study region, but they are generally meaningless outside of it.

10.7 Discussion

The theory and applications presented in this chapter represent only an introduction and a minute proportion of what spatial statistics has to offer. Many more approaches have been successfully applied to spatial problems in ecology. Correlograms and variograms belong to the family of global methods that summarize spatial patterns over the full extent of the study area. There are other global methods that now exist to generate sets of orthogonal structuring variables (spatial eigenfunctions) from regularly or irregularly spaced geostatistical data or from a connection matrix of regions (Griffith and Peres-Neto 2006). Among these the family of MEM (Moran's eigenvector maps; Dray et al. 2006) including PCNM (principal coordinates of neighbor matrices; Borcard et al. 2011) are the most commonly used. The outstanding characteristic of these structuring variables is that they allow researchers to model the spatial patterns of ecological data at multiple scales simultaneously (Borcard et al. 2004) and they can be introduced as explanatory variables in linear models (spatial eigenvector mapping, SEVM). There are also local methods that detect, describe, and map spatial patterns at individual sampling units. Spatial analysis by distance indices (SADIE) (Perry et al. 1996) is one of the most commonly used by ecologists because it was designed to analyze count data. Many ecologists also use point pattern analysis to describe and model the spatial distributions of organisms (Cressie 1993).

We have focused on linear explanatory models. In choosing between spatial linear models, the error distribution in the response variable will be an important criterion. For normal distributions, GLS, SAR, and CAR can be used efficiently. However, more flexible methods for different error distributions include spatial GLMMs and SEVM (Dormann et al. 2007; Bivand et al. 2013), not covered here. The SAR methods presented here (as well as spatial GLMMs) can be implemented within the Bayesian framework (LeSage and Pace 2009; Bivand et al. 2013). Doing so allows for a more flexible incorporation of other complications (observer bias, missing data, and different error distributions) but is much more computer-intensive than any of the methods presented here.

In brief, there is a spatial world out there to be explored! We hope that this introduction has given you a taste of the main ideas and concepts that will help you find your way to the design and analytical methods that will help you address your spatial questions. Moreover, we hope that the examples we have analyzed have convinced you that ignoring spatial correlation in data is a very bad idea. We also hope that we have scared you away from using spatial statistics as quick fixes. You should always avoid this temptation but it can be especially important with respect to spatial problems because most spatial statistical tools are descriptive and phenomenological. None of the methods presented here are dynamic or mechanistic models of ecological processes (e.g., dispersal or species interaction with the environment). An observed spatial pattern is often the end product of a combination of different processes creating patterns *that may be observationally equivalent*. Much room remains for spatial ecological research that is based on experimental work or simulated data. Such studies, in combination with the available spatial statistical tools, will make our models of the spatial distribution of ecological phenomena be more predictive.

Acknowledgments

We thank Ben Bolker and Bruce Kendall for extensive comments and discussion of an earlier draft of this chapter, and Roger Bivand for his generous advice. We also thank Rafael Ruiz Moreno for the illustrations and Isis de la Rosa for editing the figures.

## CHAPTER 11

Statistical approaches to the problem of phylogenetically correlated data

Marc J. Lajeunesse and Gordon A. Fox

11.1 Introduction to phylogenetically correlated data

Multi-species data sets violate some of the most basic assumptions of traditional statistics, and so present an important challenge for data analysis. For example, ecologists might want to use regression to explore the relationship between body mass and aerobic capacity. Given the considerable variation in body mass across mammals, from shrews to whales, there is extensive opportunity to explore trends in these two quantities. However, linear regression assumes that each observation is independent of the other (Stuart and Ord 1994). What is at risk here is that data from closely related mammals will not adequately form independent pieces of information: the shared evolutionary history of these taxa will introduce correlations, or dependencies, in data (Felsenstein 1985). A conventional regression would treat data from multiple species of canines, cats, and weasels as independent, despite the potential correlations in their characteristics due to their shared ancestry as Carnivora.

Because of these potential dependencies in data, and their effects on statistical assumptions, serious inferential errors can emerge when analyzing and comparing data from multiple species using conventional regression methods. For example, there is no longer a guarantee that statistical hypothesis tests remain valid. Standard tests, like those asking whether the slope of the regression is significantly non-zero, are no longer valid (Diaz-Uriarte and Garland 1996). To minimize this problem with interspecific (multi-species) data, techniques based on the *phylogenetic comparative method* are used to improve the reliability of inferences with regression (e.g., Felsenstein 1985; Grafen 1989; Pagel 1993). These statistics use information on phylogenetic evolutionary history to hypothesize the strength of correlations across taxa. These *phylogenetic correlations* are then applied to generalized least squares (GLS) models—a statistical framework less rigid to violations of assumptions like independence of data. GLS modeling is also used in other areas of ecology, including studies in which *serial correlations* or *spatial correlations* occur in the data (see chapters 10 and 13). Thus the principles of regression modeling used in this chapter are rather general, and we hope that readers can glean some broad lessons even if they never use multi-species data sets.

The basic principles underlying most of these phylogenetic comparative statistics are also straightforward, and share two common themes. The first is to enable a phylogenetic framework to test hypotheses on evolutionary processes (e.g., Hansen 1997; Pagel 1999). The second, and perhaps more germane to many ecologists, is to offer some assurance that inferences drawn from multi-species analyses are valid and statistically sound (e.g., Price 1997; Schluter 2000). The goals of this chapter are to introduce the basic principles of phylogenetic comparative methods, and to demonstrate *why* it is important to apply these methods when analyzing interspecific data. We emphasize *why* here because this is not often clearly addressed in introductory texts. What gets addressed typically is *how* to apply these techniques. For example, there are already several published reviews, surveys, and how-to guides on comparative methods (e.g., Harvey and Pagel 1991; Martins 1996; Martins 2000; Blomberg and Garland 2002; Felsenstein 2004; Garland et al. 2005; Nunn 2011; Paradis 2011; O'Meara 2012). By focusing here on *why* rather than *how*, we aim to provide a unique view of the risks of not applying these statistical tools, as well as insight on the limitations for what they can accomplish.

To achieve these goals, we center the chapter on a series of *Monte Carlo experiments* that aim to answer the following: *Why is it risky to use regression with multi-species data? When do you expect the greatest risk? What are phylogenetic correlations, and how are they used in regression models? What happens when the incorrect model of evolution is assumed?* Monte Carlo simulations use randomly generated data to investigate the conditions for when statistical tests, such as regression, provide reliable outcomes (Rubinstein and Kroese 2007)–or equally when they fail to provide reliable outcomes. This simulation approach has been crucial to the development of comparative phylogenetic methods and the way they are practiced (e.g., Martins and Garland 1991; Freckleton et al. 2002; Martins et al. 2002; Revell 2010; Freckleton et al. 2011). Our intention is to use simulations to: (1) reveal the underlying principles on why it is important to apply phylogenetic correlations to regression models by simulating interspecific data sets, and (2) introduce several of the powerful and diverse statistical functionalities offered in *R*. These include the widely used ape (Paradis et al. 2004) and geiger library (Harmon et al. 2008), useful for manipulating and applying phylogenies for regression modeling.

We focus exclusively on the analysis of interspecific data using linear regression, which historically has received the most attention (Felsenstein 1985; Pagel 1999), and for the purposes of this chapter, serves as an accessible introduction to more advanced statistical models and practices covered elsewhere (Martins and Hansen 1997; Pagel 1999; Revell 2010). Our aim is to channel the reader from simple to more complex topics using linear regression, assuming that the reader has little to no familiarity with comparative methods, by introducing key concepts as they emerge. We hope that this stepwise exposition helps readers gain insight on the power of these statistical tools, as well as practical information on how to interpret results from analyses with interspecific data.

11.2 Statistical assumptions and the comparative phylogenetic method

Because closely related species tend to be more similar than distantly related species, methods like regression and ANOVA—which assume normality, homogeneity of variance, and independent (uncorrelated) errors (Stuart et al. 1999)—are not generally valid for multi-species data. We use Monte Carlo experiments and regression analyses to explore the consequences of violating these assumptions for a simple reason: because we have simulated the data, we know the right answers, so we can see how different methods affect our conclusions. We begin by introducing a simple linear regression model, and then extend this model to include phylogenetic correlations. Because we will need to refer to bits of R script repeatedly, we label them somewhat like equations: the jth piece of R script we use is labeled R.j.

11.2.1 *The assumptions of conventional linear regression*

Let's start by decomposing a simple linear regression model. This will provide insight on how multi-species data sets can challenge inferences with this model. The goal of regression is to estimate the linear relationship between a dependent variable (y) and an independent (explanatory or predictor) variable (x). For example, can mass (x) predict aerobic capacity (y), or body size predict fitness? Perhaps the simplest way to ask this question is to model a straight line in $x - y$ coordinates. This line can be described in equation form as:

$$y_i = a + bx_i + \varepsilon_i. \tag{11.1}$$

Here the subscript i indexes (refers to) individual data points (or samples) of these variables, and it can take on the values $i = 1, 2, \ldots, N$ where N is the total number of observations. Equation (11.1) contains the y-intercept (a) and the slope (b) of the line. These are unknown, and what we want to estimate with regression. An important aspect of modeling the relationship between y and x is the random error (or residual) variable ε. This term is the stochastic noise in this relationship. It is often interpreted as all the variation in y that isn't explained by x. Under the simplest linear models, ε is assumed to be independent across all observations; a further assumption is that these observations will be Normally distributed (N) around a mean of zero and have a common (or homogenous) variance (σ^2). These assumptions on the distribution of ε can be summarized with:

$$\varepsilon_i \sim N(0, \sigma^2). \tag{11.2}$$

Given these assumptions, let's simulate some data that fit the relationship described in equation (11.1) and see how regression can estimate the intercept and slope of this linear model. For our simulation, and given equation (11.1), let's arbitrarily set the slope to 0.5 ($b = 0.5$), and the intercept to 1 ($a = 1.0$). What remains is to simulate ε and the explanatory variable x. Equation (11.2) describes the distribution of ε, and so for this term we will simulate random errors that are Normally distributed with a mean of zero and a variance of 1 ($\sigma^2 = 1.0$). Likewise, data for x will be generated by randomly sampling a Normal distribution with a mean of 0 and variance of 1. Finally, we will generate 30 data points for y_i and x_i ($N = 30$) and analyze these with regression. Applying equation (11.1) generates a correlation between y and x; our aim is to detect this relationship with regression. Following these parameters, we can quickly simulate random y_i and x_i in R as follows:

```
N <- 30; x_mean <- 0; x_variance <- 1; a <- 1; b <- 0.5
e <- rnorm(N, 0, 1) # sample 30 standard Normal deviates
x <- x_mean + sqrt(x_variance) * rnorm(N, 0, 1) # random x's     (R.1)
y <- a + b * x + e # see eq. (11.1)
```

Using a linear regression model in R to estimate the slope and intercept from these simulated y_i and x_i:

```
library(nlme) # load R library
# get regression results
results <- summary(gls(y ~ x, method="ML"))                          (R.2)
# print only regression coefficients
results$tTable
# print only residual error
paste("residual error = ", results$sigma)
```

we get the following R output from this `gls` function (bold our emphasis):

```
              Value Std.Error t-value p-value
(Intercept) 1.0292094 0.2212227 4.652369 0.0001
x           0.4728596 0.2209420 2.140198 0.0412
```

```
"residual error = 1.167178"
```

Given the sample size and the method used to simulate ε to add a lot of stochastic noise to the model (i.e., with large variance of $\sigma^2 = 1.0$), the regression model provides a reasonable estimation of the residual error ε (1.167 $\approx$ 1.0), and is able to detect that the intercept (1.029 $\approx$ 1.0) and slope (0.473 $\approx$ 0.5) were significantly non-zero via t-scores (i.e., p-values < 0.05). (Well, in fact, it took several runs of the R.1 and R.2 scripts to get these nice results! More on this later when we explore how sampling error can influence how well regression can detect this slope.) The variances of the regression coefficients are also close to what we simulated. For example, the regression output reported the standard error (S.E.) of the intercept to be 0.22; this translates to a variance of approximately 1 (i.e., $\sigma^2 = \text{S.E.}\sqrt{N} = 0.22\sqrt{30} = 1.21 \approx 1$). The slope's variance was also approximately 1. However, note that the expected variance of the slope in this model is 1 only because $\sigma_b^2 = \sigma_s^2/\sigma_x^2 = 1/1 = 1$. These results make sense given the way we randomly generated our data.

However, given that it took multiple runs of R.1 and R.2 to get these nice results, let's explore the error rates of this regression model in a more rigorous way. To achieve this, we will perform a Monte Carlo experiment to investigate how sampling error influences the way regression can detect a pre-defined relationship between y and x. In other words, we'll try to determine why we needed to run our previous example multiple times to get the results predicted by our linear model from equation (11.1) parameterized as: $y = 1 + 0.5x + \text{N}(0, 1)$. More specifically, we will assess the *Type II error* rates of linear regression at differing sample sizes—that is, estimate the probability of regression statistics failing to detect the intercept and slope for a given N. For this simulation, we will generate random data following the linear model in equation (11.1), analyze these with regression, and repeat this process 1,000 times with increasing sample sizes (N). For each iteration, we will count when the p-value of each t-score was greater than 0.05 (our significance level) for each regression coefficient. Finally, we divide this count by the number of simulation replications (1,000 per N). This will give us the proportion of analyses that concluded incorrectly that the regression coefficients did not differ from zero (i.e., Type II error). The R script for this simulation is found in appendix 11.A.

Our simulation of regression analyses with $N = 5$ to 50 revealed that with small sample sizes linear regression performs poorly and has large Type II error rates for detecting non-zero regression coefficients (figure 11.1). Generally, any results based on a regression with

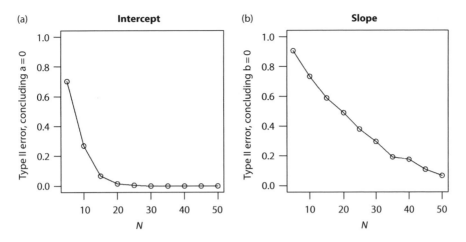

Fig. 11.1 The risk of concluding null results when few data are used in regression analyses. Presented are results from a Monte Carlo experiment exploring the Type II error rates (i.e., false negative outcomes) of ordinary least square (OLS) regression with small to large sample sizes (*N*). Error rates are based on the proportion of 1,000 regression analyses incorrectly concluding that the intercept (*a*) and slope (*b*) were zero. Here regression analyses have improved ability to detect a non-zero slope and intercept with larger sample sizes (*N*). The R script for this simulation is in appendix 11.A.

$N < 30$ should be interpreted with caution. This is because there are too few data sampled to generate enough variation for our regression analyses to properly estimate the slope and intercept of these simulated data. There is also substantial difficulty in detecting our non-zero slope; this is because we modeled the residual error (ε) to add a lot of stochastic noise to our model.

In summary, regression is a tool that aims to estimate the relationship between two variables, but the ability for regression statistics to detect this relationship is often largely dependent on the sample size (*N*). Our next goal is to repeat this simulation but with interspecific (multi-species) data and explore how these multi-species data can invalidate the assumptions of how ε is modeled, and why this can further impact the outcome of regression analyses.

11.2.2 *The assumption of independence and phylogenetic correlations*

Interspecific data sets generally violate the assumption of independence, because species form a nested hierarchy of phylogenetic relationships. This shared history introduces phylogenetic correlations among related species, and as a result, data from related species may not be statistically independent (Felsenstein 1985). Let us emphasize this in another way: data from related species may not form independent pieces of information and may be correlated—they share a common ancestor and therefore may also have common characteristics. However, we can use phylogenetic information to predict these correlations, and these predictions can then be applied to improve regression estimation and statistical inferences with interspecific data.

But how do phylogenetic correlations come into play with regression analyses? To answer this, we will first need to expand our regression model. Phylogenetic correlations are a problem for the linear model defined in equations (11.1) and (11.2) because the residual errors (ε)

are assumed to be mutually uncorrelated. This linear model is in fact a simplification of a more general way to model ε based on *ordinary least squares* (OLS):

$$\varepsilon_i^{\text{OLS}} \sim \text{MVN}(0, \sigma^2 \mathbf{I}). \tag{11.3}$$

This formulation is a different way of writing the linear regression model that we have been discussing. Writing it this way will allow us to relax assumptions about independence of data points and homoscedasticity. But first, let's understand the model in equation (11.3). Here ε has a multivariate (MV) Normal distribution, with a mean of zero and variance equal to $\sigma^2 \mathbf{I}$. The idea is that instead of a single variance σ^2 that holds for all values of y, we have a variance–covariance matrix of dimension $N \times N$. This matrix gives the variances for each value of y on its diagonal, and has important properties for linear modeling because it contains information describing the dependency between each pair of data points. These dependencies are modeled by the covariances between pairs of values in all off-diagonals of the matrix. In this case, the matrix $\mathbf{I}$ is the identity matrix (1's on the diagonal, and 0's everywhere else), so $\sigma^2 \mathbf{I}$ tells us that, indeed, every point has the same variance and they are all independent of one another.

Given this variance–covariance matrix, let's relax the assumptions of homoscedasticity and independence. We need to do this when the residual error (ε) is not distributed according to equation (11.3), such as when data are phylogenetically correlated. This is because under these conditions, OLS models may no longer provide unbiased estimates of regression coefficients, and statistical tests used for null hypothesis testing may no longer be valid (Diaz-Uriarte and Garland 1996). If phylogenetic correlations are known (or hypothesized), as is the case when we have a hypothesis on the phylogenetic history of taxa (section 11.2.3), then analyzing interspecific data now becomes a generalized least squares (GLS) problem (Pagel 1993; Revell 2010). The error term of this GLS model is defined as:

$$\varepsilon_i^{\text{GLS}} \sim \text{MVN}(0, \sigma^2 \mathbf{C}). \tag{11.4}$$

Here, $\varepsilon_i^{\text{GLS}}$ models the variance-covariance matrix ($\mathbf{C}$) to have off-diagonal covariance among data from different but related species. The next section describes exactly how we hypothesize these covariances using phylogenies.

11.2.3 *What are phylogenetic correlations and how do they affect data?*

Before exploring how regression can be modified to analyze interspecific data, we need to know a little more about phylogenies and how to extract phylogenetic correlations. Phylogenies are statistical hypotheses on the shared history of taxa (Felsenstein 2004), and for our purposes they contain information on the relative phylogenetic distances of species. The sources of phylogenies are diverse; for example, molecular or morphological information can be used to statistically group related species. The methods used to construct trees are beyond the scope of this chapter (but see Felsenstein 2004); instead we will generate a simple random tree by simulating lineages "branching-out" or diverging randomly with time. This random branching model is called a *Yule birth–death process*. Here is some *R* script that uses the `geiger` library (Harmon et al. 2008) to simulate this birth–death process to generate a small random phylogeny with 5 species (also often described as a phylogeny with five "tips"):

```
# Simulate random phylogenetic tree with 5 species using geiger
# (see Harmon 2008).
library(ape); library(geiger);
```

```
K <- 5 # five species or tips on the simulated tree
# random tree from birth-death model
tree <- sim.bdtree(b=1, d=0, stop="taxa", K)                    (R.3)
# assign letter names to tips
tree$tip.label <- paste(letters[K:1], sep="")
# plots random tree graphically plot(tree)
plot(tree)
# outputs tree in Newick text format
write.tree(tree, digits=2)
```

Running R.3, we generated the random phylogeny shown in the left of figure 11.2.

There are several characteristics of this tree that are notable in terms of predicting phylogenetic correlations. First, the branching pattern of this phylogeny, known as its *topology*, has two major lineages: one with species *a* and *b*, and a second with *c*, *d*, and *e*. One way to interpret these two lineages is to think of them as two distantly related taxonomic groups (e.g., families or orders). These broad groupings are important because they help predict which species will be correlated with one another. For example, these two lineages will translate into two clusters of phylogenetic correlations: data from species *a* and *b* will be correlated with one another, but not with *c*, *d*, and *e*. There is no correlation between these two groups because they stem from the *root* of the tree. The root is the hypothesized ancestral divergence of the entire lineage. Second, note that the nodes of the tree, which designate historic divergence or speciation events, are clustered near the tips of the tree for each group. This tight grouping will create strong correlations among species within these groups; if they were positioned closer to the root (i.e., designating more ancient divergences), then correlations would be weaker.

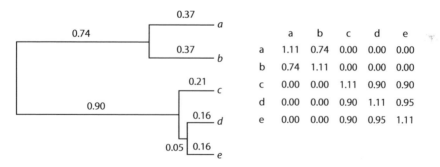

Fig. 11.2 Left, a phylogenetic tree generated from a random birth–death process (R.3). There are two major lineages: one with species *a* and *b*, and a second with *c*, *d*, and *e*. Right, the variance–covariance matrix corresponding to this phylogeny. All off-diagonals of this matrix contain the sum of the shared pairwise distance between species; for example, *a* and *b* share an internode distance of 0.74. This is not the branch-length distance from tip *a* to tip *b* (which coincidently sums to 0.74); rather it is only the internode distance shared by *a* and *b*. Also, note that the distance from root to tip for species *a* equals 1.11. This is because it shares an internode distance (common history) with *b* from the root of the tree of length 0.74, followed with a divergence period after a speciation event of length 0.37. Branch lengths calculated from this variance–covariance matrix are plotted along each branch on the left. The Newick (a computer-readable notation) version of this tree is "(((e:0.16,d:0.16):0.05,c:0.21):0.90,(b:0.37,a:0.37):0.74);".

Comparative phylogenetic methods use these characteristics of phylogenetic trees, such as the topology and the distances between each node (known as the internode distance), to predict which species will be correlated with one another and to quantify the strength of these correlations between species. Our next step is to extract these correlations from our random tree. We can quickly calculate its phylogenetic correlations using the matrix functions available in the APE library (Paradis et al. 2004), beginning with the raw phylogenetic distance matrix (*VCV*) whose elements are the sums of branch-length internode distances:

```
# Convert the phylogenetic tree into a variance-covariance matrix.
# Note that we reorder the columns of this matrix to make it easy
# to compare with the topology of the phylogeny in Figure 11.2.
# calculate matrix from phylogeny
VCV <- vcv(tree)                                                    (R.4)
# assign non-number names to tips
order <- paste(letters[1:K], sep="")
# round-down numbers, and order matrix by name
round(VCV[order,order], 2)
```

The distance matrix of our phylogeny from the left of figure 11.2 is shown in the right of figure 11.2. Note how there are essentially two submatrices (with non-zero values) in this matrix. These two submatrices designate the major lineages of our tree (i.e., group *a* and *b*, and group *c*, *d*, and *e*). Also note that all the main diagonals of this matrix equal 1.11. This is the sum of all the branch-length distances (internode distances) from the root to tip for each species.

The left of figure 11.2 also shows the internode branch-length distances. We want to emphasize that all the main diagonals of the matrix in the right of figure 11.2 are equal to 1.11 because all the tips (species, or terminal taxa) are aligned contemporaneously—that is, the distance from the root to each tip is the same. Trees with this alignment are described as having an *ultrametric* shape and are called dendrograms because they depict evolutionary time; there is a chronological ordering of nodes that hypothesize the historic divergences among lineages (Felsenstein 2004). In effect, this is how we generated our tree, by simulating random speciation events and divergences of taxa through time (via a Yule process; see Harmon et al. 2008). Trees estimated from genetic information, such as maximum likelihood trees based on nucleotide sequence data, can also generate dendrograms by assuming constant rates of random molecular change (e.g., a molecular clock). In fact, this time component of dendrograms is a crucial aspect of the comparative phylogenetic method, and later we will describe how it is used to hypothesize evolutionary processes (section 11.2.5).

The elements of the matrix in the right of figure 11.2 are still a little abstract given that they are in terms of branch-length distances; remember our goal is to use the phylogeny to estimate correlations among species. These correlations are meant to quantify the predicted relationship between interspecific variation and the phylogeny for which taxa evolved (Martins and Hansen 1997). Luckily this is straightforward, and we can quickly convert all these distance values into correlations by dividing each element in the variance–covariance matrix with the total branch-length distance from root to tip of the tree (i.e., 1.11). This is only possible because our tree is ultrametric. Dividing all the elements of the matrix by 1.11 (*R* script: `1/1.11 * vcv(tree)`) yields the correlation matrix shown in table 11.1. You can also extract this matrix by using the `cov2cor(vcv(tree))` function in *R*. Now the main diagonals in table 11.1 equal 1

Table 11.1 Correlation matrix (**C**) of our simulated phylogeny (see figure 11.2). Numbers in bold are meant to emphasize the two major subgroups *a–b* and *c–d–e* of phylogenetic correlations in this phylogeny

	a	b	c	d	e
a	**1.00**	**0.67**	0.00	0.00	0.00
b	**0.67**	**1.00**	0.00	0.00	0.00
c	0.00	0.00	**1.00**	**0.81**	**0.81**
d	0.00	0.00	**0.81**	**1.00**	**0.86**
e	0.00	0.00	**0.81**	**0.86**	**1.00**

because taxa are perfectly correlated with themselves, and off-diagonals have the pairwise correlations among taxa. For example, the correlation between *e* and *c* equals 0.81 (e.g., 0.90/1.11 = 0.81) since they only "recently" diverged—that is, recent relative to all other divergences on the tree.

Given this tree and its phylogenetic correlation matrix, our next step is to update our regression analysis and apply these correlations to model potential dependencies in interspecific data. Our matrix from table 11.1 will become the phylogenetic correlation matrix **C** used in GLS regression models [equation (11.4)]. We will also use **C** to generate random interspecific data in Monte Carlo experiments. There are many packages available to simulate correlated data in *R* (see Harmon et al. 2008), but here we will generate these directly using the *Cholesky decomposition* method (Rubinstein and Kroese 2008). The Cholesky method (described more fully following this example) takes random data and transforms it into new, correlated data. Our aim here is to generate random but correlated data with the covariance properties modeled in equation (11.4) and defined by our phylogenetic correlations in table 11.1. To start, let's first randomly generate and plot some independent (ind) and correlated (cor) *y*'s and *x* using the Cholesky method with this *R* script:

```
K <- 5; x_mean <- 0; x_variance <- 1; a <- 1.0; b <- 0.5
e_rand <- rnorm(K, 0, 1)
x_rand <- rnorm(K, 0, 1)                                        (R.5)
# independent (uncorrelated) data following the I matrix
I <- diag(K) # creates identity matrix for OLS model
e_ind <- t(chol(I)) % *% e_rand # as modeled in eq. 11.1.3
x_ind <- x_mean + sqrt(x_variance) * t(chol(I)) %*% x_rand
y_ind <- a + b * x_ind + e_ind
# correlated data following the C matrix
C <- cov2cor(vcv(tree))
e_cor <- t(chol(C)) %*% e_rand # as modeled in eq. 11.1.4
x_cor <- x_mean + sqrt(x_variance) * t(chol(C)) %*% x_rand
y_cor <- a + b * x_cor + e_cor
# now organize three scatter-plots of these data
par(mfrow=c(1,3), xpd=TRUE);
plot(x_ind, y_ind, xlim=c(-2.5,2.5), ylim=c(-1.5,3.5), main="random
data")
text(x_ind, y_ind, tree$tip.label, cex=1, pos=1, font=4)
```

```
plot(x_ind, y_ind, xlim=c(-2.5,2.5), ylim=c(-1.5,3.5), main="phylo-
transformation")
text(x_ind, y_ind, tree$tip.label, cex=1, pos=1, font=4)
arrows(x_ind, y_ind, x_cor, y_cor, length=0.05)
plot(x_cor, y_cor, xlim=c(-2.5,2.5), ylim=c(-1.5,3.5),
main="transformed data")
text(x_cor, y_cor, tree$tip.label, cex=1, pos=1, font=4)
```

Figure 11.3 contains the *R* output of two plots where each data point is labeled by its species; the left panel has independent random data and the right panel has the same data but transformed via the correlation matrix (**C**). First note that the random data, once phylogenetically transformed, are now clustered more tightly among groups *a–b* and *c–d–e*. The phylogenetic transformation had the effect of making data more similar relative to their correlations. For example, *b* and *a* are now much closer together; they are no longer independent points and therefore have some similarity due to their shared phylogenetic history.

As an aside, note that the (x, y) positioning of species *b* and *e* remained the same in both data sets (see center panel in figure 11.3). This is a property of the way we transformed our random data phylogenetically. Our transformation method finds an upper triangular matrix (**U**) or Cholesky matrix that satisfies the condition $\mathbf{C} = \mathbf{U}^{\mathrm{T}}\mathbf{U}$ (with the superscript T indicating the transposition of a matrix). Multiplying $\mathbf{U}^{\mathrm{T}}$ to a collection of random data will transform them following the correlations in **C**. However, if **C** is a proper correlation matrix, where diagonals all equal 1, and off-diagonals have correlations that range from zero to almost (but not) one, then the first (upper) element in the vector of transformed data will remain untransformed. In fact, what happens is that the phylogenetic transformation will rotate and shear the other data relative to this untransformed data point. In our case, because we have two independent groups in our correlation matrix (e.g., groups *a–b* and *c–d–e*), the transformation method will rotate and shear data relative to the way

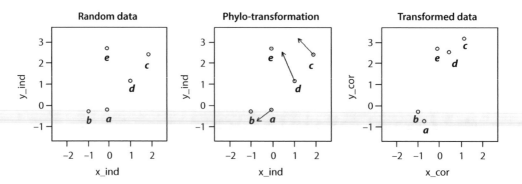

Fig. 11.3 The effects of phylogenetic correlations on randomly generated *x* and *y* data. The leftmost panel has randomly generated (independent) data for each species (*a, b, c, d, e*; see figure 11.2), the center panel depicts the direction of the phylogenetic transformation on these random data, and the rightmost panel shows the correlated random data after a phylogenetic transformation (based on **C**; see figure 11.2). The random data within groups *a–b* and *c–d–e* are sheared and rotated closer to one other because they belong to two independent groups of related species (see topology of the phylogeny in figure 11.2). The phylogenetic transformation was achieved using a Cholesky decomposition method, and the modeled relationship between *x* and *y* is defined in equation (11.1).

they are correlated with *b* and *e*. *But why species b and e?* The original correlation matrix calculated from the phylogeny with the vcv function (Paradis et al. 2004) actually had an order of *e*, *d*, *c*, *b*, *a*. We re-ordered this matrix as *a*, *b*, *c*, *d*, *e* to simplify comparisons with the phylogeny shown in the left of figure 11.2. The original un-ordered **C** had *e* and *b* occupying the first (upper) elements of each submatrix.

The random and correlated data in figure 11.3 provide a nice visualization of the effects of phylogenetic transformations. However, unless the predicted means of *y*'s and *x*'s for each species differ, or the means among groups *a–b* and *c–d–e* differ, then it is nearly impossible to predict how the phylogenetic transformation will position random data (especially with large *K*). This is because we are simulating all the *x*'s of each species to be centered around zero—what really gets affected by phylogenetic correlations is the residual error (ε) of these data, relative to *y*. For example, if we repeat R.5 30 times and plot these 30 data sets together, we can see the effects of random sampling and the unreliability of visually diagnosing phylogenetic effects in interspecific data. These results are in figure 11.4 (*R* script for this simulation is found in appendix 11.B with *N* = 30). Visually, we can barely see a positive correlation between *x* and *y*, and we can only see that because we modeled the relationship between the dependent (*y*) and predictor (*x*) variables to have a moderately strong slope (see equation (11.1)). To see why it is so difficult to visualize phylogenetic correlations in interspecific data sets, note that random data from species *a* can potentially occupy any part of that scatter plotted in figure 11.4.

Despite this large scatter in figure 11.4, the phylogenetic correlations do exist. In fact, we can recover the correlation matrix **C** and the means across all species for *x* and *y* quite

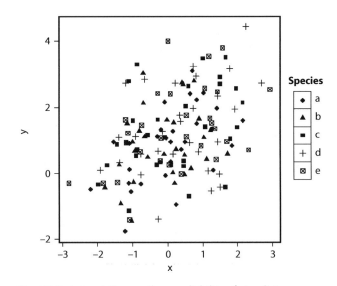

Fig. 11.4 A simulation on the unreliability of visualizing phylogenetic correlations in interspecific data. Here for each species (*a*, *b*, *c*, *d*, *e*; see figure 11.2), 30 *x* and *y* pairs were randomly generated using the Cholesky decomposition method (based on **C**, defined in table 11.1). This plot is equivalent to figure 11.2, but overlaid 30 times. Note that the random data for a single species can occupy nearly any region on the plot, and the only discernible pattern is the modeled relationship between *x* and *y* (defined in equation (11.1)). The R script for this simulation is in appendix 11.C.

easily in *R*. Again using the script found in appendix 11.B, but now with $N = 1,000$, and estimating the correlations between the random data generated for each species, we can recover the correlation matrix and means of each species for the *x*-variable; see table 11.2. With a minor modification to the script in appendix 11.B, we can also estimate the correlations and means of *y* (see table 11.2). The correlations are not perfect, but they are very close to **C** for both *x* and *y* (as in table 11.1). The means of each species are also very close—the *x*'s of all species are near 0, and all *y*'s are near 1. Had we simulated data with a larger *N*, our estimates would have converged to the expected **C** (table 11.1) and to our (expected) simulated means. This simulation emphasizes the importance of having precise species-level estimates of characteristics or traits for cross-species comparisons. Here, sampling error within species traits makes it harder for regression models to detect the underlying linear relationship between traits. However, only recently have comparative phylogenetic methods have been able to include within-species variation in regression analyses (Ives et al. 2007; Felsenstein 2008; Hansen and Bartoszek 2012).

11.2.4 *Why are phylogenetic correlations important for regression?*

Now let's return to our original regression model [equation (11.1)] and consider the case where y_i and x_i are two traits to be compared across multiple taxa—that is, the *i*th data point represents a characteristic from species *i*. In our previous simulation, the *i*th observation could be considered as *N* samples from a single species, but here let's use *K* rather than *N* to denote the total number of species analyzed. Again, *N* is the number of samples within species, and *K* is the number of species. Our goal is to repeat our previous simulation using OLS regression, but now with interspecific data—here we will assess the error rates of this regression model when the condition of independence assumed by ε is violated [equation (11.2)].

Let's start by analyzing our data set from figure 11.2 to assess how OLS and GLS models perform with our phylogenetically correlated data. Here are these data along with the GLS regression analysis including the phylogenetic correlations:

```
library(ape); library(nlme);
# raw phylogenetically correlated data from figure 11.3
x <- c(-0.07684503, 0.44569569, 1.15961757, -1.00146522, -0.71858873)
y <- c(2.7098214, 2.5464312, 3.1840059, -0.2871652, -0.7509973)
# using ape to load our Newick phylogeny; see Figure 11.2
tree <-
read.tree(text="(((e:0.16,d:0.16):0.05,c:0.21):0.90,
    (b:0.37,a:0.37):0.74);")                                      (R.6)
# The gls function of the nmle library requires a correlation
# matrix in the form of a corStruct object class.
VCV <- cov2cor(vcv(tree))
# convert matrix to corStruct object
C <- corSymm(VCV[lower.tri(VCV)], fixed=T)
# extract only coefficients
summary(gls(y ~ x, method="ML", correlation=C))$tTable
```

When a phylogenetic correlation matrix is included in a GLS model like this, it is commonly referred to as a phylogenetic generalized least squares (PGLS) regression (Martins

Table 11.2 Correlations (top) and means (bottom) for the *x* and *y* variables (left and right, respectively) estimated from random phylogenetically correlated data. Estimates are based on the script in appendix 11.B, using *N* = 1,000

Estimated correlation matrix for *x*.

	a	b	c	d	e
a	1.000	0.672	-0.004	0.007	-0.002
b	0.672	1.000	-0.009	0.001	-0.006
c	-0.004	-0.009	1.000	0.810	0.811
d	0.007	0.001	0.810	1.000	0.856
e	-0.002	-0.006	0.811	0.856	1.000

Estimated means for *x*.

	a	b	c	d	e
x	-0.007	-0.012	0.011	0.000	0.004

Estimated correlation matrix for *y*.

	a	b	c	d	e
a	1.000	0.666	-0.014	-0.012	-0.015
b	0.666	1.000	-0.022	-0.024	-0.022
c	-0.014	-0.022	1.000	0.817	0.813
d	-0.012	-0.024	0.817	1.000	0.863
e	-0.015	-0.022	0.813	0.863	1.000

Estimated means for *y*.

	a	b	c	d	e
y	0.996	0.997	0.999	1.001	1.00

and Hansen 1997; Pagel 1997, 1999; Garland et al. 1999). Our PGLS analysis estimated the following regression coefficients:

```
              Value Std.Error   t-value    p-value
(Intercept) 1.2424182 0.8446677 1.470896 0.2376872
x           0.7582988 0.5545423 1.367432 0.2649284
```

For comparison, let's also look at how a conventional OLS regression (without phylogenetic correlations) estimated the same coefficients (R script of regression without the phylogenetic correlation matrix: gls(y ~ x, method="ML")):

```
              Value Std.Error   t-value    p-value
(Intercept) 1.552139 0.4424145 3.508336 0.03924480
x           1.871734 0.5647376 3.314343 0.04524494
```

It may be useful to visualize these coefficients:

```
plot(x, y, xlim=c(-2.5,2.5), ylim=c(-1.5,3.5))
text(x, y, tree$tip.label, cex=0.7, pos=3, font=4)
abline(gls(y ~ x, method="ML")) # regression line from OLS
# regression line from PGLS
abline(gls(y ~ x, method="ML", correlation=C), lty=2)
legend(0.5, 0, c("OLS","PGLS"), cex=0.8, lty=1:2)
```

The results are shown in figure 11.5. The OLS regression line seems to have a much nicer fit to our species data than the PGLS model—it passes right through our simulated data (figure 11.5). The t-scores of the OLS estimate also concluded the slope and intercept to be non-zero (p-values are just above 0.04 for both estimates). In contrast, the regression line of the PGLS estimator does not look like a very robust fit (figure 11.5), and in fact, its slope and intercept were not significant ($p > 0.05$).

These regression results are counter-intuitive; OLS seems to provide a better fit than PGLS to the phylogenetically correlated data. However, contrasting the results from PGLS and OLS regressions underlines the importance of including phylogenetic correlations when analyzing interspecific data. Had we relied solely on the OLS regression, we would have concluded that there is a strong positive relationship between x and y. However, our PGLS analysis reveals that much of this relationship between x and y is due to their shared phylogenetic history—which is true given the way we phylogenetically transformed our data. Without the PGLS analysis, the findings of the OLS regression are at risk of making a Type II error (Diaz-Uriarte and Garland 1996; Harvey and Rambaut 1998). If we knew nothing about the underlying properties of our data, then we would have to conclude that there is no evidence for a positive linear relationship between x and y given our PGLS results. Although this is a conservative way to interpret results, it is appropriate given that only the PGLS analysis accounted for the potential phylogenetic correlations among species.

However, we simulated these data, and we know a positive relationship between x and y exists. *So what happened? Why was the OLS able to detect an effect while PGLS was not?* We do not typically have the luxury of knowing the true underlying relationships among species prior to analyses; however, our simulation approach provides us an opportunity to explore other explanations for why disparities among analyses exist. One explanation, which we will consider in section 11.2.5, is that we might have used an inappropriate model of evolution in our PGLS analysis. Another explanation, and the primary scourge of all analyses, is sampling error. It doesn't matter if you have the best phylogenetic hypothesis, or

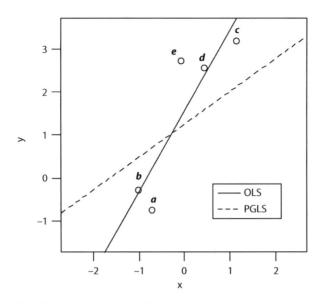

Fig. 11.5 Regression lines fit to randomly data generated with phylogenetic correlations (see figure 11.3). Regression lines were estimated with either ordinary least squares (OLS) or a phylogenetic generalized least squares (PGLS) model. These phylogenetic correlations were based on the **C** matrix from table 11.1.

the most precise trait data for your species—if you do not have enough data you will not be able to correctly estimate (with confidence) the underlying relationship (effect) with regression. In our case, this effect is the correlation (r) between x and y. Random sampling alone will generate data sets with strong positive or negative correlations—what we need is some assurance that our observed effect is true and did not emerge because of sampling error. We explored this issue of sampling error and low sample sizes previously with our simulations with conventional regression on independent data (section 11.2.1).

One way to assess the reliability of our regression analysis is to estimate its predicted false negative rate (Type II error rate, or β); that is, determine its probability of failing to detect a non-zero correlation. We can do this directly by first estimating our predicted effect, which is the expected correlation (r) between x and y, or more explicitly:

$$r = b\sqrt{\sigma_x^2/\sigma_y^2}. \tag{11.5}$$

In our simulation, $b = 0.5$ and $\sigma_x^2 = 1$ (see R.1). We also need to know the predicted variance of y, and given the way we modeled x and ε in equation (11.1), the predicted distribution of y is:

$$y_i \sim N(a + bx + \varepsilon, b^2\sigma_x^2 + \sigma_\varepsilon^2) = N(1, 1.25). \tag{11.6}$$

Thus y has a mean of 1 (i.e., $1 + 0.5 \times 0 + 0 = a + bx + \varepsilon$) and a variance of $\sigma_y^2 = b^2\sigma_x^2 + \sigma_\varepsilon^2 = 0.25 \times 1 + 1 = 1.25$. Given these values, the predicted correlation between x and y in our simulations is $0.447 \approx 0.5\sqrt{1/1.25} = r$. This is a large effect (Cohen 1988), and a strong relationship between x and y. Following our simulation conditions for x and y, we do not expect r to differ much between the raw and the phylogenetically transformed versions of x and y (Garland and Ives 2000). Finally, using the tabulated estimates of statistical power (which equal $1-\beta$) reported by Cohen (1988), we find that regression analyses with sample sizes of $K = 5$ will have a Type II error rate of 92% for detecting a correlation of approximately 0.45.

Given this large error rate, interpreting any regression results with such a small sample size is very risky. In our case, it is impossible to determine whether our OLS regression detected the true underlying effect or found a strong positive effect because of sampling error. Examining the magnitude of the estimated slope ($a = 1.87$) from OLS when the true slope equals 0.5 may provide evidence for the latter. Likewise, our PGLS regression could not detect the true effect (although the slope and intercept were very near the predicted values of 1 and 0.5, respectively). This was because the variances of regression coefficients were too large. Again, these large variances are a consequence of small sample size. This is exactly how we want our regression analyses to behave, and why the PGLS model properly estimated the variances of our random data: these variances were broad, as predicted, given our small sample size. We do not want our variances to be biased, as can potentially occur with OLS, as these will increase our chances of making wrong conclusions with our data.

Let's explore the interaction between sample size and phylogenetic correlations in more detail, and compare the Type II error rates of OLS and PGLS by simulating interspecific data. This will allow us to assess the error rates of concluding that the slope and intercept are non-zero. The R script for this simulation is in appendix 11.C. Briefly, we generated a random phylogeny with 100 species, and then randomly subsampled this phylogeny to generate subtrees of size K. We then phylogenetically transformed K random data (following R.5), and analyzed these with both OLS and PGLS. Repeating this 1,000 times for each K, we counted the number of times the p-values of t-scores for the regression coefficients were not significant (i.e., concluding that they were zero). Recall that for our linear model, the intercept and slope *are* non-zero (section 11.2.3). figure 11.6 has the simulation results of the error rates from our two regression models. Generally, increasing the sample size improves the ability to detect non-zero effects. More notably, however, the real benefits of including phylogenetic correlations in GLS models (i.e., PGLS) only emerge at larger sample sizes—given that they have significant improved ability to detect non-zero effects relative to OLS.

This is typically as far as comparative analyses can take us, and the best we can glean from a simple PGLS regression is whether non-zero effects exist given our interspecific data. But with our simulated data, we know the true underlying effects, and so we can extend our Monte Carlo experiments to investigate how close OLS and PGLS were in estimating the correct intercept and slope of our linear model. In our previous simulation, statistical tests (t-scores) assessed whether there is any evidence that $a \neq 0$ and $b \neq 0$. Now we will adjust the null hypotheses of these tests to investigate whether $a \neq 1$ and $b \neq 0.5$, and count the number of cases when t-scores incorrectly rejected our simulated regression coefficients (i.e., $a = 1$ and $b = 0.5$). This type of error is referred to as false positive, or Type I error. The simulation results are in figure 11.7, and the R script in appendix 11.D. Note that the OLS regression has a fairly high probability of incorrectly concluding that the intercept and slope were different from $a = 1$ and $b = 0.5$, and that this probability increases with larger sample sizes. This is clear evidence that the OLS estimator is not optimal for analyzing interspecific data (Martins and Garland 1991; Diaz-Uriarte and Garland 1996; Harvey and Rambaut 1998; Freckleton et al. 2002; Revell 2010). These findings also counter the seemingly amazing ability for OLS to detect a non-zero intercept (see Type II errors in figure 11.5), since OLS analyses will likely estimate significant non-zero yet erroneous

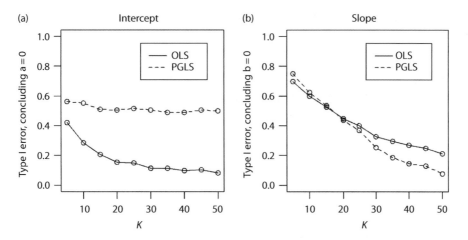

Fig. 11.6 The risk of incorrectly concluding null results with regression analyses of interspecific data. Presented are results from a Monte Carlo experiment exploring the Type I error rates (i.e., false positive outcomes) of OLS and PGLS regression with the number of species (K) varying from few to many. Error rates are based on the proportion of 1,000 regression analyses concluding that the intercept (*a*) and slope (*b*) were zero when data are phylogenetically correlated. PGLS analyses are more likely to correctly conclude that the slope is non-zero with larger K. The R script for this simulation is in appendix 11.D.

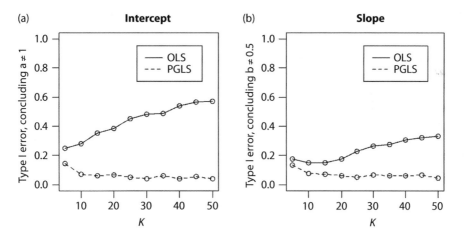

Fig. 11.7 The risk of concluding non-zero but erroneous intercept and slope estimates when using regression to analyze interspecific data. Presented are results from a Monte Carlo experiment exploring the Type I error rates (i.e., false positive outcomes) of OLS and PGLS regression with small to large number of species (K). Error rates are based on the proportion of 1,000 regression analyses concluding that the intercept (*a*) and slope (*b*) did not equal their true simulated values (i.e., $a = 0.5$ and $b = 1$). Data were simulated to have phylogenetic correlations. Here OLS analyses are more likely to incorrectly conclude significant erroneous intercept and slope values irrespective of K. The R script for this simulation is in appendix 11.D.

regression coefficients. There are also clearly issues with how PGLS estimates the intercept (see Type II errors in figure 11.5), but at least it tends to be conservative when estimating the intercept's standard error (i.e., it tends to be large). This favors the null hypothesis (see figures 11.5 and 11.6). More practically, however, evaluating whether the intercept is non-zero is typically not the focus of analyses. Generally the aim is to determine if the slope is non-zero, and PGLS seems optimal for this estimation goal with interspecific data. In fact, one of the original regression approaches to analyzing interspecific data excluded the intercept entirely from analyses (Felsenstein 1985).

11.2.5 *The assumption of homoscedasticity and evolutionary models*

Phylogenetic correlations arise because of the similarities between ancestors and their descendants, and we estimated these correlations using the pairwise phylogenetic distances between species (see table 11.1). Here, we assume that the strength of these correlations predict similarity among related taxa: the stronger the correlation, the greater the similarity of data measured between two taxa. Including these correlations in GLS models is meant to improve the way we model stochastic errors (ε) in linear regression [equation (11.4)]. However, when we apply phylogenetic correlations to GLS, we are also making an important assumption about the stochastic nature of evolution and how this process can shape variation in the characteristics of species (Martins and Hansen 1997).

For example, the way we model the residual error ε actually has an important biological interpretation regarding the variances of traits and how they are predicted to evolve along the branches of a phylogeny. Implicit in the way we quantified our phylogenetic correlations is a time component: we expect that the strength of correlations (and therefore also similarity among traits) will erode linearly with time as taxa evolve independently from a common ancestor. This type of stochastic erosion is called *Brownian motion evolution*—a model of evolutionary change where random genetic drift is the primary process resulting in the loss of similarity from ancestral characteristics (Martins and Garland 1991). As traits follow the paths along each branch of a phylogeny, random drift results in independent shifts of magnitude and direction in these characteristics, and the total change accrued is proportional to time (O'Meara et al. 2006).

Another way to think about Brownian motion (BM) evolution is that it is a hypothesis on the predicted distribution of characteristics among related species. With this in mind, we can interpret how we modeled the stochastic error (ε) in our linear model for interspecific data [equation (11.4)] as the expected variance and covariance that is proportional to shared phylogenetic history (i.e., $\sigma^2 \mathbf{C}$). Here $\mathbf{C}$ (as defined earlier in R.4), but now more precisely $\mathbf{C}^{BM}$ because we know now that it has a Brownian motion structure, quantifies the correlations among species based on the pattern and timing of their phylogenetic history. Further, σ^2 becomes the phylogenetic variance or evolutionary rate for x and y. This rate of change is an important property of BM evolution as it is a process that acts equally (i.e., has the same σ^2 rate) among the traits of evolving taxa. This satisfies the assumption of homogeneity of variances (homoscedasticity) of our GLS model as applied via PGLS.

Brownian motion is by far the most commonly assumed model of phenotypic evolution by comparative phylogenetic methods (see Edwards et al. 1963; Lynch 1991), and is the model explicitly assumed when Felsenstein's (1985) widely applied phylogenetically independent contrasts (PIC) are used to analyze interspecific data. Our PGLS analyses are a generalization of this PIC approach, as both will yield similar conclusions when assumptions of BM are met (Rohlf 2001). Interestingly, we can only recently say this confidently now, as it took nearly 30 years after the introduction of PICs to develop a mathematical proof that PGLS and PICs were equivalent under certain conditions (see Blomberg et al. 2012). Essentially, whenever phylogenetic correlations are defined by **C** (see table 11.1), and are applied to regression analyses, then the evolutionary model is a Brownian motion process.

However, random drift through Brownian motion is a rather a simplistic view of how evolution can shape the covariances among related taxa. Other processes like natural selection, along with random drift, can work together to generate very different phylogenetic correlations (Hansen 1997; O'Meara 2012). Therefore there is always the risk that the phylogenetic correlations derived from Brownian motion will not adequately model the covariances of interspecific data. Let's explore this issue by comparing the performance of PGLS when the evolutionary covariance structure differs from BM evolution. Evolution via an *Ornstein–Uhlenbeck* (OU) process is another stochastic model of evolution that is increasingly being investigated by comparative biologists (Uhlenbeck and Ornstein 1930; Lande 1976; Martins and Hansen 1997). Under the OU model, stabilizing selection acts to keep phenotypes near an optimum by removing extreme values in characters. This process works in conjunction with random genetic drift to erode phylogenetic correlations among the phenotypes of related taxa, and the process of keeping phenotypes at an optimum is what erodes phylogenetic correlations. However, because of this added selection component, phylogenetic correlations are no longer predicted to decay proportionally with time as in BM, but instead decay exponentially (i.e., at a much quicker rate) as species become more distantly related (Hansen 1997).

To visualize how stabilizing selection effects the magnitude of phylogenetic correlations, we can simulate phylogenetic correlations derived from BM and OU processes, and compare how their rates of change differ relative to the same time since divergence. The predicted phylogenetic correlations under this exponential model of evolution are estimated as:

$$\mathbf{C}^{OU} = 1/2\alpha \left\{ e^{-2\alpha \left[\mathrm{diag}(\mathbf{C}^{BM}) - \mathbf{C}^{BM} \right]} - e^{-2\alpha \left[\mathrm{diag}(\mathbf{C}^{BM}) \right]} \right\}, \tag{11.7}$$

where α is the stabilizing selection parameter that can range from zero (no selection) to infinity (very high selection), and where "diag" indicates a vector containing only the main diagonals of $\mathbf{C}^{BM}$. By manipulating the strength of stabilizing selection (α) in $\mathbf{C}^{OU}$, we can visualize the effects of selection eroding phylogenetic correlations by simulating a random tree and plotting the divergence time versus the correlations found in $\mathbf{C}^{BM}$ and $\mathbf{C}^{OU}$. The script for this simulation is in appendix 11.E, and the phylogenetic correlations derived from these two models of phenotypic evolution are shown in figure 11.8. As expected under BM, the phylogenetic correlations are linearly proportional with the time since divergence (figure 11.8); they form a straight line between the time of divergence (i.e., shared phylogenetic branch-length distance) and the correlations. However, under the OU model, increasing intensity of stabilizing selection (i.e., larger values of α), the magnitudes of correlations erode exponentially. Taxa far apart in a phylogeny quickly achieve evolutionary independence relative to those under BM. When α is near 0, the phylogenetic correlations of an OU model ($\mathbf{C}^{OU}$) are equivalent to $\mathbf{C}^{BM}$. However, as selection (α) increases in intensity, $\mathbf{C}^{OU}$ approaches **I**; species become nearly independent (section 11.2.2; also see Lajeunesse 2009). In the latter case, selection is so strong that it quickly erases all phylogenetic correlations among related taxa.

11.2.6 *What happens when the incorrect model of evolution is assumed?*

With the predicted phylogenetic correlations $\mathbf{C}^{OU}$ and $\mathbf{C}^{BM}$ described above, we can simulate interspecific data derived from these two models of phenotypic evolution, and then compare the performance of our PGLS analyses using interspecific data that do not fit the BM model of evolution. Thus, we will assess the Type II error rates of PGLS analyses assuming a variance–covariance structure of ε based on BM evolution. This will provide some insight as how PGLS performs when an incorrect model of evolution is assumed with interspecific data. We simulated interspecific data evolving via an OU model with strong selection ($\alpha = 3$; see figure 11.8) using the R script in appendix 11.E. We then analyzed these data using PGLS assuming BM evolution [equation (11.6)]. The results (figure 11.9), revealed a slight loss of efficiency when estimating the slope with a *PGLS assuming BM* with data simulated under an *OU* model. There is also a lack of efficiency when estimating the intercept. In terms of the slope parameter, this may seem like a trivial amount of error when the incorrect model of evolution is applied to interspecific data. However, these results more likely reflect the relative similarity between BM and OU models, rather an apparent robustness of BM when analyzing data from a different evolutionary model. For example, even though the OU data were modeled with strong stabilizing selection ($\alpha = 3$), these two models still preserve groups of correlations based on the topology of the tree; such as within the two major groups $a-b$ and $c-d-e$ described earlier (although at smaller magnitudes; figure 11.4). Assuming BM under our simulation conditions provides (albeit

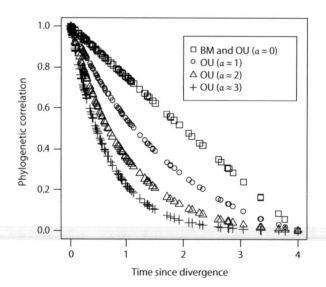

Fig. 11.8 The change in magnitude among phylogenetic correlations when they are based on different models of evolution. Brownian motion (BM) assumes a linear decay of correlations with time; whereas the Ornstein–Uhlenbeck (OU) model assumes an increasingly exponential decay with rising intensities of stabilizing selection, $\alpha = \{1, 2, 3\}$. Note that when stabilizing selection is near zero, the OU model converges to a BM model. The R script of this simulation is in appendix 11.E.

very coarsely) some useful correlational structure to assist the linear regression with *OU* data. Nonetheless, the potential risk of incorrectly concluding a null result still exists, and fitting the appropriate model of evolution to your data will help minimize this risk—even if in our case with simulated data the improvement was only about 5%.

11.3 Establishing confidence with the comparative phylogenetic method

Using Monte Carlo experiments, we were able to investigate the challenges of analyzing interspecific data and ask how applications of the comparative phylogenetic method can improve inferences with these data. An advantage of our simulation approach is that we knew the underlying relationships in our simulated data. With real (observed) interspecific data, little to no information will be known with any certainty about such underlying processes. This suggests that one should make a great deal of effort to approach interspecific data with a robust statistical framework, and to present results in a way that provides confidence that the observed relationships are biologically meaningful and not statistical artifacts. Below, we sketch a few guidelines on how to approach and present your comparative analyses to achieve these goals (for more extensive guidelines see also Garland et al. 2005; Freckleton 2009).

Nearly all comparative phylogenetic methods assume that the phylogenies used to estimate phylogenetic correlations are known without error (Rohlf 2001). However, phylogenies are only statistical hypotheses on the evolutionary history of taxa. They vary

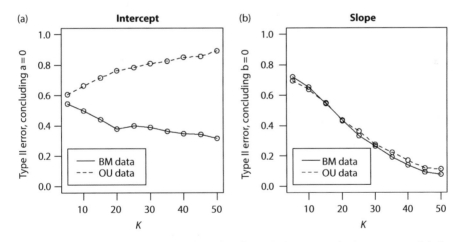

Fig. 11.9 The risk of concluding null results when PGLS assumes the incorrect model of evolution. Presented are results from a Monte Carlo experiment exploring the Type II error rates (i.e., false negative outcomes) of PGLS with the number of species (*K*) varying from few to many. Error rates are based on the proportion of 1,000 regression analyses incorrectly concluding that the intercept (*a*) and slope (*b*) were zero. Data were simulated to have phylogenetic correlations derived from an Ornstein–Uhlenbeck (OU) model of evolution with a stabilizing selection parameter set to $\alpha = 3$; these data were analyzed with a PGLS assuming a model of Brownian motion (BM) evolution. The *R* script for this simulation is in appendix 11.F.

tremendously in availability and uncertainty for distinguishing both deep and recent divergences, as well as their relative timing (Felsenstein 2004). Our phylogenetic correlation matrix (**C**) is at best a hypothesis of the expected true correlations (i.e., variance–covariances) that may or may not exist among the traits of related taxa. It is therefore always important to ask how uncertainty in **C** can influence the performance and statistical outcomes of PGLS analyses. Incorrectly specifying **C**, either by using an incorrect tree topology or an incorrect model of evolution, can result in PGLS models performing more poorly than OLS models (Mittelhammer et al. 2000). Deep topological errors near the root of the tree are also expected to have more strongly negative effects on the performance of PGLS, compared with errors in the positioning of nodes near the tips of phylogenetic trees (Martins and Housworth 2002). Therefore, if multiple (alternative) hypotheses of divergences are available for a collection of taxa, it is good practice to incorporate each of these phylogenetic hypotheses in PGLS analyses (see Donoghue and Ackerly 1996). Doing so allows one to compare regression results based on alternative phylogenetic hypotheses if they are biologically meaningful. Alternatively, multiple separate regression analyses can be averaged to provide an aggregate view based on different phylogenetic hypotheses. Model selection criteria (e.g., AIC scores) may also be useful in assessing the relative fit of competing phylogenetic hypotheses (Lajeunesse et al. 2013).

A more common challenge with phylogenetic trees is a lack of information needed to connect the divergences among taxa. Complete phylogenetic information (e.g., a phylogenetic tree that is completely bifurcated) can help minimize the Type I error rates of comparative analyses (see Purvis and Garland 1993). Several solutions to this problem of missing topologies within trees are available. For example, a sophisticated approach applies birth–death models to simulate random divergences among taxa with missing phylogenetic information (Kuhn et al. 2011). This imputation approach (see chapter 4) is not too different from the way we simulated our random phylogenic tree (section 11.2.3). The aim of these imputations are to fill gaps of information about the topology (and therefore correlations in **C**) by randomly resolving polytomies (nodes that specify unresolved divergences among lineages or taxa (Maddison 1989). Models of evolution can also be assumed to make the internode branch-length distances (i.e., simulated divergence times) less arbitrary (Kuhn et al. 2011). Analyses are then repeated several times with these randomly resolved topologies to minimize the risk of the method itself introducing bias to PGLS results. Alternatively, a coarse hypothesis on phylogenetic history, such as estimating **C** with a tree based on Linnaean rankings (e.g., grouped by class or order), can also help improve the performance of PGLS models, as long as the overall topology is correct and matches the true major divergence events (e.g., Freckleton et al. 2002). The disadvantage of this coarse approach is a lack of information about relative divergence times; these are useful for making predictions regarding the evolutionary basis for phenotypic change and their predicted phylogenetic correlations (i.e., **c**). Several online resources are also available that can help supplement phylogenetic information to generate fully bifurcated trees. For example, the widely used *phylomatic* by Webb and Donoghue (2005) is an important tool for generating phylogenetic trees for PGLS. The massive tree of life project called *timetree* by Hedges et al. (2006) is very helpful for determining the divergence time between distantly related taxonomic groups.

Another problem with PGLS analyses is the risk of over-fitting phylogenetic correlations to data that are weakly or not phylogenetically correlated. One common solution to this problem is to apply a transformation parameter such as Pagel's λ to the phylogenetic correlations in **C** (Pagel 1999). The idea is to make the correlation structure of analyses flexible relative to the observed phylogenetic signal (λ) of the interspecific data. A phylogenetic signal is a measure that quantifies the overall statistical dependence among species traits, relative to their phylogenetic relationships. Applying this transformation is meant to relax the assumption of Brownian motion as the primary evolutionary model for phenotypic change, and therefore help minimize the potential of over-specifying C for interspecific data that are not actually phylogenetically correlated (Garland et al. 2005). For example, λ can first be estimated via maximum likelihood with a PGLS model, and then λ multiplies all the off-diagonals of C (i.e., all the correlations). If λ is estimated to be near zero, then λC approaches I. Therefore, when no phylogenetic signal is detected, the PGLS will converge to OLS, which is more efficient at estimating regression coefficients if the data are independent (Revell 2010). Likewise, other evolutionary models can be used to adjust phylogenetic correlations following different models of evolution (e.g., the selection parameter of the OU model; Hansen 1997). It is also important to note that the accuracy of estimating evolutionary parameters like Pagel's λ is largely dependent on the number of species included in the analysis. Generally, phylogenies with fewer than 30 species will provide unreliable estimates of λ (Revell 2010).

11.4 Conclusions

By focusing solely on simple linear regression and Monte Carlo experiments, we hope that this chapter provides some clarity to why it is important to apply this statistical framework to interspecific data. However, it is important to note that the same statistical issues and interpretive problems outlined here are equally relevant to *any* analyses using phylogenetic correlations to model dependencies in interspecific data. These include more elaborate phylogenetic analogues such as GLS modeling to perform ANOVA or ANCOVA, principle component analysis (Revell 2009), and meta-analysis (Lajeunesse et al. 2013). Finally, we urge readers interested in applying these methods to think beyond treating phylogenetic correlations as a nuisance to be controlled in analyses. Phylogenetic dependence in our data is not just another pitfall to avoid, like pseudoreplication. Much more can be gleaned from these analyses if one adopts an evolutionary framework and compares multiple evolutionary models (e.g., BM vs. OU) with an aim of providing insight into how and why phenotypic and ecological data are phylogenetically correlated (Butler and King 2004). Dobzhansky (1973) famously commented that "Nothing in biology makes sense except in the light of evolution;" this also applies to ecological problems!

Acknowledgments

Support for this work was funded by the College of Arts and Sciences, University of South Florida.

CHAPTER 12

Mixture models for overdispersed data

Jonathan R. Rhodes

12.1 Introduction to mixture models for overdispersed data

Much of ecological statistics, including most of the methods in this book, rely on parametric statistics. Parametric statistics make specific assumptions about the nature of the probability distributions that our data arise from, in contrast to non-parametric (distribution-free) statistics that make far fewer such assumptions. These assumptions can be an advantage because they allow us to make clearly defined and transparent assumptions about the processes generating our data (Royle and Dorazio 2008). These assumptions, in turn, allow us to test explicit hypotheses about the ecological processes that led to observed data. However, a disadvantage is that ecological data can fail to meet the assumptions of the standard probability distributions (e.g., Normal, Poisson, and binomial distributions) used in parametric statistics. A particularly common problem in this context is a phenomenon known as overdispersion that arises when data are more variable than can be accommodated by the parametric distribution being used to describe them (McCullagh and Nelder 1989). This chapter is about how to deal with overdispersion when using parametric statistics for ecological inference. In particular, I show how a class of models known as mixture models (Mengersen et al. 2011) can be used to help ensure that our statistical tests are valid when overdispersion is present and to better understand the drivers of overdispersion for improved ecological inference.

To illustrate the idea of overdispersion, imagine you go out to a number of randomly selected sites and count the number of individuals of a species at each site. What would these data look like if the distribution of the species was highly spatially aggregated, occurring at high densities at locations where habitat is suitable, but absent from other areas where habitat is unsuitable? If this were the case, we would expect the data to consist predominantly of high values in sites where habitat is suitable and zero values elsewhere. That is, we would tend to observe data at the two extremes of the distribution, with values in between being much less common. A consequence of the data lying at the two extremes is that their variance will be higher than the theoretical variance for the standard parametric distribution used to model count data, which is the Poisson distribution. This happens because the Poisson distribution assumes that the data lie predominantly around the center of the distribution, rather than the extremes. In this example, it is the spatial aggregation process that leads to overdispersion, but overdispersion in ecological

Ecological Statistics: Contemporary Theory and Application. First Edition. Edited by Gordon A. Fox, Simoneta Negrete-Yankelevich, and Vinicio J. Sosa. © Oxford University Press 2015. Published in 2015 by Oxford University Press.

data can be caused by a range of ecological, observation, and modeling processes (Haining et al. 2009; Lindén and Mäntyniemi 2011).

A major issue with overdispersion is that it generates bias in statistical tests. Overdispersion means that the true variances of the data are larger than the theoretical variances assumed by parametric distributions. This leads to incorrect model likelihoods and, because variances are underestimated, we will tend to incorrectly reject the null hypothesis (i.e., make Type I errors) more often than we should (also see chapters 2 and 3). Consequently, we need tools that allow us to account for overdispersion in our statistical models so that bias in our statistical tests is reduced or eliminated. Anderson et al. (1994) illustrate one way to do this using quasi-likelihood methods (see chapter 6) to adjust Akaike's Information Criterion (AIC) values to correct for overdispersion. However, this approach tells us little about the nature of the overdispersion itself—it just accounts for it—but the nature of overdispersion can also provide important information about underlying ecological processes and/or observation processes (Martin et al. 2005b). For example, Rhodes et al. (2008b) model the effects of marine pollution on fecundity in the copepod *Tigriopus japonicus* and use a mixture model that explicitly models the processes driving an excess of zeroes (and therefore overdispersion) in their data. By being explicit about the processes driving overdispersion they were able to make inferences about the effect of pollution on two different processes: the number of individuals that entirely fail to breed (causing an excess of zeroes), and the number of young per successful breeder. In this case, overdispersion is not just a nuisance that we want to control for, but reflects a key ecological process of interest. Consequently, we often want methods to explicitly model the overdispersion process when these processes themselves are of direct interest.

Mixture models are a particular class of statistical model that allow us to control both for overdispersion in our statistical tests and to model explicitly the processes that drive overdispersion, resulting in improved ecological inference. These models allow for greater variability than standard distributions by allowing the parameters (and sometimes the structure) of standard statistical models to vary randomly, rather than being fixed. To illustrate the idea of a mixture model, let us go back to our hypothetical example of the spatially aggregated count data that we looked at above. The standard way to model this type of data would be using a Poisson distribution that has a single parameter, λ, representing the mean. However, since our count data have a high frequency of zeroes and a high frequency of high values, the true variance of the data will be greater than the theoretical variance of the Poisson distribution. To deal with this, we could use a mixture model that assumes that the λ parameter can vary randomly and take one of two values: either zero, or a fixed value greater than zero, with either case occurring with a given probability. This model explicitly accounts for the high frequencies of zeroes and high frequencies of high values by allowing the mean to be either zero or a fixed value greater than zero respectively. This mixture model is known as a zero-inflated Poisson distribution (Lambert 1992). Importantly, by allowing λ to vary randomly we characterize sites as suitable ($\lambda > 0$) and unsuitable ($\lambda = 0$), and therefore the process that leads to the overdispersion is explicitly characterized. A nice property of a mixture of this type, therefore, is that not only do we control for overdispersion, but we can make inferences about the processes that lead to that overdispersion, such as estimating the proportion of sites where habitat is suitable versus unsuitable.

Although I have used count data and the problem of estimating the distribution and abundance of a species to illustrate the idea of a mixture model, we can apply mixture models to other classes of problems. In fact, mixture models represent a highly flexible approach for dealing with overdispersion across a very wide range of classes of statistical problems (Mengersen et al. 2011). In ecology, mixture models have been successfully applied to deal with overdispersion and heterogeneity in a range of applications, including: modeling species' distributions and abundance (Tyre et al. 2003; Royle 2004; Wenger and Freeman 2008), survival analysis (Pledger and Schwarz 2002), population dynamics (Kendall and Wittmann 2010); disease ecology and parasitology (Calabrese et al. 2011); community ecology (Colwell et al. 2004); and dispersal ecology (Clark et al. 1999). However, there are three common problems in ecology where mixture models are particularly useful: (1) accounting for an excess of zeroes in data (arising either due to ecological or observation processes); (2) accounting for heterogeneity among sampling units (e.g., individuals or social groups); and (3) making explicit inferences about two or more ecological or observation processes that jointly give rise to overdispersed data (e.g., short- and long-distance dispersal processes that both contribute to the distribution of dispersal distances).

In this chapter I present mixture models as a powerful and flexible way to deal with overdispersion in ecological data and discuss how this approach can be used to account for overdispersion and facilitate improved inference by understanding the overdispersion process itself. Although mixture models are not new and some mixture models (e.g., the negative binomial distribution) are commonly used in ecology, the routine consideration of mixture models as a flexible approach for modeling complex ecological data is rare. The discipline can therefore benefit greatly from a better-informed use of mixture models that will lead to improved ecological inference. Further, faster computers and new computational methods now make it possible for most ecologists to routinely fit complex statistical models to ecological data. Ecologists are therefore in an ideal position to extend their toolkit to the more general use of mixture models.

The chapter is divided into four sections. In the first section I define overdispersion and provide guidance on how it can be detected. In the second section I discuss mixture models in more detail and highlight the main types of mixture models used in ecology. I then present two empirical examples. The first example is a survival analysis problem where I use a mixture model to deal with heterogeneity among groups. In the second example I present a problem where the aim is to estimate species' abundance from count data that are zero-inflated. In this example, I show how mixture models can be used to model both ecological and observational sources of the zero inflation. I end with a discussion of the benefits and challenges of using mixture models for ecological inference, especially in comparison to alternative approaches, and highlight key things to consider when using mixture models.

12.2 Overdispersion

12.2.1 *What is overdispersion and what causes it?*

Data are defined as overdispersed if the variance of the data is greater than the theoretical variance of the probability distribution being used to describe the data generation process (Hinde and Demetrio 1998). In other words, overdispersion is always relative to a specified probability distribution. Overdispersion is often most apparent in count and presence/absence data because the variance of a standard Poisson or binomial distribution is a function of the mean, rather than estimated independently from the data. In the case of the Poisson distribution, the variance is equal to the mean (i.e., $\sigma^2 = \lambda$) and, in the case of the binomial distribution, the variance is equal to the number of trials multiplied by the success probability multiplied by the failure probability (i.e., $\sigma^2 = np[1-p]$). If the data fail to conform to these characteristics of the variance, then the true variance of the data can be higher than the theoretical variance and therefore overdispersed. Such overdispersion in ecological data can arise from ecological processes, observation processes and/or mis-specification of the mean (table 12.1; Haining et al. 2009; Lindén and Mäntyniemi 2011). We will look at these sources of overdispersion next.

Two important ways in which ecological mechanisms can lead to overdispersion are: (1) causing spatial/temporal clustering or aggregation (see also chapter 10) and/or (2) introducing heterogeneity among sampling units. As I highlighted in section 12.1, spatial

Table 12.1 Main causes of overdispersion and, for each cause, an ecological example illustrated by abundance data (counts), together with typically what a histogram of the overdispersed data would look like relative to a Poisson distribution fitted to the data and what a quantile–quantile plot of the overdispersed data would look like relative to a Poisson distribution fitted to the data (the Poisson distribution is represented by the solid line in both the histogram and the quantile-quantile plots). See Haining et al. (2009) and Lindén and Mäntyniemi (2011) for further discussion of the causes of overdispersion.

Cause of overdispersion	Ecological example	Histograms of overdispersed count data relative to the Poisson	Quantile–quantile plots of overdispersed count data relative to the Poisson
Spatial/temporal clustering	A species only occurs in a small part of the landscape sampled		
Heterogeneity among sampling units	The expected abundance of a species varies randomly across a landscape based on variation in habitat quality		
Measurement error	Individuals that are truly present sometimes fail to be detected		
Misspecification of the mean	The relationship between abundance and habitat quality is actually non-linear but we model it as linear, resulting in underestimation of the mean for some habitat qualities		

clustering tends to generate data with too many high counts (from locations where the species is present) and/or too many zero counts (from locations where the species is absent), resulting in a variance that is greater than the theoretical variance of the Poisson distribution. Cunningham and Lindenmayer (2005), for example, show that, in the Central Highlands of Victoria, Australia, counts of the threatened Leadbeater's possum (*Gymnobelideus leadbeateri*) are overdispersed primarily due to an excess of zero values. This is driven by the species' distribution being highly spatially clustered in only a few areas of the landscape where its habitat occurs. In a similar way, temporal clustering can also lead to overdispersion. For example, disturbance events that impact on ecosystems such as cyclones (also known as hurricanes or typhoons, for those not fortunate enough to live in Australia) can be highly temporally clustered and therefore overdispersed (Mumby et al. 2011). Heterogeneity among sampling units (e.g., genetic or phenotypic variation among individuals) can also result in an excess of high and/or low values and therefore overdispersion relative to the Poisson or binomial distributions. A typical example of this is where breeding success varies among individuals, leading to highly variable reproductive output and overdispersion in data on reproduction (Quintero et al. 2007; Kendall and Wittmann 2010).

Observation processes commonly result in data inaccuracies that can also lead to overdispersion by increasing variability in the data. For example, presence/absence data collected where there is imperfect detection (which is almost always the case) can lead to an excess of zeroes and overdispersion relative to the binomial distribution (Tyre et al. 2003). However, although zero-inflation caused by detection error can appear similar to zero-inflation caused by ecological processes, the inferences we make from the data are normally quite different. This is because, in the presence of detection error, we are usually interested in making ecological inferences after stripping out the process (detection error) causing overdispersion. On the other hand, when ecological processes are the cause of zero-inflation, we are commonly interested in making inferences about overdispersion as a component of the ecological processes of interest.

The final way overdispersion can arise is when the mean is misspecified. Most statistical models specify the mean of the appropriate distribution as a function of covariates (e.g., in linear regression, the mean of the Normal distribution is specified as a function of covariates). In the case of the Poisson and binomial distributions, if the function that links the mean to the covariates is misspecified in a way that results in the mean being underestimated (e.g., due to important covariates, or non-linear terms being missed), then the variance will also be underestimated. This leads to overdispersion because the estimated variance is lower than the true variance of the data. Missing covariates are likely to be common in ecology, particularly in applications such as modeling the distribution of species, where the factors driving distributions are often not well understood, or, even when they are understood, often cannot be directly measured (Barry and Elith 2006).

12.2.2 *Detecting overdispersion*

Prior to and during the development of statistical models for data that may be overdispersed, it is important to be able to identify whether the data are in fact overdispersed or not. There are three primary ways in which we can detect overdispersion: (1) inspect histograms of the raw data; (2) inspect quantile–quantile plots of the residuals of the model; and/or (3) conduct formal hypothesis tests, or model selection. Often, simply inspecting a histogram of the raw data and comparing this against expected frequencies based on the relevant standard distribution can reveal important information on whether data are overdispersed or not. Table 12.1 illustrates what histograms look like relative to the Poisson distribution for different causes of overdispersion in count data, but we can construct similar plots for any distribution. However, overdispersion in the raw data can sometimes be accounted for by the relationship with a covariate included in the model, rendering residuals that are not overdispersed. Therefore, a preferred, and more sophisticated, approach is to inspect a quantile–quantile plot of the residuals; this is a standard method for visually comparing the distribution of data versus the expected distribution. Quantile–quantile plots show the actual ordered residuals from the model against the expected ordered residuals of the model; a plot lying close to the 1:1 line represents good agreement between the distribution of the data and the expected distribution. Quantile–quantile plots of overdispersed data will tend to lie below the 1:1 line at the lower end of the distribution and/or lie above the 1:1 line at the higher end of the distribution. This reflects the tendency for overdispersed data to contain more extreme values than expected, but the exact pattern will depend upon the nature of the overdispersion present. Table 12.1 illustrates what quantile–quantile plots look like relative to the Poisson distribution for different causes of overdispersion. Landwehr et al. (1984) develop a useful simulation approach for constructing quantile–quantile plots for logistic regression, but the approach is flexible enough to be applied to any model. The inspection of histograms of the raw data and quantile–quantile plots of the residuals represent qualitative approaches for detecting overdispersion. One advantage of these approaches is that they allow a visual representation of the distribution of the data relative to the expected distribution that can help in pinpointing how overdispersion arises in the data. However, it is also possible to take a more formal approach and explicitly test for overdispersion by conducting hypothesis (score) tests or to use multi-model selection methods (see also chapter 3; Dean 1992; Richards 2008). I will expand on and illustrate these approaches in the empirical examples later in the chapter.

12.3 Mixture models

In the previous section I discussed the nature of overdispersion and described how to identify whether your data are overdispersed or not. Now we are going to turn our attention to mixture models as a way of dealing with overdispersion in our statistical models. In this section I define what a mixture model is, identify some typical mixture models used in ecology and briefly mention the different ways in which mixture models can be fit to data. A complete technical treatment of mixture models is not possible within a single book chapter, but McLachlan and Peel (2000), Johnson et al. (2005), and Mengersen et al. (2011) provide more comprehensive and technical treatments of mixture models.

12.3.1 *What is a mixture model?*

To illustrate the idea of a mixture model, let us go back to our hypothetical example of the surveys of the highly spatially aggregated species that I described in section 12.1. There I discussed the idea that we could formulate a mixture model in a way that allowed the mean to vary randomly between having a value of zero and having a value greater than zero. That is to say, some data points will come from a distribution with a mean of zero and some will come from a distribution with a mean greater than zero. But let us now look at this more formally. First, assume that, regardless of whether the mean is zero or greater than zero, the data are Poisson distributed (see book appendix for the definition of the probability density function for the Poisson distribution). Then, when the mean, λ, is greater than zero, the probability density function is

$$f(y|\lambda > 0) = \frac{e^{-\lambda}\lambda^y}{y!}, \qquad (12.1)$$

and, by substituting zero for the mean, λ, when the mean equals zero, the probability density function is

$$f(y|\lambda = 0) = \frac{e^{-0}0^y}{y!}$$

$$= \begin{cases} 1 \text{ when } y = 0 \\ 0 \text{ when } y > 0. \end{cases} \qquad (12.2)$$

If we let p be the probability that the mean is zero, so the probability that the mean is greater than zero is $1 - p$, then the probability density function is

$$g(y|\lambda) = pf(y|\lambda = 0) + (1 - p)f(y|\lambda > 0)$$

$$= p\frac{e^{-0}0^y}{y!} + (1 - p)\frac{e^{-\lambda}\lambda^y}{y!}$$

$$= \begin{cases} p + (1 - p)e^{-\lambda} \text{ when } y = 0 \\ (1 - p)\dfrac{e^{-\lambda}\lambda^y}{y!} \text{ when } y > 0. \end{cases} \qquad (12.3)$$

This is the probability density function for the zero-inflated Poisson distribution (Lambert 1992); a mixture model that has a mean equal to $(1-p)\lambda$ and variance equal to $(1-p)(\lambda + p\lambda^2)$. The variance of the zero-inflated Poisson distribution is always greater than the mean (in contrast to the Poisson), since $(1-p)(\lambda + p\lambda^2) > (1-p)\lambda$ when $p > 0$. Therefore, if we were to use a Poisson distribution to model these data we would underestimate the variance. Using, instead, a zero-inflated Poisson distribution corrects this problem and allows for overdispersion to be accommodated in our model (in this case, in the form of zero-inflation).

Equation 12.3 shows that the mixture model is essentially a weighted sum of two probability density functions (the mixture components), with weights p and $1-p$ (the mixture weights). This is what is known as a *finite mixture distribution* because it is a finite sum of distributions. We can generalize this idea to a K-component finite mixture model, $g(y|\Theta)$, which is any convex combination of K probability density functions such that

$$g(y|\Theta) = \sum_{i=1}^{K} \omega_i f_i(y|\theta_i) \text{ subject to } \sum_{i=1}^{K} \omega_i = 1, \qquad (12.4)$$

where $f_i(y|\theta_i)$ is a probability density function, with parameters θ_i, representing mixture component i, ω_i is the mixture weight for component i, and $\Theta = (\omega_1, \ldots, \omega_k, \theta_1, \ldots, \theta_k)$. In ecology, mixtures of more than two distributions may be appropriate when we want to capture more than two processes generating the data. For example, Kendall and Wittmann (2010) use a finite mixture model with more than two components to model multiple processes that drive reproductive output in birds, mammals, and reptiles. Many of the processes they consider (including nest building success, number of eggs laid or births, chance of nest destruction, and offspring survival) can result in overdispersion in reproduction data, but they explicitly account for them using a finite mixture model with more than two components. In general, finite mixture models provide a highly flexible approach for modeling non-standard distributions and are well suited to account for many kinds of overdispersion where model parameters can take a finite number of discrete values (Mengersen et al. 2011).

Another kind of mixture model arises when one or more of the parameters of a probability distribution varies randomly and can take an infinite number of values. In this case, the number of mixture components (the number of different possible probability distributions) is not discrete anymore and becomes infinite, reflecting the infinite number of possible values for the parameter(s). In the example above we assumed that the mean, λ, could take one of two discrete values: (1) a value of zero; or (2) a fixed value greater than zero. But, what if the mean can actually take any random value between zero and infinity? In this case, rather than just being able to take two discrete values, λ could be an infinite number of values, with the appropriate model changing from a finite to an *infinite mixture*. This could arise, for example, if mean abundance or density varies randomly across a landscape due to some ecological processes, such as variation in habitat quality. This would likely result in a pattern different from zero-inflation, but still cause overdispersion in the data.

If the parameter with random variation (such as the mean) varies according to a discrete distribution (e.g., Poisson), the resulting mixture is known as a *countable mixture*, but if it varies randomly according to a continuous distribution (e.g., Normal or gamma), the resulting mixture is known as a *continuous mixture*. An example of a continuous mixture is modeling the spatial distribution of a species' abundance, in the presence of continuous random variation in the *mean* abundance of the species across a landscape. Here, random variation in mean abundance would best be described by a continuous distribution, so a continuous mixture would be appropriate. An example of a countable mixture is modeling the proportion of individuals of a species calling across a landscape, but where the number of individuals present at sites varies randomly. Here, variation in the number of individuals must be described by a discrete distribution (since the number of individuals must be an integer) and so a countable mixture would be used.

To more formally define these types of mixture models, consider a distribution that has only one parameter, θ, and that this parameter varies randomly according to some probability density function, $h(\theta|\psi)$, where ψ is a vector of the parameters. If the distribution of θ is discrete, then a countable mixture results and the probability density function is

$$g(y|\psi) = \sum_i f(y|\theta_i)h(\theta_i|\psi), \tag{12.5}$$

where θ_i are the discrete values of θ and the sum is over all possible values of θ_i. This can be thought of as analogous to a finite mixture model, except the mixture weights, ω_i, are replaced by $h(\theta_i|\psi)$, and there are an infinite number of possible mixture components, $f(y|\theta_i)$. If, on the other hand, the distribution of θ is continuous, then a continuous mixture results and the probability density function is

$$g(y|\psi) = \int f(y|\theta)h(\theta|\psi)d\theta, \tag{12.6}$$

where the integration is over all possible values of θ. This is similar to a countable mixture, but we integrate over continuous values of θ rather than summing over discrete values of θ. Countable and continuous mixture models are important because they provide an explicit and flexible framework for modeling heterogeneity across sampling units (Johnson et al. 2005). In the empirical examples later in the chapter I will illustrate the use of finite and infinite (countable and continuous) mixture models in an ecological context.

12.3.2 Mixture models used in ecology

Most standard distributions used in ecology have corresponding overdispersed versions based on mixture distributions (table 12.2). The Poisson distribution, which is typically used to model count data, has an overdispersed version known as the negative binomial. This distribution is a Poisson–gamma continuous mixture that explicitly models heterogeneity (e.g., variation among individuals) in the Poisson rate parameter, through a gamma distribution. The zero-inflated Poisson distribution that I discussed above is another overdispersed version of the Poisson distribution that is increasingly being used to model excess zeroes and observation error in ecological data (Lambert 1992; Martin et al. 2005b). The binomial distribution, which is typically used to model presence/absence data, has an overdispersed version known as the beta-binomial that accounts for heterogeneity in the binomial probability among sampling units. This distribution is a continuous mixture that models heterogeneity in the binomial probability through a beta distribution. Similarly to the Poisson distribution, the binomial distribution also has a zero-inflated version; the zero-inflated binomial distribution (Hall 2000; Martin et al. 2005b). The beta-binomial distribution can also be generalized to an overdispersed multinomial distribution; the Dirichlet–multinomial distribution.

One of the most commonly used of these mixture models in ecology is the negative binomial distribution, which tends to be the default strategy for dealing with overdispersion in count data (Lindén and Mäntyniemi 2011). However, the negative binomial distribution does not deal with overdispersion that arises due to zero-inflation that is common in ecological data (Martin et al. 2005b). This is because, although the negative binomial models variation in the mean of the Poisson distribution, it does not model any process that specifically leads to zero values. Zero-inflated models are more appropriate in this case, but the fact that zero-inflation can arise through either ecological and/or observation processes complicates the choice of mixture model and inference. Recognition of the particular problem of observation error has led to a rapidly growing area of statistical ecology that uses mixture models to explicitly account for observation error in both count and presence/absence data (Mackenzie et al. 2006; Royle and Dorazio 2008).

Table 12.2 Commonly used probability distributions and some overdispersed mixture distribution equivalents. See book appendix for formal definitions of the likelihood functions for some of these distributions.

Distribution	Equivalent overdispersed mixture distributions
Normal	**Student's *t*-distribution** - *Normal distribution with variance following an inverse gamma distribution (continuous mixture)*
Poisson	**Negative binomial distribution** - *Poisson distribution with the rate parameter following a gamma distribution (continuous mixture)*
	Zero-inflated Poisson distribution - *Poisson distribution with a Bernoulli distribution determining value of rate parameter of either zero or greater than zero (finite mixture)*
Binomial	**Beta-binomial distribution** - *binomial distribution with binomial probability parameter following a beta distribution (continuous mixture)*
	Zero-inflated binomial distribution - *binomial distribution with a Bernoulli distribution determining value of binomial probability parameter of either zero or greater than zero (finite mixture)*
Multinomial	**Dirichlet-multinomial distribution** - *multinomial distribution with multinomial probability parameters following a Dirichlet distribution (continuous mixture)*

12.4 Empirical examples

In this section I present two empirical examples to illustrate the use of mixture models to account for overdispersion that arises from both ecological processes and observation processes. Throughout the examples I illustrate the process of detecting overdispersion and choosing appropriate models. I emphasize how mixture models allow us to reduce bias in our statistical tests, but also, importantly, that they allow us to make inferences that would otherwise not be possible. In the first example, I apply mixture models to data from koala dung decay trials to illustrate how these models can be used to disentangle the role of two distinct decay processes that lead to overdispersion in the data. In the second example, I use data on lemur counts from eastern Madagascar for two species, to illustrate how mixture models can be used to distinguish between, and control for, overdispersion that arises from ecological and observation processes in modeling abundance.

There are a number of libraries available in R for fitting mixture models, including: Flexmix; mixtools; and mclust (mixtures of Normal distributions only). However, in the empirical examples I present here I use R to construct likelihoods for the appropriate mixture models and fit these models to data by maximizing the likelihood using the libraries bbmle. One advantage of writing your own likelihood functions is that it allows for greater flexibility, but here it also illustrates how we can fit mixture models by maximum likelihood in a way that expands on the concepts developed in chapter 3. However, for many applications, existing mixture model libraries available in R may be perfectly sufficient.

12.4.1 *Using binomial mixtures to model dung decay*

For many species that are highly cryptic, the only realistic way we can determine their presence/absence is indirectly through signs that they leave. However, some signs, such as dung and snow tracks, decay and disappear over time and the rates at which they decay can vary substantially both spatially and temporally. This causes problems when using these types of indirect signs to estimate presence or absence for two main reasons. First, if signs decay very slowly we may detect old signs in locations where the species is no longer present and we will make false-positive errors. On the other hand, if signs decay very quickly, we may fail to detect signs in areas where the species is still present, and we will make false-negative errors. If there are high levels of spatial and temporal variation in decay rates, this will likely introduce bias into our estimates of presence/absence, and thus will bias our estimates of how presence/absence depends on habitat variables. One way to try to deal with this is to model decay rates of the signs and then adjust for those decay rates in our estimates of presence/absence.

Koala dung is a primary means by which koala distributions are estimated since our ability to detect koalas using direct observations is poor (McAlpine et al. 2008; Rhodes et al. 2008a). This is because the species is highly cryptic and occurs at low densities over much of its range. Nonetheless, dung decay rates can vary both spatially and temporally, thus introducing biases into estimates of presence and absence, so there is a need to understand how spatially and temporally variable dung decay rates are. Here, I illustrate how we can use overdispersed binomial mixture models to analyze data from koala dung decay trials that were designed to estimate decay rates under a range of habitat and climatic conditions (Rhodes et al. 2011).

The decay trial data were collected in Coffs Harbour, New South Wales, Australia between April 1996 and March 1997 (online appendix 12A). The trials were conducted at five sites, with three plots nested within each site, and at each site, plots were located in different topographic positions: one on a ridge; one on a mid-slope; and one in a gully. Each month, a group of 10 fresh koala dung pellets were laid out in each plot and pellets that had disappeared were counted at approximately fortnightly intervals. Data were also available on daily rainfall, daily average humidity, and daily average temperatures for the study area. The data therefore consisted of the number of pellets that had disappeared (decayed) in each recording interval and a series of possible covariates for predicting the probability of dung decay, namely: site, topography, pellet age, rainfall, humidity, and temperature.

The raw decay data are overdispersed, with many more low values and more high values than would be expected under the standard binomial distribution (figure 12.1). This is confirmed by looking at the relative variances of the raw data versus the expected values based on the binomial distribution (0.064 for the raw data vs. 0.044 for the binomial distribution). In particular, there are many more zeroes in the data (data points where no pellets disappeared in a time interval) than we would expect under the binomial distribution.

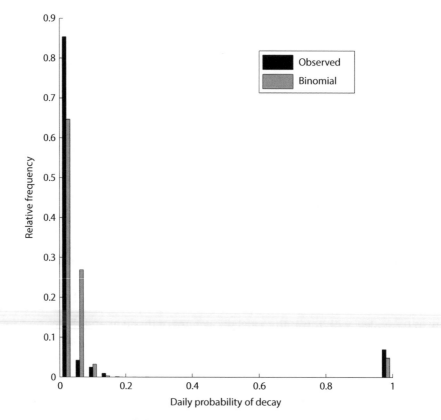

Fig. 12.1 Histogram of observed daily probabilities of a koala pellet decay (black bars) versus expected values based on a binomial distribution with the same mean as the observed data (gray bars). Daily probabilities of decay relating to each recording interval were calculated as $1 - (s/n)^{1/t}$, where $s =$ the number of pellets that survived the interval (observed or expected), $n =$ number of pellets at the start of the interval, and $t =$ number of days in the interval.

There is also a slightly greater frequency of high values (where all pellets disappeared in a time interval), than would be expected under the binomial distribution. However, this difference is small relative to the zero-inflation.

One way to model the overdispersion would be to use the overdispersed version of the binomial distribution, the beta-binomial, which is a continuous mixture model that assumes the binomial probability varies randomly according to the beta distribution (*sensu* Rhodes et al. 2011). This would account for extra-binomial variation that may have arisen due to inherent variation in decay rates among groups, possibly due to random variation in environmental conditions or pellet susceptibility to decay. However, this approach may fail to adequately account for the high levels of zero-inflation, which could have arisen via an entirely different process. It is possible, based on our understanding of the decay process, that there is another mechanism operating, related to whether the agents that cause pellet decay (e.g., insect, animal, and/or bacterial activity) are present or not. If agents are not present, then no decay will occur, resulting in zero values in the data. On the other hand, if agents are present, decay will occur, but the rate of decay may then still depend on environmental variation or variation in susceptibility to decay. One way to proceed to deal with this is to use a zero-inflated binomial or zero-inflated beta-binomial mixture model, rather than the standard binomial or beta-binomial models to account for the extra process of agents being present or absent.

One thing to note at this point is that here we are paying careful attention to the possible processes that may drive patterns in the data, and this is critical in informing the choice of mixture model and its interpretation. I will return to this issue in the discussion (section 12.5), but I mention it here to stress that this is an important habit to get into, to ensure that your choices of models are ecologically sensible.

The first thing that we will do now is explore the relative support from the data for four possible decay models: (1) binomial; (2) beta-binomial; (3) zero-inflated binomial; and (4) zero-inflated beta-binomial. The first model is the standard binomial model, with the other three models being mixture models representing different mechanisms through which overdispersion may arise, as discussed above. The binomial model has one parameter, s, representing the expected daily pellet survival probability (with $1-s$ being the decay rate). The beta-binomial model has an additional parameter, γ, that controls the level of overdispersion, with low values of γ representing high levels of overdispersion and high values of γ representing low levels of overdispersion. The two zero-inflated models then have a further parameter, q, representing the level of zero-inflation, with low values of q representing high levels of zero-inflation and high values of q representing low levels of zero-inflation. The formulations of the likelihoods of these models are described in online appendix 12B. In the context of this empirical example, we interpret the parameter q as the probability that agents causing decay (e.g., insect, animal, and/or bacterial activity) are present.

In addition to estimating the support for each of these models, we are also interested in whether pellet survival rates, s, and the probability that decay agents are present, q, vary with environmental factors or remain roughly constant. To incorporate covariates we model s and q as functions of covariates, rather than assuming they are constant. Specifically, we model them using the standard logit link function, such that the daily survival probability, s, of a pellet in group i on day t of interval j is

$$s_{ijt} = \frac{\exp\left(\alpha + \boldsymbol{\beta}^{\mathrm{T}}\mathbf{X}_{ijt}\right)}{1 + \exp\left(\alpha + \boldsymbol{\beta}^{\mathrm{T}}\mathbf{X}_{ijt}\right)}, \tag{12.7}$$

where α is an intercept, $\boldsymbol{\beta}$ is a vector of regression coefficients, and $\mathbf{X}_{ijt}$ is a vector of covariates for pellet group i in interval j on day t. We model the probability that decay agents, q, are present in pellet group i in interval j as

$$q_{ij} = \frac{\exp\left(\eta + \boldsymbol{\upsilon}^{\mathrm{T}}\mathbf{Y}_{ij}\right)}{1 + \exp\left(\eta + \boldsymbol{\upsilon}^{\mathrm{T}}\mathbf{Y}_{ij}\right)}, \tag{12.8}$$

where η is an intercept, $\boldsymbol{\upsilon}$ is a vector of regression coefficients, and $\mathbf{Y}_{ij}$ is a vector of covariates for pellet group i in interval j.

We can construct expressions for the likelihoods for each of these models (see online appendix 12B for details) and then use the function `mle2` from the libraries `bbmle` in R to find the maximum likelihood parameter estimates. The `mle2` function accepts, as one of its arguments, a function for the negative log-likelihood and then finds the parameter values that minimize this function using numerical optimization (note that minimizing the negative log-likelihood is identical to maximizing the log-likelihood, so this finds the maximum likelihood estimates of the model parameters). Although this can also be achieved by using the `optim` function (see chapter 3), the `mle2` function provides additional functionality, such as the generation of standard errors for the parameter estimates that is very useful. In appendix 12C, I provide R code for the likelihood-functions and example code for fitting the models using the `mle2` function.

If we fit the models described above with all covariates (site, topography, pellet age, rainfall, humidity, and temperature) as predictors of s_{ijt}, but with q constant initially (i.e., assuming that environmental variables determine decay rates, but not the presence or absence of decay agents), the best-supported model (based on Akaike's Information Criterion (AIC)—see chapter 3) is the beta-binomial model. But the zero-inflated beta-binomial model also has considerable support (having an AIC only 1.2 units larger than the beta-binomial model) (table 12.3). On the other hand, the binomial and zero-inflated binomial models have almost no support from the data, with AIC values much greater than either of the two best models. This provides strong evidence that random variation in pellet survival is a key process driving overdispersion (i.e., both of the top two models contain the beta-binomial mixture representing random variation in pellet survival). However, there is some evidence that the presence or absence of decay agents may operate together with random variation in survival rates (i.e., the zero-inflated beta-binomial model also has support relative to the beta-binomial model). If we plot the quantile–quantile plots of residuals for the binomial and beta-binomial models (figure 12.2), we can see that the beta-binomial model adequately accounts for overdispersion in the data, but the binomial model does not (i.e., figure 12.2A shows the characteristic quantile–quantile plot shape for overdispersed data). Moreover, the standard errors of the coefficient estimates for the beta-binomial models are larger than for the binomial model, which is as expected because it is accounting for overdispersion, and standard errors are no longer underestimated (table 12.3).

So far we have been able to account for overdispersion and say something about the relative support for each hypothesized mechanism driving overdispersion. But now let us look at predictors of pellet survival rates and the presence of decay agents. We will do this for the two best supported models: the beta-binomial and the zero-inflated beta-binomial models. The variables hypothesized to be potentially important drivers of pellet decay include spatial variables (site and topography) and temporal variables (pellet age, rainfall, humidity, and temperature). A sensible question to ask, therefore, might be, "What do the data tell us about the importance of spatial versus temporal variables?" For the beta-binomial model we can ask this by constructing models (using equation 12.7) that contain

Table 12.3 Akaike's Information Criterion (AIC) and coefficient estimates for the binomial (Bin), beta-binomial (BBin), zero-inflated binomial (ZIBin), and the zero-inflated beta-binomial (ZIBBin) models fitted to the koala dung decay data (standard errors in parentheses)

Value	Model			
	Bin	BBin	ZIBin	ZIBBin
AIC	3,215.7	2,360.6	2,771.4	2,361.8
ΔAIC	855.1	0.0	410.8	1.2
α	5.30 (0.110)	5.07 (0.174)	4.18 (0.125)	5.04 (0.181)
β_{age}	0.02 (0.001)	0.01 (0.001)	0.01 (0.001)	0.01 (0.001)
β_{site}	−1.19 (0.096)	−1.19 (0.162)	−0.89 (0.103)	−1.21 (0.164)
	0.14 (0.104)	0.31 (0.168)	−0.04 (0.113)	0.31 (0.170)
	0.71 (0.083)	0.59 (0.13)	0.58 (0.094)	0.60 (0.136)
β_{mid}	−0.59 (0.088)	−0.53 (0.146)	−0.32 (0.102)	−0.53 (0.148)
β_{gully}	−0.52 (0.088)	−0.59 (0.144)	−0.14 (0.099)	−0.59 (0.146)
β_{rain}	−0.01 (0.001)	−0.01 (0.002)	−0.01 (0.002)	−0.01 (0.002)
β_{hum}	−0.08 (0.008)	−0.08 (0.013)	−0.06 (0.008)	−0.08 (0.0127)
β_{temp}	−0.12 (0.014)	−0.08 (0.021)	−0.13 (0.014)	−0.09 (0.022)
$\log(\gamma)$		0.52 (0.098)		0.57 (0.119)
$\text{logit}(q)$			0.13 (0.099)	3.31 (1.316)

ΔAIC = difference between model AIC and model with the lowest AIC; α = intercept; β_{age} = coefficient for pellet age; β_{site} = coefficients for the sites; β_{mid} = coefficient for mid-slope; β_{gully} = coefficient for gully; β_{rain} = coefficient for rainfall; β_{hum} = coefficient for humidity; β_{temp} coefficient for temperature; $\log(\gamma)$ = logarithm of the overdispersion parameter in the beta-binomial distribution; and $\text{logit}(q)$ = logit of the probability that decay agents are present for the zero-inflated models. Continuous covariates were centered based on: median age = 71 days, median rainfall = 0 mm, median humidity = 71.50%, and median temperature = 19.48 °C.

none of the variables, either the spatial or temporal variables, or both, and comparing the four resulting models using AIC. For the zero-inflated beta-binomial model, however, there is the possibility that the variables may determine the expected survival rate, s, and/or the probability that decay agents are present, q. In this case, there are sixteen possible combinations of models representing the different ways in which the spatial and temporal variables could influence s and q (using equations 12.7 and 12.8), and the support for each of these models can also be explored using AIC. Note that, in constructing these models, I always include pellet age as a covariate for s, but never include it as a covariate for q (since there is no reason to expect pellet age to determine whether decay agents are present or not), and the temporal covariates for q are quantified based on their mean values within each time interval.

So what does this tell us? For the beta-binomial model, there is very strong indication that both temporal and spatial variables drive pellet decay (table 12.4). For the zero-inflated beta-binomial model there is still strong support for both spatial and temporal variables driving pellet decay (variables for s), but there is also strong evidence that the temporal climatic variables are important determinants of whether decay agents are present or not (variables for q; tables 12.4 and 12.5). Interestingly, in this case, the best zero-inflated beta-binomial model has a considerably lower AIC than the best beta-binomial model (2,345.64 vs. 2,360.60). Hence, once we include covariates for both s and q, there is compelling evidence for two different processes operating to drive pellet decay; one that determines whether decay agents are present and one that determines the decay rate if decay agents are present.

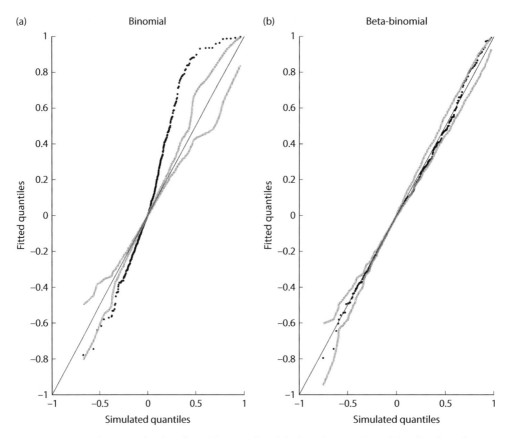

Fig. 12.2 Quantile–quantile plots for: (a) binomial and (b) beta-binomial models of koala pellet decay. Dots represent the quantile–quantile plot, with the solid black line and gray lines representing the expected (1:1) relationship and 95% point-wise confidence intervals respectively. The quantile–quantile plot for the binomial model shows a pattern characteristic of overdispersion, with the lower end of the distribution lying below the confidence intervals and the upper end of the distribution lying above the confidence intervals. On the other hand, the quantile–quantile plot for the beta-binomial model lies close to the 1:1 line and within the confidence intervals, indicating that overdispersion has been accounted for.

Table 12.4 Akaike's Information Criterion (AIC) values for each alternative beta-binomial model fitted to the koala dung decay data

Model rank	Spatial variables	Temporal variables	AIC	$\triangle$AIC
1	X	X	2,360.60	0.00
2	X		2,436.30	102.70
3		X	2,499.94	139.34
4			2,583.33	222.73

X = variables present in the model and $\triangle$AIC = difference between model AIC and model with the lowest AIC. Continuous covariates were centered based on: median age = 71 days, median rainfall = 0 mm, median humidity = 71.50%, and median temperature = 19.48 °C.

Table 12.5 Akaike's Information Criterion (AIC) values for each alternative zero-inflated beta-binomial model fitted to the koala dung decay data

Model Rank	Covariates for s		Covariates for q		AIC	ΔAIC
	Spatial	Temporal	Spatial	Temporal		
1	X	X		X	2,345.64	0.00
2	X	X	X	X	2,354.72	9.08
3	X	X			2,361.77	16.13
4	X	X	X		2,366.30	20.66
5	X		X	X	2,414.09	68.45
6		X	X		2,430.38	84.74
7		X	X	X	2,431.38	85.74
8	X			X	2,442.54	96.90
9	X		X		2,455.50	109.86
10	X				2,465.30	119.66
11			X	X	2,472.17	126.53
12		X		X	2,484.54	138.90
13			X		2,519.61	173.97
14				X	2,551.20	205.56
15			X		2,558.11	212.47
16					2,585.34	239.70

X = variables present in the model (Covariates for s represent explanatory variables for the survival rates and Covariates for q represent explanatory variables for the probability that decay agents are present) and ΔAIC = difference between model AIC and model with the lowest AIC. Continuous covariates for p were centered based on: median age = 71 days, median rainfall = 0 mm, median humidity = 71.5%, and median temperature = 19.48 °C. Continuous covariates for q were centered based on: median rainfall = 2.87 mm, median humidity = 71.02%, and median temperature = 19.79 °C.

This example provides an illustration of the power of mixture models to account for overdispersion and to allow inferences about the processes that drive that overdispersion. By grounding our model construction explicitly in terms of hypotheses about the ecological mechanisms that drive overdispersion, we are able to say something useful about the support for each of those mechanisms. This may be particularly important here since it appears that the drivers of whether decay agents (e.g., insect, animal, and/or bacterial activity) are present may be different from the drivers of decay rates if decay agents are present. The development of approaches for using these types of models to reliably calibrate surveys of indirect signs will depend on being able to correctly identify and quantify the processes that drive the decay process. Mixture models are an important tool for helping us to do this.

12.4.2 *Using Poisson mixtures to model lemur abundance*

Count data are one of the most commonly collected types of data for estimating species' distributions and abundance. However, as I have already pointed out, these data commonly exhibit overdispersion that precludes analysis based on standard distributions. Although both ecological processes and observation errors can lead to overdispersion in these types of data (table 12.1), correct inference relies on distinguishing between these two sources of overdispersion. In this context, Royle (2004) demonstrates how to account for detection errors in count data by using so called *N*-mixture models. These are countable mixture models that are explicit about zero-inflation arising from detection errors and the distribution of the true underlying abundances, which can in turn also be represented by a mixture model if necessary. For example, Royle (2004) adopts the Poisson distribution to describe the true underlying abundances, but also illustrates how the negative binomial may be used instead so that the model accounts for both zero-inflation that arises due to detection errors and overdispersion in the underlying abundances. Wenger and Freeman (2008) extend the approach to allow the true underlying abundances to be described by zero-inflated models. This allows for the possibility of simultaneously representing zero-inflation that arises from observation errors and zero-inflation that arises from ecological processes in the same model.

In this second empirical example, I illustrate the use of N-mixture models to make inferences about the abundance of two lemur species (the common brown lemur *Eulemur fulvus fulvus* and the black and white ruffed lemur *Varecia variegata variegata*) at two sites in the Zahamena reserve in eastern Madagascar. I use this example to illustrate how we can construct mixture models to account for and make inferences about overdispersion that arises from both observational and ecological sources. You will see as we go through the example that, once again, thinking carefully about the sources of overdispersion is central to successful model construction. I will use mixture models to try to distinguish between two sources of overdispersion; one that relates to overdispersion arising from observation error and one that relates to overdispersion arising from an ecological process. I will show that the source of overdispersion has profound implications for ecological inference. This is because, in the case of the observation error process, we actually want to "strip out" the effect of that process so as to reduce bias in ecological inference, while in the case of the ecological process, we are interested in the process itself and it is therefore retained as a component for ecological inference.

The data I use were collected in 1999 and 2000 at two sites in the Zahamena reserve in eastern Madagascar; one in mid-altitude rainforest (Antenina; elevation 900 m) and one in lowland rainforest (Namarafana; elevation 450 m). The data consist of direct group counts of lemurs along 300 m or 400 m long transect sections at each site. Although data were collected on all lemur species at the sites, I will only focus here on counts of groups of *Eulemur f. fulvus* and *Varecia v. variegata* (appendix 12D) and we will aim to quantify differences in abundance for these species between the two sites. Once again, let us start by looking at histograms of the data (figure 12.3). Histograms of the raw data reveal substantial zero-inflation, but there is also some suggestion of an excess of high values too. Overdispersion is also indicated by the variance of the data relative to the expected variance based on the Poisson distribution (0.056 vs. 0.016 for *Eulemur f. fulvus* and 0.091 vs. 0.018 for *Varecia v. variegata*). However, it is unclear on inspection of the histograms whether the overdispersion occurs primarily due to observation error, or as a result of an ecological process, such as a highly clumped spatial distribution. Understanding this is critical because it will likely make a major difference to our interpretation. We will now begin to explore these issues starting with a simple model and then adding complexity.

The simplest way to model these data is to ignore any observation error and overdispersion and use a standard Poisson distribution with one parameter, λ, representing the mean. However, it would make sense to try to account for the zero-inflation in some way. A straightforward way to accommodate the zero-inflation is to use a zero-inflated Poisson model with two parameters, λ and q. If we assume that there is no observation error then we can interpret q as the probability that habitat is suitable and then λ is the mean abundance, given that habitat is suitable (remember we discussed this idea earlier in the chapter in section 12.3.1). However, since we also seem to have an excess of

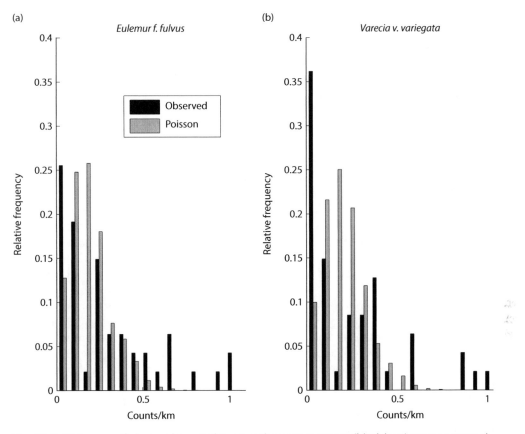

Fig. 12.3 Histogram of observed counts/km on each transect section (black bars) versus expected values based on a Poisson distribution with the same mean as the observed data (gray bars) for: (a) *Eulemur f. fulvus* and (b) *Varecia v. variegata*.

high values in the data, as well as zero-inflation, it could also make sense to extend this to the zero-inflated negative binomial model which has one further parameter, κ, that represents the level of overdispersion in the negative binomial component of the mixture (high values of κ imply high levels of overdispersion and low values of κ imply low levels of overdispersion).

The likelihoods for the Poisson, zero-inflated Poisson, and zero-inflated negative binomial models are described in online appendix 12E and can, once again, be fitted to the data using the function `mle2`. As in the first empirical example we can make the model parameters functions of covariates. We are interested in the difference in abundance between sites so it would make sense to introduce a covariate for site. However, because total survey effort varies between transect sections, we need to control for this by incorporating survey effort as a covariate too. Survey effort can be controlled for in a simple way: let λ (mean abundance [given suitable habitat in the case of the zero-inflated models]) depend on survey effort, with $\lambda = \gamma S$, where $S > 0$ is the survey effort and $\gamma > 0$ is the expected count per unit of survey effort (given suitable habitat in the case of the zero-inflated models). Then, to introduce the site covariate, we model γ and q (the probability that habitat is suitable) as functions of the site using the standard log and logit link functions respectively, such that

$$\gamma_i = \exp(\alpha + \beta X_i),$$ (12.9)

where γ_i is the expected count per unit of survey effort (our index of abundance) for transect section i, α is an intercept, β is a regression coefficient, and X_i is a categorical covariate, taking values of zero or one, representing the site within which transect section i is located. Finally, let

$$q_i = \frac{\exp(\eta + \upsilon X_i)}{1 + \exp(\eta + \upsilon X_i)},$$ (12.10)

where q_i is the probability that the habitat in transect section i is suitable, η is an intercept, and υ is a regression coefficient. In appendix 12F, I provide R code for the likelihood-functions and example code for fitting the models using the `mle2` function.

Fitting the Poisson model to the data and inspecting the quantile–quantile plots of the residuals reveals high levels of overdispersion in the data for both species, with the characteristic pattern of too many low values and too many high values in the data (figures 12.4A and 12.4C). The zero-inflated Poisson model reduces the level of overdispersion, but some points in the quantile–quantile plot still lie outside the 95% confidence intervals, suggesting some remaining overdispersion. On the other hand, the zero-inflated negative binomial model adequately accounts for overdispersion, with the quantile–quantile plot lying close to the expected 1:1 line and within the 95% confidence intervals (figures 12.4B and 12.4D). Therefore, a model whereby overdispersion is represented by both zero-inflation (representing whether habitat is suitable or not) and heterogeneity among transect sections (represented by the negative binomial component of the mixture) appears to be adequate for accounting for the overdispersion.

Although we have accounted for overdispersion here through both zero-inflation and heterogeneity among sections and the model seems to fit well, we have not considered the possibility that the zero-inflation may arise due to detection error (i.e., where the probability of detecting an individual that is present is less than one) rather than through the processes of habitat being suitable or not. If detection errors are present, then our estimates of abundance will be biased if not accounted for, especially if detection errors vary between the two sites. In recent years there has been substantial progress made in the development of methods for dealing with detection errors in ecological data (Mackenzie et al. 2006; Royle and Dorazio 2008). In general, these methods have, at their core a mixture model that enables the explicit representation of zero-inflation or missed counts, arising from the failure to detect individuals that are actually present. For example, Mackenzie et al. (2002) and Tyre et al. (2003) use zero-inflated binomial models to estimate occupancy while accounting for a failure to detect occupancy, thus reducing bias in occupancy estimates. However, to distinguish false-negatives (i.e., a failure to detect individuals that are actually present) from true-negatives (i.e., true absences) requires repeat surveys of sites within a short enough time period that the true occupancy or abundance state can be assumed to be unchanged. Fortunately, the lemur data consist of repeat surveys of each transect sections and therefore we can take advantage of this to explicitly account for detection error and reduce bias in abundance estimates.

Earlier I mentioned Royle's (2004) N-mixture model for dealing with detection errors in count data and we are going to use this model to examine the implications of detection error on our inferences about abundance at the two sites. The details of the likelihood for an N-mixture model are given in appendix 12E, but I will describe the model briefly here. The model is a countable mixture model based on an observation process defined by a binomial distribution that represents the probability of detecting an individual given that it is present. In the binomial distribution, the binomial probability, p, represents

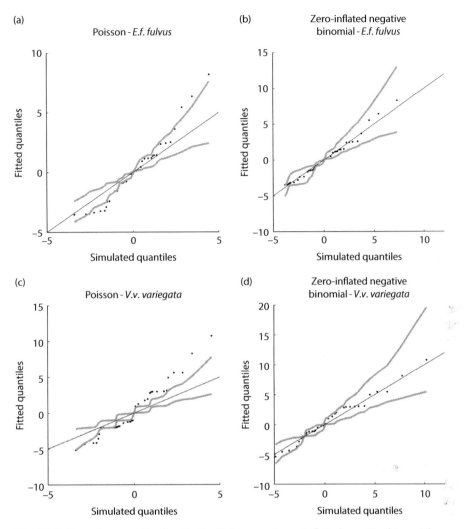

Fig. 12.4 Quantile–quantile plots for the Poisson and zero-inflated negative binomial models for counts of *Eulemur f. fulvus* (a, b), and for the Poisson and zero-inflated negative binomial models for counts of *Varecia v. variegata* (c, d). Dots represent the quantile–quantile plot, with solid black line and gray lines representing the expected 1:1 relationship and 95% point-wise confidence intervals respectively. The quantile–quantile plots for the Poisson models show a pattern characteristic of overdispersion, with the lower end of the distribution lying below the confidence intervals and the upper end of the distribution lying above the confidence intervals. On the other hand, the quantile–quantile plots for the zero-inflated negative binomial models lie close to the 1:1 line and with points lying within the confidence intervals, indicating that overdispersion has been accounted for.

the probability of detection, while the number of trials, N, represents the true number of groups present at a site. The number of trials, N, is assumed to vary randomly according to a Poisson distribution with mean λ (although other distributions, such as the negative binomial are also possible). Covariates for q and λ can be included in a similar way to equations 12.9 and 12.10. In these mixture models we are explicit about the observation process, via the binomial distribution, and explicit about the true underlying abundance, via the Poisson or negative binomial distribution. I provide R code for the likelihood functions of the N-mixture models that I use here, and example code for fitting these models to the lemur count data using the `mle2` function, in appendix 12F.

The interpretation of N and p is worth a note here before moving on. We interpret N for this case study as the number of groups that use a transect section, rather than the usual interpretation that would be the number of groups present in a transect section at the time of survey. Due to the mobile nature of the species, the number of groups present on a transect section may be different from day to day. An important assumption of these models is that the state of the system does not change between repeat surveys (an assumption known as the closure assumption) and this is likely to be broken here possibly leading to bias (Kendall and White 2009; Rota et al. 2009). This is because detection errors can occur for two reasons that are confounded in the estimate of p: (1) a group may be present at the time of a survey, but not observed; and (2) a group that uses the section may not be present at the time of the survey. However, if we interpret p as the probability that we detect a group that uses a transect section, rather than the probability that we detect a group that is present at the time of the survey, this issue is resolved. This is because the number of groups that used a transect section over the study period would have been relatively constant and so by making inferences at this level, the closure assumption holds and the confounding of sources of detection error does not matter. This also means, however, that we must interpret N as the number of groups using a transect section over the study period, rather than being the number of groups present in a transect section at the time of the survey.

We will now explore the extent to which the use of N-mixture models (i.e., assuming that the zero-inflation arises due to observation error), as opposed to using the zero-inflated Poisson or zero-inflated negative binomial models (i.e., assuming that the zero-inflation arises due to the availability of habitat), modifies our conclusions about differences in abundance between the two sites. If we fit both Poisson and negative binomial N-mixture models to the lemur data, assuming that both detection errors and abundance can vary between sites (i.e., we include a site covariate on p and λ), and compare these models to the zero-inflated models, we see a number of key differences (table 12.6). The first thing to note is that, although the negative binomial distribution has better support than the Poisson distribution for the zero-inflated models (based on AIC), this is not necessarily the case for the N-mixture models. The second thing to note is that, although for *Eulemur f. fulvus* the zero-inflated models suggest that abundance is lower at the lowland site than at the mid-elevation site (although not significantly so, based on the standard error estimate), the N-mixture models suggest that abundance is greater at the lowland site than at the mid-elevation site. This is because, although sighting rates are lower at the lowland site, the N-mixture model estimates that the probability of detection is much lower at the lowland site than at the mid-elevation site. The lower probability of detection more than compensates for the lower sighting rates at the lowland site, resulting in a higher estimate of abundance. For *Varecia v. variegata* the two types of model are in agreement, with abundance estimated to be higher in the lowland than in the mid-elevation site, but the probability of detection is similarly estimated to be lower in the lowland than in the mid-elevation site.

This example shows that our assumptions about sources of overdispersion can have profound implications for the inferences we make. In developing our inferences in this case, we need to make a decision about whether we believe that zero-inflation arises through observation error, or through some ecological processes related to the availability

Table 12.6 Model coefficients, Akaike's Information Criterion (AIC) values, and the estimated difference in abundance between the lowland and mid-elevation sites for each of the zero-inflated and N-mixture models fitted to the lemur count data (standard errors) in parentheses

Model	Coefficients for p or q		Coefficients for γ		$\ln(\kappa)$	AIC	$\hat{N}_{lowland} - \hat{N}_{mid\text{-}elevation}$ (groups/km)
	Intercept	Lowland site	Intercept	Lowland site			
Eulemur f. fulvus							
Zero-inflated Poisson	1.522 (0.493)	19.616 (2×10^{-10})	-0.986 (0.010)	-1.632 (0.028)		197.77	-0.233 (0.029)
Zero-inflated negative binomial	1.770 (0.659)	11.935 (715.179)	-1.030 (0.141)	-1.575 (0.323)	1.624 (0.728)	194.86	-0.175 (0.168)
N-mixture (Poisson)	-2.335 (0.297)	-2.767 (0.741)	1.253 (0.284)	1.244 (0.682)		738.02	8.645 (2.629)
N-mixture (negative binomial)	-2.336 (0.297)	-2.767 (0.741)	1.253 (0.284)	1.244 (0.682)	31.193 (2×10^{-10})	740.02	8.645 (2.629)
Varecia v. variegata							
Zero-inflated Poisson	-0.028 (0.366)	22.375 (2×10^{-10})	-1.044 (0.129)	0.143 (0.171)		222.72	0.233 (0.156)
Zero-inflated negative binomial	0.157 (0.423)	15.460 (615.348)	-1.132 (0.221)	0.267 (0.295)	0.937 (0.454)	204.94	0.247 (0.374)
N-mixture (Poisson)	-1.805 (0.230)	-0.550 (0.450)	0.202 (0.263)	1.375 (0.444)		746.50	3.617 (1.172)
N-mixture (negative binomial)	-1.892 (0.274)	-0.994 (0.514)	0.279 (0.302)	1.798 (0.523)	1.457 (0.828)	747.09	6.659 (1.741)

p = probability of suitable habitat in the zero-inflated models, q = detection probability in the N-mixture models, κ = the overdispersion parameter for the negative binomial distribution, and $\hat{N}_{lowland} - \hat{N}_{mid\text{-}elevation}$ = the difference in estimated abundance between the lowland site and mid-elevation site in units of groups/km (standard errors for the differences were estimated using the delta method).

of habitat. It is unlikely in this example that observation error is zero; it is almost certainly the case that groups that are present on a transect section could have been missed and the mobile nature of the species means that a group that uses a transect section may not be present at the time of survey. There are two possible reasons for detection errors being higher in the lowland site than the mid-elevation site. The first reason is that groups present on the transect sections are more often not detected at the lowland than at the mid-elevation site. The lowland site has a more dense understory and higher canopy than the mid-elevation site (J. Rhodes, personal observation), which would tend to make lemur observations more difficult, so this is consistent with the N-mixture models. However, this could also be driven by differences in field personnel between the two sites. The second reason is that groups that use a transect section are more often absent from a transect section at the lowland site than at the mid-elevation site. This could occur, for example, if groups tend to move more frequently at the lowland site than at the mid-elevation site. Although we have no information about the relative movement frequencies at the two sites, the N-mixture models make sense in terms of the likely presence of detection errors and variation in forest structure and personnel between the two sites. Importantly, in this example, a mixture modeling approach has allowed us to be explicit about the mechanisms driving overdispersion and to understand the implications of the assumptions we have made.

12.5 Discussion

Mixture models should become a critical component of the ecologist's statistical tool box. Ecological data arise from complex interacting processes; they rarely conform nicely to the assumptions of standard statistical distributions. When they do not, this often manifests itself as overdispersion, playing havoc with our statistical tests and inference. Fortunately, mixture models provide a flexible way to deal with overdispersion, but they are useful for much more than simply controlling for overdispersion. This is because they allow us to make inferences about the causes of overdispersion, leading to greatly improved ecological inference. In particular, it allows us to have a much more mechanistic understanding of the processes that lead to the observed data. In this chapter I have outlined what mixture models are and illustrated their use in two quite different applications. The applications demonstrate how careful consideration of the mechanisms driving overdispersion in the data can lead to a much richer understanding of the underlying ecological and observation processes. Although the applications I have presented come from survival analysis and abundance estimation, mixture models are applicable to almost any area of ecology. As such, they are an important and widely applicable approach in ecological statistics.

In this chapter I have focused on some of the more typical and standard mixture models. However, it is possible to more generally construct complex mixtures of distributions to represent a wide range of ecological mechanisms that may be hypothesized to generate any observed data. For example, I mentioned earlier Kendall and Wittmann's (2010) stochastic model of breeding success that explicitly models the probability of laying eggs, nest survival, clutch size, and offspring survival as mechanisms leading to the observed data on reproductive output. They apply it to 53 vertebrate species, and, to achieve this, they model the number of offspring as a finite mixture distribution, with mixing weights defined by the probability that eggs are laid and then model the probability of nest survival, given eggs are laid, using a distribution that is itself a mixture model. The model for nest survival reflects the contribution of clutch size and offspring survival to the number of offspring, and is assumed to be a countable mixture, with offspring survival defined by a binomial distribution and the number of trials specified by a Poisson distribution (in a similar way to an *N*-mixture model). This provides inference about these separate component processes that would otherwise not be possible without the use of a mixture model. More broadly, flexible mixture models form the basis of so-called state–space models that aim to represent ecological and observation processes in a mechanistic fashion (for nice examples see Buckland et al. 2004; Patterson et al. 2008). Specifying these models often results in complex mixtures, but because they are explicit about the ecological and observation processes that generate the observed data, they provide a powerful and flexible framework for ecological inference that is becoming increasingly popular.

I have demonstrated how we should ground our choice of mixture model in mechanistic hypotheses about the processes that may have led to the data. An alternative is to adopt a, so-called, quasi-likelihood approach. Rather than characterizing the full likelihood of the data, quasi-likelihood approaches characterize a quasi-likelihood function that depends only on the mean and variance, but behaves in a similar way to the full likelihood (see chapter 6; McCullagh and Nelder 1989; Burnham and Anderson 2002). Essentially, what this means is that the quasi-likelihood does not characterize the full distribution of the data, but simply adjusts the variance to account for overdispersion. In contrast, mixture models use information about the full distribution of the data and this is what allows us to make more mechanistic inferences that would not necessarily be possible using quasi-likelihood methods. For example, in the koala dung decay example, our mixture model uses the amount of zero-inflation in the data to distinguish between the processes driving the presence of decay agents versus processes driving decay rates when decay agents are present. This type of analysis would not be possible with a quasi-likelihood; inferences about the presence of decay agents would not be possible, although it would still control for overdispersion allowing us to perform correct statistical tests.

Although inference about mechanisms is a major strength of the mixture modeling approach it can also be problematic if we have no, or little, a priori information about potential causes of overdispersion. In cases where we are not specifically interested in the causes of overdispersion, or are unable to develop sensible mixture models, then quasi-likelihood approaches are often a suitable alternative for dealing with overdispersion. In fact, Richards (2008) shows that quasi-likelihood approaches to model selection produce very similar results to the negative binomial mixture model based on Akaike's Information Criterion (AIC). Nonetheless, we still need to think critically about our choice of model and model assumptions because this can have important bearing on inference. For example, Ver Hoef and Boveng (2007) show that quasi-Poisson (a quasi-likelihood version of the Poisson distribution) and negative binomial models can produce quite different parameter estimates. Using an example of harbor seals in Alaska they show that regression coefficient estimates are affected by the choice of model because they make different assumptions about how the variance of the data changes with abundance. The quasi-Poisson model assumes that the variance increases linearly with abundance, while the negative binomial model assumes that the variance increases quadratically with abundance. In their case, they find that the quasi-Poisson is the better model for their data, and suggest plotting abundance versus variance of data to get an idea of which model may be most appropriate. The important take-home message here is that, even if we use a quasi-likelihood approach, we should ensure that the assumed variance–mean relationship is sensible for our data.

Despite the great promise of mixture models, they should be used with some caution. One issue is that they make strong assumptions about the distribution of the data and if these assumptions do not hold this could result in biased parameter estimates. Therefore, I recommend that careful a priori consideration should be given to the choice of assumed mechanisms as I have done in this chapter. First, you should think carefully about what the implications of failing to meet those assumptions might be. For example, in the lemur example, inference is strongly dependent upon whether you assume that zero-inflation arises from observation or ecological processes. Second, because mixture models often contain a large number of unobserved (latent) variables, model parameters can often fail to be identifiable. Model parameters are not identifiable when the data are insufficient to distinguish between the values for two or more parameters because their values are confounded. For example, in the lemur case study I was unable to fit an N-mixture model based on a zero-inflated Poisson distribution because there was insufficient information in the data to separate zero-inflation that arises due to detection error from zero-inflation that arises due to the availability of suitable habitat. This issue can limit the extent to which mixture models are able to be applied to specific ecological questions.

Mixture models are closely related to mixed-effects (or random-effects) models that are commonly used in ecology (see chapter 13). Mixed-effects models introduce random variation in model parameters (through the specification of random effects) that can account for additional variation in the data in a similar way to mixture models. However, whereas mixture models introduce random variation at the level of individual data points, variation in mixed-effects models is usually specified at a hierarchical level above that of the individual data points. For example, in a mixed-effects model we may have random variation among sites, but not among data points within sites as in a mixture model. For this reason, mixed-effects models are most often used to account for hierarchical structure, or dependencies, in the data rather than overdispersion. For example, Thomas et al. (2006) use individual-level random effects to model variation in habitat selection among individuals in their analysis of caribou location data. The purpose for doing so was to account for dependencies in the data within individuals and variation among individuals, rather than dealing with and understanding overdispersion per se. Therefore, despite the close links between the two approaches, their use in ecology is quite different.

What should you report in a paper using mixture models? One of the most critical aspects is to be clear about how you constructed your mixture models, the mechanisms you hypothesize that the different components of the mixture represent, and what assumptions you have made. In describing your models this is critical so that the reader understands what your models represent. In this context, Kendall and Wittmann (2010) provide an excellent example where the rationale for the model is very clearly described. You should also present evidence that your models have dealt adequately with overdispersion in the data using techniques such as quantile–quantile plots as I have used in this chapter. Finally, providing inference in terms of the different components of the mixture model is important so that readers can relate this inference back to the proposed mechanisms. For example, if you use a negative binomial distribution to deal with random variation in habitat quality, then, in addition to reporting the regression parameters, report and interpret the overdispersion parameter in terms of the hypothesized source of overdispersion. This will provide readers with a richer understanding of the ecological processes that would not be possible if the overdispersion parameter was not interpreted in this way.

Acknowledgments

The writing of this chapter was funded in part by the Australian Research Council Centre of Excellence for Environmental Decisions. Thank you to Dan Lunney for providing access to the koala dung decay trial data.

## CHAPTER 13

Linear and generalized linear mixed models

Benjamin M. Bolker

13.1 Introduction to generalized linear mixed models

Generalized linear mixed models (GLMMs) are a powerful class of statistical models that combine the characteristics of generalized linear models (GLMs: chapter 6) and *mixed models* (models with both fixed and random predictor variables). They handle a wide range of types of response variables, and a wide range of scenarios where observations have been sampled in some kind of groups rather than completely independently. While they can't do everything—an expert might sometimes choose custom-built models for greater flexibility (Bolker et al. 2013)—GLMMs are fast, powerful, can be extended to handle additional complexities such as zero-inflated responses, and can often be fitted with off-the-shelf software. The only real downsides of GLMMs are due to their generality: (1) some standard recipes for model testing and inference do not apply, and (2) it's easy to build plausible models that are too complex for your data to support. GLMMs are still part of the statistical frontier, and even experts don't know all of the answers about how to use them, but this chapter will try to provide practical solutions to allow you to use GLMMs with your data.

GLMs allow modeling of many kinds of response variables, particularly those with binomial and Poisson distributions; you should definitely feel comfortable with GLMs before attempting the methods described in this chapter. In contrast, you may be unfamiliar with the mixed models, and with the central distinction between *fixed effects* (the typical way to compare differences between treatments or the effects of continuous predictor variables) and *random effects* (roughly speaking, experimental or observational blocks within which you have several observations). Models with normally distributed responses that incorporate some kind of random effects are called *linear mixed models* (LMMs); they are a special, slightly easier case of GLMMs. This chapter will review the basic idea of experimental blocks (for a reminder see Gotelli and Ellison (2004) or Quinn and Keough (2002)). If you are already well-versed in classic ANOVA approaches to blocked experimental designs, you may actually have to unlearn some things, as modern approaches to random effects are quite different from the classical approaches taught in most statistics courses.

As well as using different conceptual definitions of random effects (section 13.3.1), modern mixed models are more flexible than classic ANOVAs, allowing, for example, non-Normal responses, unbalanced experimental designs, and more complex grouping

Ecological Statistics: Contemporary Theory and Application. First Edition. Edited by Gordon A. Fox, Simoneta Negrete-Yankelevich, and Vinicio J. Sosa. © Oxford University Press 2015. Published in 2015 by Oxford University Press.

structures. Equally important is a new philosophy: modern approaches use a model-building rather than a hypothesis-testing approach (chapter 3). You can still test hypotheses, but instead of a list of F statistics and p-values the primary outputs of the analysis are quantitative parameter estimates describing (1) how the response variable changes as a function of the fixed predictor variables, and (2) the variability among the levels of the random effects.

Random effects such as variation among experimental blocks are often neglected in model-based analyses because they are relatively difficult to incorporate in custom-built statistical models. While one can use software such as WinBUGS, AD Model Builder, or SAS PROC NLMIXED to incorporate such components in a general model (Bolker et al. 2013), generalized linear mixed models are general enough to encompass the most common statistical problems in ecology, yet can be fitted with off-the-shelf software.

Section 13.2 (Running examples) introduces several case studies from the literature, and from my own work, for which the data are freely accessible. Section 13.3 (Concepts) gets philosophical, exploring different definitions of random effects; related concepts like pooling, shrinkage, and nested vs. crossed experimental designs; the statistical issues of overdispersion and variable correlation within groups; and the extended definitions of likelihood required for mixed models (see chapter 3 for the basic definition). Section 13.4 (Setting up a GLMM) is practical but short; once you understand the ins and outs of random effects, and the concepts of GLMs from chapter 6, writing the code to define a GLMM is actually quite straightforward. Sections 13.5 (Estimation) and 13.6 (Inference) go into nitty-gritty detail about the choices you have when fitting a GLMM and translating the results back from statistical to scientific answers.

13.2 Running examples

- *Tundra carbon dynamics*: Belshe et al. (2013) did a meta-analysis of previous studies of carbon uptake and release in tundra ecosystems. They asked how the CO_2 flux, or net ecosystem exchange at measured experimental sites, was changing over time. The residual variation of the primary response variable (GS.NEE, net carbon flux during the growing season) was assumed to be Normal, so the model is a linear mixed model. Sites were treated as random effects, meaning that site was a *grouping variable* (a categorical predictor across which effects are assumed to vary randomly), with the baseline CO_2 flux (i.e., the intercept term of the model) varying across sites. Time (Year) was the primary fixed effect, although the paper also considered the effects of mean annual temperature and precipitation, as well as the additional response variables of winter and total annual carbon flux.

- *Coral symbiont defense*: McKeon et al. (2012) ran a field experiment with coral (*Pocillopora* spp.) inhabited by invertebrate symbionts (crabs [*Trapezia* spp.] and shrimp [*Alpheus* spp.]) and exposed to predation by sea stars (*Culcita* spp.). They asked whether combinations of symbionts from different species were more, less, or equally effective in defending corals from predators, compared to expectations based on the symbionts' independent protective effects. The design is a randomized complete block design with a small amount of replication: 2 replications per treatment per block; 4 treatments (no symbionts, crabs only, shrimp only, both symbionts), with each of these units of 8 repeated in 10 blocks. The response (predation) is binomial with a single trial per unit (also called Bernoulli or binary, see book appendix); treatment (ttt), a categorical variable, is the only fixed-effect input variable; block is the only grouping variable, with intercepts (i.e., baseline predation probability) varying among blocks.

- *Gopher tortoise shells*: Ozgul et al. (2009) analyzed the numbers of gopher tortoise shells found at different sites to estimate whether shells were more common (implying a higher mortality rate) at sites with higher prevalence of a mycoplasmal pathogen (prev). The response is the count of fresh shells (shells), for which we will consider Poisson and negative binomial distributions (book appendix); seroprevalence of mycoplasma (prev: i.e., the fraction of tortoises carrying antibodies against the disease) is a continuous, fixed predictor variable. We initially considered year and site as crossed grouping variables (section 13.3.1) with variation in baseline shell counts (intercepts) among them; we also included the logarithm of the site area (Area) as an *offset* term to account for variation in site area, effectively modeling shell density rather than shell numbers.
- *Red grouse ticks*: Elston et al. (2001) used data on numbers of ticks sampled from the heads of red grouse chicks in Scotland to explore patterns of aggregation. Ticks have potentially large fitness and demographic consequences on red grouse individuals and populations, but Elston et al.'s goal was just to decompose patterns of variation into different scales (within-brood, within-site, by altitude and year). The response is the tick count (TICKS, again Poisson or negative binomial); altitude (HEIGHT, treated as continuous) and year (YEAR, treated as categorical) are fixed predictor variables. Individual within brood (INDEX) and brood within location are nested random-effect grouping variables, with the baseline expected number of ticks (intercept) varying among groups.

All of these case studies include some kind of grouping (sites in the tundra carbon example; experimental blocks for the sea star example; areas and years for the gopher tortoise example; and individuals within broods within sites for the tick example), requiring mixed models. The first has Normal responses, requiring a LMM, while the latter three have non-Normal response variables, requiring GLMMs.

13.3 Concepts

13.3.1 *Model definition*

The complete specification of a GLMM includes the distribution of the response variable; the link function; the definition of categorical and continuous fixed-effect predictors; and the definition of the random effects, which specify how some model parameters vary randomly across groups. Here we focus on random effects, the only one of these components that is not already familiar from chapter 6.

Random effects

The traditional view of random effects is as a way to do correct statistical tests when some observations are correlated. When samples are collected in groups (within sites in the tundra example above, or within experimental blocks of any kind), we violate the assumption of independent observations that is part of most statistical models. There will be some variation within groups (σ^2_{within}) and some among groups (σ^2_{among}); the total variance is $\sigma^2_{total} = \sigma^2_{within} + \sigma^2_{among}$; and therefore the correlation between any two observations in the same group is $\rho = \sqrt{\sigma^2_{among}/\sigma^2_{total}}$ (observations that come from *different* groups are uncorrelated). Sometimes one can solve this problem easily by taking group averages.

For example, if we are testing for differences between deciduous and evergreen trees, where every member of a species has the same leaf habit, we could simply calculate species' average responses, throwing away the variation within species, and do a t-test between the deciduous and evergreen species means. If the data are balanced (i.e., if we sample the same number of trees for each species), this procedure is exactly equivalent to testing the fixed effect in a classical mixed model ANOVA with a fixed effect of leaf habit and a random effect of species. This approach correctly incorporates the facts that (1) repeated sampling within species reduces the uncertainty associated with within-group variance, but (2) we have fewer *independent* data points than observations—in this case, as many as we have groups (species) in our study.

These basic ideas underlie all classical mixed-model ANOVA analyses, although the formulas get more complex when treatments vary within grouping variables, or when different fixed effects vary at the levels of different grouping variables (e.g., randomized-block and split-plot designs). For simple nested designs, simpler approaches like the averaging procedure described above are usually best (Murtaugh 2007). However, mixed-model ANOVA is still extremely useful for a wide range of more complicated designs, and as discussed below, traditional mixed-model ANOVA itself falls short for cases such as unbalanced designs or non-Normal data.

We can also think of random effects as a way to combine information from different levels within a grouping variable. Consider the tundra ecosystem example, where we want to estimate linear trends (slopes) across time for many sites. If we had only a few years sampled from a few sites, we might have to *pool* the data, ignoring the differences in trend among sites. Pooling assumes that σ^2_{among} (the variance in slopes among sites) is effectively zero, so that the individual observations are uncorrelated ($\rho = 0$).

On the other hand, if we had many years sampled from each site, and especially if we had a small number of sites, we might want to estimate the slope for each site individually, or in other words to estimate a fixed effect of time for each site. Treating the grouping factor (site) as a fixed effect assumes that information about one site gives us no information about the slope at any other site; this is equivalent, for the purposes of parameter estimation, to treating σ^2_{among} as infinite. Treating site as a random effect compromises between the extremes of pooling and estimating separate (fixed) estimates; we acknowledge, and try to quantify, the variability in slope among sites. Because the trends are assumed to come from a population (of slopes) with a well-defined mean, the predicted slopes in CO_2 flux for each site are a weighted average between the trend for that site and the overall mean trend across all sites; the smaller and noisier the sample for a particular site, the more its slope is compressed toward the population mean (figure 13.1).

For technical reasons, these values (the deviation of each site's value from the population average) are called *conditional modes*, rather than *estimates*. The conditional modes are also sometimes called *random effects*, but this could also refer to the grouping variables (the sites themselves, in the tundra example). Confusingly, both the conditional modes and the estimates of the among-site variances can be considered parameters of the random effects part of the model. For example, if we had independently estimated the trend at one site (i.e., as a fixed effect) as −5 grams $C/m^2/year$, with an estimated variance of 1, while the mean rate of all the sites was −8 g $C/m^2/year$ with an among-site variance of 3, then our predicted value for that site would be $(\mu_{site}/\sigma^2_{within} + \mu_{overall}/\sigma^2_{among})/(1/\sigma^2_{within} + 1/\sigma^2_{among}) = (-5/1 + -8/3)/(1/1 + 1/3) = -5.75$ g $C/m^2/year$. Because $\sigma^2_{within} < \sigma^2_{among}$—the trend estimate for the site is relatively precise compared to the variance among sites—the random-effects prediction is closer to the site-specific value than to the overall mean. (Stop and plug in a few different values of among-site variance to convince yourself that

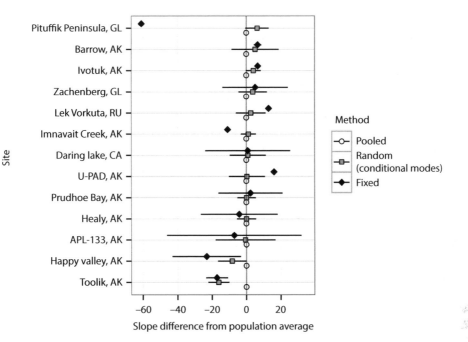

Fig. 13.1 Estimated differences in slope (annual change in growing season NEE) among sites, with 95% confidence intervals. The conditional modes are (mostly) intermediate between the fixed estimates and the pooled estimate of zero (the two exceptions, Pituffik Peninsula and Imnaivit Creek, have compensating differences in their intercept estimates); sites with only one year's data, for which a fixed-effect slope cannot be estimated, are not shown. The confidence intervals are generally much narrower for the conditional modes than for the fixed-effect estimates (the four fixed-effect estimates with error bars not shown have 95% CIs that extend beyond the limits of the plot).

this formula agrees with verbal description above of how variance-weighted averaging works when σ^2_{among} is either very small or very large relative to σ^2_{within}.)

Random effects are especially useful when we have (1) lots of levels (e.g., many species or blocks), (2) relatively little data on each level (although we need multiple samples from most of the levels), and (3) uneven sampling across levels (box 13.1).

Frequentists and Bayesians define random effects somewhat differently, which affects the way they use them. Frequentists define random effects as categorical variables whose levels are chosen *at random from a larger population*, e.g., species chosen at random from a list of endemic species. Bayesians define random effects as sets of variables whose parameters are drawn from a distribution. The frequentist definition is philosophically coherent, and you will encounter researchers (including reviewers and supervisors) who insist on it, but it can be practically problematic. For example, it implies that you can't use species as random effect when you have observed *all* of the species at your field site—since the list of species is not a sample from a larger population—or use year as a random effect, since researchers rarely run an experiment in randomly sampled years—they usually use either a series of consecutive years, or the haphazard set of years when they could get into the field. This problem applies to both the gopher tortoise and tick examples, each of which use data from consecutive years.

Box 13.1 WHEN TO TREAT A PREDICTOR VARIABLE AS A RANDOM EFFECT

You may want to treat a predictor variable as a random effect if you:

- don't want to test hypotheses about differences between responses at particular levels of the grouping variable;
- do want to quantify the variability among levels of the grouping variable;
- do want to make predictions about unobserved levels of the grouping variable;
- do want to combine information across levels of the grouping variable;
- have variation in information per level (number of samples or noisiness);
- have levels that are randomly sampled from/representative of a larger population;
- have a categorical predictor that is a nuisance variable (i.e., it is not of direct interest, but should be controlled for).

Cf. Crawley (2002); Gelman (2005)

If you have sampled fewer than five levels of the grouping variable, you should strongly consider treating it as a fixed effect even if one or more of the criteria above apply.

Random effects can also be described as predictor variables where you are interested in making inferences about the distribution of values (i.e., the variance among the values of the response at different levels) rather than in testing the differences of values between particular levels. Choosing a random effect trades the ability to test hypotheses about differences among particular levels (low vs. high nitrogen, 2001 vs. 2002 vs. 2003) for the ability to (1) quantify the variance among levels (variability among sites, among species, etc.) and (2) generalize to levels that were not measured in your experiment. If you treat species as a fixed effect, you can't say anything about an unmeasured species; if you use it as a random effect, then you can guess that an unmeasured species will have a value equal to the population mean estimated from the species you did measure. Of course, as with all statistical generalization, your levels (e.g., years) must be chosen in some way that, if not random, is at least *representative* of the population you want to generalize to.

People sometimes say that random effects are "factors that you aren't interested in." This is not always true. While it is often the case in ecological experiments (where variation among sites is usually just a nuisance), it is sometimes of great interest, for example in evolutionary studies where the variation among genotypes is the raw material for natural selection, or in demographic studies where among-year variation lowers long-term growth rates. In some cases fixed effects are also used to control for uninteresting variation, e.g., using mass as a covariate to control for effects of body size.

You will also hear that "you can't say anything about the (predicted) value of a conditional mode." This is not true either—you can't formally test a null hypothesis that the value is equal to zero, or that the values of two different levels are equal, but it is still perfectly sensible to look at the predicted value, and even to compute a standard error of the predicted value (e.g., see the error bars around the conditional modes in figure 13.1). Particularly in management contexts, researchers may care very much about *which* sites are particularly good or bad relative to the population average, and how much better or worse they are than the average. Even though it's difficult to compute formal inferential summaries such as *p*-values, you can still make common-sense statements about the conditional modes and their uncertainties.

The Bayesian framework has a simpler definition of random effects. Under a Bayesian approach, a fixed effect is one where we estimate each parameter (e.g., the mean for each species within a genus) independently (with independently specified priors), while for a random effect the parameters for each level are modeled as being drawn from a distribution (usually Normal); in standard statistical notation, species_mean $\sim$ Normal(genus_mean, $\sigma^2_{species}$).

I said above that random effects are most useful when the grouping variable has many measured levels. Conversely, random effects are generally ineffective when the grouping variable has too few levels. You usually can't use random effects when the grouping variable has fewer than five levels, and random effects variance estimates are unstable with fewer than eight levels, because you are trying to estimate a variance from a very small sample. In the classic ANOVA approach, where all of the variance estimates are derived from simple sums-of-squares calculations, random-effects calculations work as long as you have at least two samples (although their power will be very low, and sometimes you can get negative variance estimates). In the modern mixed-modeling approach, you tend to get warnings and errors from the software instead, or estimates of zero variance, but in any case the results will be unreliable (section 13.5 offers a few tricks for handling this case). Both the gopher tortoise and grouse tick examples have year as a categorical variable that would ideally be treated as random, but we treat it as fixed because there are only three years sampled: treating years as a random effect would most likely estimate the among-year variance as zero.

Simple vs. complex random effects

The most common type of random effect quantifies the variability in the baseline values of the response variable among levels of a categorical grouping variable (e.g., baseline numbers of ticks in different locations). Although technically location is the *grouping variable* in this case, and the thing that varies among levels is the intercept term of a statistical model, we would often call this simply a random effect of location. This is a random intercept model, which is also a scalar random effect (i.e., there is only one value per level of the grouping variable). In R it would be specified within a modeling formula as ~group or ~(1):group (MCMCglmm library), ~1|group (nlme or glmmADMB libraries), or (1|group) (lme4 or glmmADMB libraries) (the 1 specifies an intercept effect; it is implicit in the first example).

More generally, we might have observed the effects of a treatment or covariate within each level, and want to know how these effects (described by either a categorical or a continuous predictor) vary across levels; this is the case for slopes (i.e., the effect of time) in the tundra example. Since the intercept as well as all of the parameters describing the treatment would vary across levels, this would be called a non-scalar or a vector random effect. This could be specified as ~1+x|group in nlme or glmmADMB, (1+x|group) in lme4 or glmmADMB), or ~us(1+x):group in MCMCglmm. In many cases the 1 is optional—(x|group) would also work—but I include it here for concreteness. The us in the third specification refers to an unstructured variance–covariance matrix: MCMCglmm offers several other options (see the *Course Notes* vignette that comes with the library).

For example, the coral symbiont data follow a randomized block design, with replicates of all treatments within each block. So we could in principle use the random effects model (1+ttt|block) (equivalent to (ttt|block) because the intercept is implicitly included) to ask how the effects of symbionts varied among different blocks, with four random parameters per block (intercept and three treatment parameters), where the intercept parameter describes the variation among control treatments across blocks and the treatment parameters describe the variation in the effects of symbionts (crab vs. control, shrimp vs. control, and crab + shrimp vs. control) among blocks. However, this is another case where the ideal and the practical differ; in practice this approach is not feasible because we have too little information—there are only two binary samples per treatment per block—so we would likely proceed with an intercept-only (scalar) random effect of blocks.

Non-scalar effects represent *interactions* between the random effect of block and the fixed effect (symbionts), and are themselves random—we assume, for example, that the difference in predation rate between corals with and without symbionts is drawn from a distribution of (differences in) predation rates. The interaction between a random effect and a continuous predictor is also random; the tundra carbon example includes a site × year interaction which describes the variation in temporal trends among sites. This type of interaction is the only case in which it makes sense to consider a random effect of a continuous variable; a continuous variable (year in this example) cannot itself be a *grouping* variable, but can vary across grouping variables (sites). One should in general consider the random × fixed effect interactions whenever it is feasible, i.e., for *all* treatments that are applied within levels of a random effect; doing otherwise assumes a priori that there is no variation among groups in the treatment effect, which is rarely warranted biologically (Schielzeth and Forstmeier 2009; Barr et al. 2013). It is often impossible or logistically infeasible to apply treatments within groups: in the gopher tortoise example the prevalence of disease is fundamentally a site-level variable, and can't vary within sites. Or, as in the coral symbiont example, we may have so little statistical power to quantify the among-group variation that our models don't work, or that we estimate the variation as exactly zero. In these cases we have to accept that there probably is a real interaction that we are ignoring, and temper our conclusions accordingly.

Nesting and crossing

What about the interaction between two random effects? Here we have to specify whether the two effects are *nested* or *crossed*. If at least one of the levels of each effect is represented in multiple levels of the other effect, then the random effects are crossed; otherwise, one is nested in the other. In the gopher tortoise example, each site is measured in multiple years, and multiple sites are measured in each year, so site and year are crossed (although as pointed out above we don't actually have data for enough years to treat them as random); this would be specified as (1|site) +(1|year). On the other hand, in the tick example each chick occurs in only one brood, and each brood occurs in only one site: the model specification is (1|SITE/BROOD/ INDEX), read as "chick (INDEX) nested within brood nested within site," or equivalently (1|SITE) + (1|SITE:BROOD) + (1|SITE:BROOD:INDEX). If the broods and chicks are uniquely labeled, so that the software can detect the nesting, (1|SITE) + (1|BROOD) + (1|INDEX) will also work (do *not* use (1|SITE) + (1|SITE/ BROOD) + (1|SITE/BROOD/INDEX); it will lead to redundant terms in the model). Another way of thinking about the problem is that, in the gopher tortoise example, there is variation among sites that applies across years, variation among years that applies across all sites, and variation among site-by-year combinations. In the tick example, there is variation among broods and variation among chicks within broods, but there is no sensible way to define variation among chicks *across* broods. In this sense a nested model is a special case of crossed random effects that sets one of the variance terms to zero.

Crossed random effects are more challenging computationally than nested effects (they are largely outside the scope of classical ANOVAs), and so this distinction is often ignored in older textbooks. Most of the software that can handle both crossed and nested random effects can automatically detect when a nested model is appropriate, provided that the levels of the nested factor are uniquely labeled. That is, the software can only tell individuals are nested if they are labeled as A1, A2, ..., A10, B1, B2, ... B10, ... If individuals are instead identified only as 1, 2, ... 10 in each of species A, B, and C, the software can't tell that individual #1 of species A is not related to individual #1 of species B. In this case you can specify nesting explicitly, but it is safer to label the nested individuals uniquely.

You should usually treat interactions between two or more fixed effects as crossed, because the levels of fixed effects are generalizable across levels of other fixed effects ("high nitrogen" means the same thing whether we are in a low- or high-phosphorus treatment). Random effects can be nested in fixed effects, but fixed effects would only be nested in random effects if we really wanted (for example) to estimate different effects of nitrogen in each plot.

Overdispersion and observation-level random effects

Linear mixed models assume the observations to be normally distributed conditional on the fixed-effect parameters and the conditional modes. Thus, they need to estimate the residual variance at the level of observations. If there is only one observation for each level of a grouping variable, the variance of the corresponding random effect will be confounded with the residual variance—we say that the variance of the observation-level random effect is *unidentifiable*. For example, if we decided to treat year as categorical variable in the tundra ecosystem analysis, and included a random effect of the site × year interaction, we would have exactly one observation for each site-by-year combination, and this random effect variance would be confounded with the residual variance. Many libraries (e.g., nlme) will fail to detect this problem, and will give arbitrary answers for the residual variance and the confounded random-effect variance. The same situation applies for any GLMM where the scale parameter determining the variance is estimated rather than fixed, such as Gamma GLMMs or quasi-likelihood models.

Most GLMMs, in contrast, assume distributions such as the binomial or Poisson where the scale parameter determining the residual variance is fixed to 1—that is, if we know the mean then we assume we also know the variance (equal to the mean for Poisson distributions, or to $Np(1 - p)$ for binomial distributions, see book appendix). However, as discussed in chapters 3, 6, and 12, we frequently observe *overdispersion*—residual variances higher than would be predicted from the model, due to missing predictors or among-individual heterogeneity. Overdispersion does not occur in LMMs or in GLMMs with an estimated scale parameter, because the scale or residual variance parameter adjusts the model to match the residual variance. Overdispersion occurs, but is not identifiable, with binary/Bernoulli responses, unless the data are grouped so that there are multiple observations with the same sets of predictor variables (e.g., in the coral predation data there are two replicates in each site/treatment combination). If so, the data can be collapsed to a binomial response, in this case by computing the number of predation events (out of a maximum of 2) for each site/treatment combination, and then overdispersion will be identifiable.

You can allow for overdispersion in GLMMs in some of the same ways as in regular GLMs—use quasi-likelihood estimation to inflate the size of the confidence intervals appropriately, or use an overdispersed distribution such as a negative binomial. These options may not be available in your GLMM software: at present, none of the libraries discussed here offers quasi-likelihood estimation, and only glmmADMB has a well-tested negative binomial option.

A GLMM-specific solution to overdispersion is to add *observation-level* random effects, i.e., to add a new grouping variable with a separate level for every observation in the data set. This seems like magic—how can we estimate a separate parameter for every observation in the data set?—but it is just a way to add more variance to the data distribution. For Poisson distributions, the resulting *lognormal-Poisson* distribution is similar to a negative binomial distribution (sometimes called a *Gamma-Poisson* distribution because it represents a Poisson-distributed variable with underlying Gamma-distributed heterogeneity). Most GLMM packages allow observation-level random effects: for technical reasons, MCMCglmm *always* adds an observation-level random effect to the model, so you can *only* fit overdispersed models. Another advantage of using observation-level random effects is that this variability is directly comparable to the among-group variation in the model; Elston et al. (2001), the source of the grouse tick data, exploit this principle (see also Agresti 2002, section 13.5).

Correlation within groups (R-side effects)

As described above, grouping structure induces a correlation $\rho = \sqrt{\sigma^2_{among}/\sigma^2_{total}}$ between every pair of observations within a group. Observations can also be differentially correlated within groups; that is, an observation can be strongly correlated with some of the observations in its group, but more weakly correlated with other observations in its group. These effects are sometimes called *R-side effects* because they enter the model in terms of correlations of residuals (in contrast with correlations due to group membership, which are called *G-side effects*). The key feature of R-side effects is that the correlation between pairs of observations within a group typically decreases with increasing distance between observations. As well as physical distance in space or time, pairs of observations can be separated by their amount of genetic relatedness (distance along the branches of a pedigree or phylogeny). To include R-side effects in a model, one typically needs to specify both the distance between any two observations (or some sort of coordinates—observation time, spatial location, or position in a phylogeny—from which distance can be computed), as well as a model for the rate at which correlation decreases with distance. While incorporating R-side effects in *linear* mixed models is relatively straightforward—Belshe et al. (2013) included temporal autocorrelation in their model, and chapter 10 gives other examples—putting them into GLMMs is, alas, rather challenging at present.

Fixed effects and families

For a complete model, you need to specify the fixed effects part of your model, and the family (distribution and link function) as well as the random effects. These are both specified in the usual way as for standard (non-mixed, fixed-effect-only) GLMs (chapter 6).

Depending on the package you are using, the fixed effects may be specified separately or in the same formula as the random effects; typically the fixed-effect formula is also where you specify the response variable (the model has only one response variable, which is shared by both the fixed and the random effects). In the tundra ecosystem example, time (year) is the only fixed effect. In the coral symbiont example, the fixed effect is the categorical treatment variable ttt (control/shrimp/crabs/both). In the gopher tortoise example we have the effects of both disease prevalence and, because we didn't have enough levels to treat it as random, year (treated as a categorical variable); we also have an offset term that specifies that the number of shells is proportional to the site area (i.e., we add a log(area) term to the predicted log number of shells). Finally, the grouse tick example uses fixed effects of YEAR and HEIGHT.

13.3.2 *Conditional, marginal, and restricted likelihood*

Once you have defined your GLMM, specifying (1) the conditional distribution of the response variable (`family`) and link function (chapter 6); (2) the categorical and continuous predictors and their interactions (chapter 6); and (3) the random effects and their pattern of crossing and nesting (table 13.1), you are ready to try to fit the model. Chapter 3 describes the process of maximum likelihood estimation, which we extend here to allow for random effects.

Conditional likelihood

If we somehow knew the values of the conditional modes of the random effects for each level (e.g., the predation rates for each block), we could use standard numerical procedures to find the maximum likelihood estimates for the fixed-effect parameters, and all of the associated things we might like to know: confidence intervals, AIC values, and *p*-values for hypothesis tests against null hypotheses that parameters or combinations of parameters were equal to zero. The likelihood we obtain this way is called a conditional likelihood, because it depends (is conditioned on) a particular set of values of the conditional modes. If x is an observation, β is a vector of one or more fixed effects parameters, and b is a vector of the conditional modes of a random effect, then the conditional likelihood for x would be expressed as $L(x|\beta, b)$. If b were a regular fixed effect parameter, then we could go ahead and find the values of β and b that jointly gave the maximum likelihood, but that would ignore the fact that the conditional modes are random variables that are drawn from a distribution. In order to account for this extra variability, we need to define the marginal likelihood.

Marginal likelihood

The marginal likelihood is the modified form of the likelihood that allows for the randomness of the conditional modes. It compromises between the goodness of fit of the conditional modes to their overall distribution and the goodness of fit of the data within grouping variable levels. For example, a large number of attacks on a coral defended by both crabs and shrimp, which would be typically expected to be well protected, could be explained either by saying that the coral was an unlucky individual within its (perfectly typical) block or by saying that the coral was not unlucky but that the block was unusual, i.e., subject to higher-than-average attack rates. Because the block effect is treated as a random variable, in order to get the likelihood we have to average the likelihood over *all possible values* of the block effect, weighted by their probabilities of being drawn from the Normal distribution of blocks. The result is called the marginal likelihood, and we can generally treat it the same way as an ordinary likelihood. In mathematical terms, this average is expressed as an integral. If we take the definitions of x (observation), b (conditional mode), and β (fixed effect parameter) given above, and abbreviate the among-group variance introduced above (σ^2_{among}) as σ^2, then the likelihood of a given value of b is $L(b|\sigma^2)$ (the b values are defined as having a mean of zero) and the marginal likelihood of x is the integral of the conditional likelihood weighted by the likelihood of b:

$$L(x|\beta, \sigma^2) = \int L(x|b, \beta) \cdot L(b|\sigma^2)\, db.$$

Figure 13.2 shows the conditional likelihood $L(x|b, \beta)$ as a dashed line; the likelihood of the conditional mode $L(b|\sigma^2)$ as a dotted line; and the marginal likelihood as the gray area under the product curve. The marginal likelihood is a function of β and σ^2, which are the parameters we want to estimate. In a more complex model, σ^2 would be replaced

Table 13.1 Model specifications in R syntax for the examples

	nlme/glmmADMB	lme4/glmmADMB	MCMCglmm
tundra CO$_2$	fixed = NEE ~ year, random = ~ year | Site (family not specified for LMMs)	formula = NEE ~ year + (year | Site) (family not specified for LMMs)	fixed = NEE ~ year, random = ~ us (year) : Site, family = "gaussian"
coral symbiont	fixed = pred ~ ttt, random = ~ 1 | block, family = "binomial"	formula = pred ~ ttt + (1 | block), family = "binomial"	fixed = pred ~ ttt, random = ~ block, family = "categorical"
gopher tortoise	fixed = shells ~ factor (year) + prev + offset (log (Area)), random = ~ 1 | Site, family = "poisson"	formula = shells ~ factor (year) + prev + offset (log (Area)) + (1 | Site), family = "poisson"	fixed = shells ~ factor (year) + prev + offset (log (Area)), random = ~ Site, family = "poisson"
grouse ticks	fixed = ticks ~ 1 + factor (year) + height, random = ~ (1 | location / brood / index), family = "poisson"	formula= ticks ~ 1 + factor (year) + height + (1 | location / brood / index), family = "poisson"	fixed = ticks ~ 1 + factor (year) + height, random = ~ location + brood + index, family = "poisson"

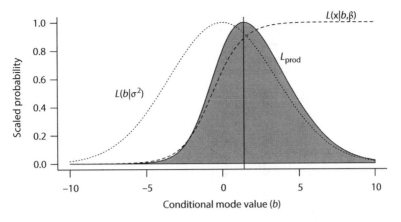

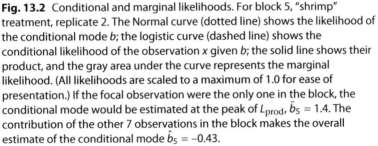

Fig. 13.2 Conditional and marginal likelihoods. For block 5, "shrimp" treatment, replicate 2. The Normal curve (dotted line) shows the likelihood of the conditional mode b; the logistic curve (dashed line) shows the conditional likelihood of the observation x given b; the solid line shows their product, and the gray area under the curve represents the marginal likelihood. (All likelihoods are scaled to a maximum of 1.0 for ease of presentation.) If the focal observation were the only one in the block, the conditional mode would be estimated at the peak of L_{prod}, $\bar{b}_5 = 1.4$. The contribution of the other 7 observations in the block makes the overall estimate of the conditional mode $\hat{b}_5 = -0.43$.

by a vector of parameters, representing the variances of all of the random effects and the covariances among them.

Restricted likelihood

Many of the useful properties of maximum likelihood estimates, such as efficiency and lack of bias, only hold *asymptotically*—that is, when the data set is large. In particular, maximum likelihood estimates of variances are biased downward because they ignore uncertainty in the sample means. You may remember that the usual formula for estimating sample variance is $\sum(x - \bar{x})^2/(n - 1)$, rather than $\sum(x - \bar{x})^2/n$ (the latter is the maximum likelihood estimate), for exactly this reason: dividing by a smaller number ($n - 1$ rather than n) increases the estimate just enough to account for the uncertainty in $\bar{x}$. Restricted maximum likelihood (REML) generalizes this idea to allow for less biased estimates of the variances in mixed models. Technically, it is based on finding some way to combine the observations that factors out the fixed effects. For example, in a pairwise *t*-test the average difference between the two observations in a pair is equal to the difference between treatments, which is the fixed effect. Since we are usually interested in the difference between the treatments, we compute the difference between treatments in each pair. If instead we took the average of each pair, we would cancel out the fixed effect, and could then compute an unbiased estimate of the variance among the pairs. A broader way of thinking about REML is that it applies to any statistical method where we integrate over the fixed effects when estimating the variances. When using REML, you *cannot* compare the restricted likelihoods of two models with different sets of fixed effects, because they are likelihoods of completely different models for the variance. While REML in principle applies to GLMMs as well as LMMs, they are more easily defined and more accessible in software for LMMs than for GLMMs (Bellio and Brazzale 2011; Millar 2011). It's generally good to use REML, if it is available, when you are interested in the magnitude of the random effects variances, but *never* when you are comparing models with different fixed effects via hypothesis tests or information-theoretic criteria such as AIC.

13.4 Setting up a GLMM: practical considerations

13.4.1 *Response distribution*

The conditional distribution of the response variable, which we often abbreviate to "the response distribution" or "the distribution of the data," is the expected distribution of each observed response around its predicted mean, given the values of all of the fixed and random effects for that observation. That is, when we collect a data set of (for example) counts, we don't expect the overall (marginal) distribution of the data to be Poisson; we expect each point to be drawn from a Poisson distribution with its own mean that depends on the predictors for that point (chapter 6). In the gopher tortoise example, the distribution of number of shells S in a given site s (with infection seroprevalence $P(s)$) and year y is $S_{sy} \sim \text{Poisson}(\beta_0 + \beta_y + \beta_P P(s) + b_s)$, where β_0 is the baseline (year-0, 0-prevalence, average site) expectation; β_y is the difference between year y and the baseline; β_P is the effect of an additional percentage of seroprevalence; and b_s gives the difference between site s and the overall average.

If the conditional distribution is Gaussian, or can sensibly be transformed to be Gaussian (e.g., by log transformation), as in the tundra ecosystem example, then we have a *linear* mixed model, and several aspects of the modeling process are simpler (we can more easily define R-side effects and restricted maximum likelihood; statistical tests are also easier: see section 13.6). As with GLMs (chapter 6), binomial (including binary or Bernoulli, i.e., 0/1 responses) and Poisson responses comprise the vast majority of GLMMs. The Gamma distribution is the other common distribution handled by GL(M)Ms; it is useful for continuous, skewed distributions, but treating such data as lognormal (i.e., log-transforming and then using a linear mixed model) is easier and usually gives very similar results.

In addition to these standard distributions, there are other useful distributions that do not technically fall within the scope of GLMMs, but can sometimes be handled using simple extensions. These include the negative binomial distribution for overdispersed count data; zero-inflated distributions for count data with excess zeros (chapter 12); the beta distribution for proportional data that are not proportions out of a known total count; and the Tweedie distribution for continuous data with a spike at zero (glmmADMB handles the first three cases). Ordinal responses (i.e., categorical responses that have more than two ordered categories) and multinomial responses (categorical responses with more than two categories, but without ordering) can be handled by extensions of binomial GLMMs, implemented in the clmm function in the ordinal library. These extensions are often useful, but using them will generally make it harder to analyze your model (you are more likely to run into computational difficulties, which will manifest themselves as warnings and errors from software), and restrict your choice of software more than if you stick to the simpler (Normal, binomial, Poisson) distributions.

As is typical in ecological applications, the examples for this chapter all use either Normal (tundra ecosystem), binary (coral symbiont), or Poisson (gopher tortoise, grouse tick) conditional distributions (table 13.1). The family is specified almost exactly as in standard GLMs, with a few quirks. For linear mixed models, you should use the `lme` (in the `nlme` library) or `lmer` (in the `lme4` library), or `family="gaussian"` in MCMCglmm or glmmADMB. MCMCglmm and glmmADMB require the family argument to be given as a quoted string (e.g., `family="poisson"`), in contrast to `lme4`, which allows more flexibility (e.g., `family="poisson"` or `poisson()`). MCMCglmm has different names from the standard R conventions for binary/logit (`family="categorical"`) and binomial (`family="multinomial2"`) models.

13.4.2 *Link function*

As with GLMs, we also have to choose a link function to describe the shape of the response curve as a function of continuous predictor variables. The rules for picking a link function are the same as for GLMs: when in doubt, use the default (canonical) link for the response distribution you have chosen. We will follow this rule in the examples, using the default logit link for the coral symbiont (binary) example and a log link for the gopher tortoise and grouse tick (Poisson) examples (table 13.1), although we did also consider a log link for the coral symbiont example. In `lme4` links are specified along with the family as for standard GLMs in R, e.g., `family=binomial(link="logit")` or `binomial(link="log")`; in glmmADMB they are specified as a separate string (`link="logit"`); and MCMCglmm uses alternative family names where alternate links are available (e.g., `family="ordinal"` for a binary/probit link model).

13.4.3 *Number and type of random effects*

As discussed in section 13.3.1, it is not always easy to decide which variables to treat as random effects. The more random effects a model includes, the more likely you are to run into computational problems. It is also more likely that the fit will be *singular*: some random effects variances will be estimated as exactly zero, or some pairs of random effects will be estimated as perfectly correlated. While this does not necessarily invalidate a particular model, it may break model-fitting software in either an obvious way (errors) or a non-obvious way (the model is more likely to get stuck and give an incorrect result, without warning you). Model complexities interact: for example, some of the software available to fit models with non-standard distributions can only handle a single random effect. In general you should avoid: (1) fitting random effects to categorical variables with fewer than five levels, and, unless you have very large data sets and a fast computer, (2) fitting more than two or three random effects in a single model or (3) fitting vector-valued random effects (i.e., among-group variation of responses to categorical variables) for categorical predictors with more than two or three levels.

13.5 Estimation

Once the model is set up, you need to estimate the parameters—the fixed-effect parameters that describe overall changes in the response, the conditional modes of the random effects that describe the predicted differences of each level of the grouping variable from the population average, and the variances of, and covariances among, the random effects. This isn't always easy; there are a variety of methods, with trade-offs in speed and availability.

13.5.1 *Avoiding mixed models*

Sometimes fitting a mixed model is difficult: for example, if you have too few levels of your random effect, or repeated measurements within just a few blocks. In this case fitting a mixed model doesn't have many advantages, and you may be able to take a shortcut instead.

- For data from a nested experimental design, taking the average of each block and doing a one-way ANOVA on the results will give you exactly the same results for the fixed effects as you would get from a mixed model (Murtaugh 2007); if your data are unbalanced you can do a weighted ANOVA with weights of $1/n_i$ (where n_i is the number of observations in the ith block). If you want to allow (for example) varying slopes across blocks, you can fit a *two-stage model*, where you fit a linear regression for each block separately and then do a one-way ANOVA on the slopes. This works best with Normal data, but if you have many points per block the block averages will be approximately Normal—although you may still need to deal with heteroscedasticity, e.g., by transforming the data appropriately.
- You can try showing that random effects are ignorable by fitting a model that ignores random effects, and then using a one-way ANOVA on the residuals of the model by block to show that they do not vary significantly across blocks.
- If you need to compute the among-block variance when there are too few levels (< 5), you can fit the blocks as a fixed effect, with "sum to zero" contrasts set, and compute the mean of the squared coefficients $(\sum (\beta_i - \text{mean}(\beta))^2/(n-1))$.
- If you have paired comparisons (i.e., you are testing the difference between two fixed effect levels, such as treatment vs. control within each block) for normally distributed responses, you can replace the test of the fixed effect with a paired t-test, and estimate the among-block variance by computing the variance of ((control + treatment)/2) across blocks.

For many situations (e.g., randomized block or crossed designs, or pairwise comparisons of non-Normal data), you may not be able to use these shortcuts and will have to proceed with a mixed model.

13.5.2 *Method of moments*

The traditional way to fit a mixed ANOVA model is to compute appropriate sums of squares (e.g., the sum of squares of the deviations of the group means from the grand mean, or the deviations of observations from their individual group means) and dividing them by the appropriate degrees of freedom to obtain mean squares, which are estimates of the variances. This approach is called the method of moments because it relies on the correspondence between the sample moments (mean squares) and the theoretical parameters of the model (i.e., the random effects variances). This approach is simple, fast, always gives an answer—and is extremely limited, applying only to Normal responses (i.e., linear mixed models), in balanced or nearly balanced designs, with nested random effects only (see Gotelli and Ellison 2004 or Quinn and Keough 2002).

13.5.3 *Deterministic/frequentist algorithms*

Instead of computing sums of squares, modern estimation approaches try to find efficient and accurate ways to compute the marginal likelihood (section 13.3.2), which can be challenging. The first class of approaches for estimating mixed models, which I call deterministic approaches (note that this is not standard terminology), are typically used in a frequentist statistical framework to find the maximum likelihood estimates and confidence intervals.

- *Penalized quasi-likelihood* (PQL, Breslow 2004) is a quick but inaccurate method for approximating the marginal likelihood. While it is fast and flexible, it has two important limitations. (1) It gives biased estimates of random-effects variances, especially with binary data or count data with low means (e.g., Poisson with mean < 5). More accurate versions of PQL exist, but are not available in R. The bias in random-effect variances may be unimportant if your questions focus on the fixed effects, but it's hard to be sure. (2) PQL computes a quantity called the "quasi-likelihood" rather than the likelihood, which means that inference with PQL is usually limited to less-accurate Wald tests (section 13.6.2).
- *Laplace approximation* is a more accurate, but slower and less flexible, procedure for approximating the marginal likelihood.
- *Gauss–Hermite quadrature* (GHQ) is a more accurate, but still slower and less flexible approach. Where Laplace approximation uses one point to integrate the marginal likelihood, GHQ uses multiple points. You can specify how many points to use; using more is slower but more accurate. The default is usually around eight; `lme4` allows up to 25, which is usually overkill. Many software packages restrict GHQ to models with a single random effect.

You should use the most accurate algorithm available that is fast enough to be practical. If possible, spot-check your results with more accurate algorithms. For example, if Laplace approximation takes a few minutes to fit your models and GHQ takes a few hours, compare Laplace and GHQ for a few cases to see if Laplace is adequate (i.e., whether the difference between the coefficient values between the two methods is small relative to their standard errors).

13.5.4 *Stochastic/Bayesian algorithms*

Another approach to GLMM parameter estimation uses *Markov chain Monte Carlo* (MCMC), a stochastic estimation algorithm. There's not nearly enough room in this chapter to give a proper explanation of MCMC; you can just think of it as a general computational recipe for sampling values from the probability distribution of model parameters. Stochastic algorithms are usually much slower than deterministic algorithms, although a single run of the algorithm provides both the coefficients and the confidence intervals, in contrast to deterministic algorithms, where computing reliable confidence intervals may take several times longer than just finding the coefficients.

Although there is at least one "black box" R library (`MCMCglmm`) that allows the user to define the fixed and random effects via the sorts of formulas shown in table 13.1, many researchers who opt for stochastic GLMM parameter estimation use the BUGS language instead (i.e., the WinBUGS package or one of its variants such as OpenBUGS or JAGS) to fit their models. BUGS is a flexible, powerful framework for fitting ecological models to data in a Bayesian context (McCarthy 2007; Kéry 2010), not just GLMMs, but it comes with its own steep learning curve.

For technical reasons, most stochastic algorithms use a Bayesian framework, usually with weak priors (chapter 1); except when you have parameters that are very uncertain, this distinction doesn't make a huge practical difference. Bayesian inference, and stochastic algorithms in general, make it much easier to compute confidence intervals that incorporate all the relevant sources of uncertainty (section 13.6.2). If you want to use stochastic algorithms but avoid Bayesian methods, you can use a stochastic algorithm that works within a frequentist framework, such as *data cloning* (Ponciano et al. 2009; Sólymos 2010).

13.5.5 *Model diagnostics and troubleshooting*

Model checking for GLMMs overlaps a lot with the procedures for GLMs (chapter 6). You should plot appropriately scaled residuals (i.e., deviance or Pearson residuals) against the fitted values and against the input variables, looking for unexplained patterns in the mean and variance; look for outliers and/or points with large influence (leverage); and check that the distribution of the residuals is reasonably close to what you assumed. For Poisson or binomial GLMMs with $N > 1$, you should compare the sum of the squared Pearson residuals to the residual degrees of freedom (number of observations minus number of fitted parameters) to check for overdispersion (unless your data are binary, or the model already contains an observation-level random effect; appendix 13A).

The first GLMM-specific check is to see whether the model is singular: that is, whether non-zero variances (and non-perfect correlations among random effects, i.e., $|\rho| < 1$) could be estimated for all the random effects in the model. If some of the variances are zero or some correlations are ± 1, it indicates that not only was the among-group variation not significantly different from zero, the best estimate was zero. Your model is probably too complex for the data: the best way to avoid this problem in general is to try to simplify the model *in advance* to a level of complexity that you think the data can support, by leaving out random-effects terms or by converting them to fixed effects. It does take some practice to calibrate your sense of what models can be fitted. For example, in the coral symbiont example I left out the block × treatment interaction, successfully fitting a non-singular model, but in the gopher tortoise example the model with site and observation-level random effects was singular even though I had tried to be conservative by treating year as a fixed effect.

Although in principle the results from a singular model fit will be the same as if you had just left the zero-estimate terms out of the model in the first place, you should probably refit the model without them to make sure this is true (i.e., that the software hasn't run into computational problems because the model was too complicated). Another possible solution to this problem is to impose a Bayesian prior on the variances to push them away from zero, which you can do using the `blme` (Chung et al. 2013) or `MCMCglmm` libraries. Although some researchers advocate simply picking a reasonable model and sticking with it (i.e., not looking for a more parsimonious reduced model: Barr et al. 2013), you can also use information-theoretic approaches (AIC or BIC) to choose among possible candidate random-effects models (see chapter 3 and section 13.6.2), especially if you are interested in prediction rather than in testing hypotheses.

Another diagnostic specific to (G)LMMs is checking the estimates of the conditional modes. In theory these should be Normally distributed (you can check this using a *quantile–quantile* ("q–q") plot). You should only worry about extreme deviations: no-one really knows how badly a non-Normal distribution of conditional modes will compromise a (G)LMM, and fitting models with non-Normal modes is difficult. Look for extreme conditional modes and treat them as you would typically handle outliers; for example, figure out whether there is something wrong with the data for those groups, or try fitting the model with these groups excluded and see whether the results change very much.

For MCMC analyses (e.g., via MCMCglmm), you should use the usual diagnostics for convergence and mixing (read more about these in McCarthy (2007) and Kéry (2010)), check quantitative diagnostics such as the Gelman–Rubin statistic and effective sample size, and examine graphical diagnostics (trace and density plots) for both the fixed and random effects parameters. With small data sets, the variance–covariance parameters often mix badly, sticking close to zero much of the time and occasionally spiking near zero; the corresponding density plots typically show a spike at zero with a long tail of larger values. There are no really simple fixes for this problem, but some reasonable strategies include (1) running much longer chains; (2) adding an informative prior to push the variance away from zero; (3) taking the results with a grain of salt (appendix 13A).

As you try to troubleshoot the random-effects component of your analysis, you should keep an eye on the fixed-effect estimates and confidence intervals associated with models with different random effects structures; the fixed-effect estimates often stay pretty much the same among models with different random effects. This can be comforting if your main interest is in the fixed effects, although you should be careful since fitting multiple models also allows some scope for cherry-picking the results you like.

13.5.6 *Examples*

Some technical issues that arose as I fitted and diagnosed models for the examples above were (appendix 13A for more details):

- *Tundra CO_2 flux*: Overall the fits were well-behaved, but one site (Toolik) differed from the others; its observations were poorly fitted by the full model (they had large residuals with high variance) and its conditional mode for the slope was an outlier. Including the Toolik data made the model harder to fit, and generated autocorrelation in the residuals that could not be completely accounted for. However, the primary estimate of the population-level rate of increasing CO_2 flux remained qualitatively similar whether we included the Toolik data or not.

- *Coral symbionts*: One observation in the data set was poorly predicted—it was a coral that escaped predation although it had a high expected predation risk (it was in the no-symbiont treatment in a frequently attacked block). Refitting the model without this observation led to nearly complete separation (chapter 6), making the estimates even more extreme. In the end we retained this data point, since including it seemed to give conservative estimates. Other aspects of the model looked OK—the distribution of conditional modes was sensible, and using GHQ instead of the Laplace approximation changed the estimates only slightly.

- *Gopher tortoise shells*: Poisson sampling accounted for nearly all the variation in the data—the estimated variances both among observations and among sites were very close to zero. Thus, the conditional modes were also all near zero. In other words, we would have obtained similar results from a simple Poisson GLM. The residuals looked reasonable, with similar variation in each site. The MCMCglmm fit, which included both among-site and among-observation variation, showed unstable estimates of the random-effects variance, as described in section 13.5.5. We couldn't simplify the model, but instead used a stronger prior on the among-site variance to stabilize it. This didn't change the estimated effect of disease prevalence, but did increase its uncertainty. In Ozgul et al. (2009) we fitted the full model with WinBUGS; if I ran the analysis again today I would either fit a simple Poisson model or use MCMCglmm or blme to fit the full model with stabilizing priors.

- *Grouse ticks*: The residuals and estimated conditional modes all looked reasonable. We didn't test for overdispersion since the model includes observation-level random effects. Deterministic algorithms (lme4 and glmmADMB) gave positive estimates for all of the variances, but MCMCglmm disagreed; unless we added a prior, it estimated the among-location variance as nearly zero, suggesting that the separation of variation into among-brood vs. among-location components is unstable.

13.6 Inference

13.6.1 *Approximations for inference*

Estimates of parameters are useless without confidence intervals, or hypothesis tests (*p*-values), or information criteria such as AIC, that say how much we really know. Inference for GLMMs inherits several assumptions from GLMs and linear mixed models that do not apply exactly for GLMMs, and which (as with the estimation methods for GLMMs) require trade-offs between accuracy, computation time, and convenience or availability in software. As with estimation (section 13.5), you should generally use the slowest but most accurate method that is practical, double-checking your results with a slower and more accurate method if possible. Inference for GLMMs involves three separate types of approximation, which we will discuss in general before discussing specific methods for inference in section 13.6.2.

Shape: the fastest but least accurate approaches to GLMM inference (Wald intervals and tests) make strong assumptions about the shape of the likelihood curve, or surface, that are exactly true for linear models (ANOVA/regression), but only approximately true for GLMs, LMMs, and GLMMs. These approximations are more problematic for smaller data sets, or for data with high sampling variance (binary data or Poisson or binomial data with small observed counts or numbers of successes/failures).

Finite-size effects: When the data set is not very large (e.g., < 40 observations, or < 40 levels for the smallest random-effect grouping variable) we have to make further assumptions about the shapes of distributions of summaries such as the likelihood ratio or *F* statistic. In the classical ANOVA or regression framework, these assumptions are taken care of by specifying the "denominator degrees of freedom," that is, specifying the effective number of independent observations. For GLMs, for better or worse, people usually ignore these issues completely.

- For LMMs that don't fit into the classical ANOVA framework (i.e., with unbalanced designs, crossed random effects, or R-side effects), the degrees of freedom for the *t* distribution (for testing individual parameters), or the denominator degrees of freedom for the *F* distribution (for testing effects), are hard to compute and are at best approximate. If your experimental/observational design *is* nested and balanced, you can use a software package that computes the denominator degrees of freedom for you or you can look the experimental design up in a standard textbook (e.g., Gotelli and Ellison 2004 or Quinn and Keough 2002). If not, then you will need to use the approximation methods implemented in the lmerTest and pbkrtest libraries (Kenward and Roger 1997; Halekoh and Højsgaard 2013), or use a resampling-based approach (section 13.6.2).
- GLMMs involve a different finite-size approximation (the distribution of the likelihood ratio test statistic is approximate rather than exact). Stroup (2014) states that the Kenward–Roger approximation procedure developed for LMMs works reasonably well for GLMMs, but neither it nor Bartlett corrections (another approximation method

described in McCullagh and Nelder 1989) are implemented for GLMMs in R; you will need to use stochastic sampling methods (section 13.6.2) if you are concerned about finite-size inference for GLMMs.

Boundary effects: statistical tests for linear models, including GLMMs, typically assume that estimated parameters could be either above or below their null value (e.g., slopes and intercepts can be either positive or negative). This is not true for the random effect variances in a (G)LMM—they must be positive—which causes problems with standard hypothesis tests and confidence interval calculations (Pinheiro and Bates 2000). In the simplest case of testing whether a single random-effect variance is zero, the *p*-value derived from standard theory is twice as large as it should be, leading to a conservative test (you're more likely to conclude that you can't reject the null hypothesis). To test the null hypothesis that the sole random-effect variance in a model is equal to zero you can just divide the *p*-value by 2. If you want to test hypotheses about random effects in a model with more than one random effect you will need to simulate the null hypothesis (section 13.6.2).

13.6.2 *Methods of inference*

Wald tests

The standard errors and *p*-values that R prints out when you summarize a statistical model (*Wald* standard errors and tests) are subject to artifacts in GLM or GLMM modeling. They're especially bad for binomial data where some categories in the data have responses that are mostly (or all) successes or failures (*complete separation*: the related inference problems are called the *Hauck–Donner effect*: Venables and Ripley 2002). The typical symptom of these problems is large parameter estimates (e.g., absolute value>10) in conjunction with huge standard errors and very large ($p \approx 1$) *p*-values: sometimes, but not always, you will also get warnings from the software. More generally, Wald statistics are less accurate than the other methods described below. However, they are quick to compute, can be useful for a rapid assessment of parameter uncertainty, and are reasonably accurate for large data sets. If you can guess the appropriate residual degrees of freedom, then you may try to use appropriate *t* statistics rather than *Z* statistics for the *p*-values and confidence interval widths in order to account for finite sample sizes, but this is a crude approximation in the case of GLMMs.

Likelihood ratio tests

Likelihood ratio tests and profile confidence intervals are an improvement over Wald statistics, but come at a computational cost that may be significant for large data sets. You can use the likelihood ratio test to compare nested models (via the anova command in R, or by computing the *p*-value yourself based on the χ^2 distribution); this provides a significance test for the factors that differ between the two models. The corresponding confidence intervals for a parameter are called profile confidence intervals. Profile confidence intervals are computationally challenging—they may take dozens of times as long to compute as the original model fit. Furthermore, because profile likelihood calculations have to evaluate the likelihood for extreme parameter values, they are much more subject to computational problems than the original model fit.

Finally, although likelihood-based comparisons are more reliable than Wald statistics, they still assume infinite denominator degrees of freedom. If your effective sample size is large enough (e.g., the smallest number of levels of any grouping variable in your model is > 40), you don't need to worry. Otherwise you may need to use a stochastic resampling method such as parametric bootstrapping or Markov chain Monte Carlo for accurate inference.

Bootstrapping

Bootstrapping means resampling data with replacement to derive new pseudo-data sets, from which you can estimate confidence intervals (chapter 1). Parametric bootstrapping (PB) instead simulates pseudo-data from the fitted model (or from *reduced* models that omit a parameter you are interested in making inferences about). You can then refit your model to these pseudo-data sets to get reliable *p*-values or confidence intervals.

PB is very slow (taking hundreds or thousands of times as long as fitting the original model), and it does make assumptions—that the model structure is appropriate, and that the estimated parameters are close to the true parameters—but it is the most accurate way we know to compute *p*-values and confidence intervals for GLMMs.

Specialized forms of PB are faster. For example, the `RLRsim` library in R (Scheipl et al. 2008) does a kind of PB to compute *p*-values for random-effect terms in LMMs, orders of magnitude faster than standard PB.

You can also use non-parametric bootstrapping—resampling the original data values—but you must respect the grouping structure of the data. For example, for a model with a single grouping variable you could do two-stage bootstrapping (Field and Welsh 2007), first sampling with replacement from the levels of the grouping variable, then sampling with replacement from the observations within each sampled group. For more complex models (with crossed random effects, or R-side effects), appropriate resampling may be difficult.

MCMC

The results of an MCMC fit (section 13.5.4) give estimates and confidence intervals on parameters; you can also get *p*-values from MCMC, although it is unusual (since most MCMC is based in a Bayesian framework). MCMC is very powerful—it automatically allows for finite size effects, and incorporates the uncertainty in all the components of the model, which is otherwise difficult. It's so powerful, in fact, that some frequentist tools (such as AD Model Builder) use a variant of MCMC to compute confidence intervals. This pseudo-Bayesian approach is convenient, but may have problems when the information in the data is weak. For small, noisy data sets the distribution of the variance parameters is often composed of a spike at zero along with a second component with a mode away from zero. In this case, many MCMC algorithms can get stuck sampling either the spike or the non-zero component, and thus give poor results.

Information-theoretic approaches

Many ecological researchers use information-theoretic approaches to select models and generate parameter importance weights or weighted multimodel averages of parameters and predictions (chapter 3; Burnham and Anderson 2002). In principle, AIC (and other indices like BIC) do apply to mixed models, but several of the theoretical difficulties discussed in section 13.6.1 affect information criteria (Greven and Kneib 2010; Müller et al. 2013).

- AIC comparisons among models with different variance parameters have the same problem as null-hypothesis tests of variances (section 13.6.1)—they tend to understate the importance of variance terms.

- When comparing models with different random-effects terms, or when using a finite-size corrected criterion such as AICc, the proper way to compute the model complexity (number of parameters) associated with a random effect depends on whether you are trying to predict at the population level (predicting the average value of a response across all random-effects levels) or at the individual level (a *conditional* prediction, i.e., making predictions for specific levels of the random effect). For population-level prediction, you should count one parameter for each random-effects variance or covariance/correlation. For conditional prediction, the correct number of parameters is somewhere between 1 and $n-1$, where n is the number of random-effects levels: methods for computing appropriate AIC values in this case (Vaida and Blanchard 2005) are not widely implemented. Academic ecologists typically want to know about effects at the level of the whole population, which allows them to use the easier one-parameter-per-variance-parameter rule; applied ecologists might be more interested in predictions for specific groups. Bayesian MCMC has an information-theoretic metric called the *deviance information criterion* (DIC: Spiegelhalter et al. 2002), for which the so-called level of focus must be defined similarly (O'Hara 2007).
- Finite-size-corrected criteria such as AICc are poorly understood in the mixed model context. For example, for n in the denominator of the AICc correction term ($n-k-1$: chapter 3), should one count the total number of observations in a nested design, or the number of groups? For better or worse, most ecologists use AICc for model selection with GLMMs without worrying about these issues, but this may change as statisticians come to understand AICc better (Shang and Cavanaugh 2008; Peng and Lu 2012).

In general you should *pick a single approach to modeling and inference in advance*, or after brief exploration of the feasibility of different approaches, in order to avoid the ever-present temptation to pick the results you like best.

13.6.3 *Reporting the GLMM results*

Graphical summaries of statistical analyses that display the model coefficients and their uncertainty, or that overlay model predictions and their uncertainties on the original

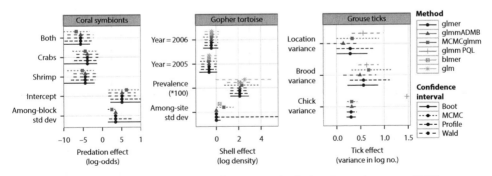

Fig. 13.3 Comparisons of all estimation/inference methods showing estimates and 95% confidence intervals for all three GLMM examples. For the most part all methods give similar results; the biggest differences are in the MCMCglmm estimates (which represent posterior means rather than maximum likelihood estimates) and in the estimates and confidence intervals for random-effect standard deviations or variances.

data, are important (Gelman et al. 2002). However, you also need to summarize the results in words. This summary should include the magnitudes and confidence intervals of the fixed effects; the magnitude of the among-group variation for each random effect, whether it is of primary interest or not; and possibly the confidence intervals of the among-group variation (if the random effects are included because they are part of the design, you should *not* test the null hypothesis that they are zero). If you are interested in the partitioning of variance across levels, report among-group variation as random-effect variances, or proportions of variance (see the grouse tick example below). If you are more interested in the fixed effects, report among-group variation as random-effect standard deviations, as these are directly comparable to the corresponding fixed effects. The following are sample reports for the four worked examples; appendix 13A shows the technical details of deriving these results. The results from all the combinations of estimation and inference methods in this chapter are summarized in Figure 13.3.

- *Tundra carbon*: The main effect of interest is the across-site average change in growing-season carbon flux per year; the estimated slopes are negative because the rate of carbon loss is increasing. Our conclusion from the fitted model with the year variable centered (i.e., setting Year=0 to the overall mean of the years in the data) would be something like: "the overall rate of change of growing season NEE was –3.84 g C/m^2/season/year (t_{23} = –2.55, p = 0.018, 95% CI = {–6.86, –0.82}). We estimated a first-order autocorrelation within sites of ρ = 0.39; among-site variation in the intercept was negligible, while the among-site standard deviation in slope was 5.07 g C/m^2/season/year, with a residual standard deviation of 58.9 g C/m^2/season."
- *Coral symbionts*: For the analysis done here (logit link, one-way comparison of crab/shrimp/both to control) we could quote either the fixed-effect parameter estimates (clarifying to the reader that these are differences between treatments and the baseline control treatment, on the logit or log-odds scale), or the changes in predation probability from one group to another. Taking the first approach: "Crab and shrimp treatments had similar effects (–3.8 log-odds decrease in predation probability for crab, –4.4 for shrimp); the dual-symbiont treatment had an even larger effect (–5.5 units), but although the presence of any symbiont caused a significant drop in predation probability relative to the control (Wald p-value 0.0013; parametric bootstrap p-value < 0.003), none of the symbiont treatments differed significantly from each other (likelihood ratio test p = 0.27, parametric bootstrap test (N = 220) p = 0.23); in particular, two symbionts did not have significantly greater protective effects than one (Wald and PB p-values both $\approx$0.15). The among-block standard deviation in log-odds of predation was 3.4, nearly as large as the symbiont effect." (McKeon et al. (2012) present slightly different conclusions based on a model with a log rather than a logit link.) Alternately, one could quote the predicted predation probabilities for each group, which might be more understandable for an ecological audience.
- *Gopher tortoise*: The main point of interest here is the effect of prevalence on the (per-area) density of fresh shells. This makes reporting easy, since we can focus on the estimated effect of prevalence. Because the model is fitted on a log scale and the parameter estimate is small, it can be interpreted as a proportional effect. For example: "A 1% increase in seroprevalence was associated with an approximately 2.1% increase (log effect estimate = 0.021) in the density of fresh shells (95% CI = {0.013, 0.031} by parametric bootstrap [PB]). Both of the years subsequent to 2004 had lower shell densities (log-difference = –0.64 (2005), –0.43 (2006)), but the differences were not statistically significant (95% PB CI: 2005 = {– 1.34, 0.05}, 2006 = {– 1.04, 0.18}). There was no

detectable overdispersion (Pearson squared residuals/residual df = 0.85; estimated variance of an among-observation random effect was zero). The best estimate of among-site standard deviation was zero, indicating no discernible variation among sites, with a 95% PB CI of {0, 0.38}."

- *Grouse ticks*: In this case the random-effects variation is the primary focus, and we report the among-group variance rather than standard deviation because we are interested in variance partitioning. "Approximately equal amounts of variability occurred at the among-chick, among-brood, and among-location levels (MCMCglmm, 95% credible intervals: σ^2_{chick} = 0.31 [95% CI {0.2, 0.43}, σ^2_{brood} = 0.59 {0.36, 0.93}, $\sigma^2_{location}$ = 0.57 {0.29, 1.0}]. The among-brood variance is estimated to be approximately twice the among-chick and among-location variances, but there is considerable uncertainty in the brood/chick variance ratio ($\sigma^2_{brood}/\sigma^2_{chick}$ = 2.01 {1.007, 3.37}), and estimates of the among-location variance are unstable. Year and altitude also have strong effects. In 1996, tick density increased by a factor of 3.3 relative to 1995 (1.18 {0.72, 1.6} log units); in 1997 density decreased by 38% (–0.98 {–1.49, 0.46} log units) relative to 1995. Tick density increased by approximately 2% per meter above sea level (–0.024 {–0.03, –0.017} log-units), decreasing by half for every 30 (log(2)/0.024) m of altitude."

13.7 Conclusions

I hope you are convinced by now that GLMMs are a widely useful tool for the statistical exploration of ecological data. Once you get your head around the multi-faceted concept of random effects, you can see how handy it is to have a modeling framework that naturally combines flexibility in the response distribution (GLMs) with the ability to handle data with a variety of sampling units with uneven and sometimes small sample sizes (mixed models).

GLMMs cannot do everything; especially for very small data sets, they may be overkill (Murtaugh 2007). Ecologists will nearly always have too little data to fit as sophisticated a model as they would like, but one can often find a sensible middle ground.

In this chapter I have neglected the other end of the spectrum, very large data sets. Ecologists dealing with Big Data from remote sensing, telemetry, citizen science, or genomics may have tens or hundreds of thousands of observations rather than the dozens to hundreds represented in the examples here. However, telemetry and genomic data often contain huge amounts of detail about a small number of individuals; in this case a fixed-effect or two-stage (Murtaugh 2007) model may work as well as a GLMM. The good news is that some of the computational techniques described here scale well to very large data sets, and some of the most computationally intensive analyses become unnecessary when all the grouping variables have more than 40 levels.

I have also neglected a variety of useful GLM extensions such as non-standard link functions (for fitting specific non-linear models such as the Beverton–Holt or Ricker functions); methods for handling multinomial or ordinal data; and zero-inflation. The good news is that most of these tricks are at least in principle extendable to GLMMs, but your choice of software may be more limited (Bolker et al. 2013).

Unfortunately, GLMMs do come with considerable terminological, philosophical, and technical baggage, which I have tried in this chapter to clarify as much as possible. As GLMM software, and computational power, continue to improve, many of the technical difficulties will fade, and GLMMs will continue their growth in popularity; a firm grasp of the *conceptual* basis of GLMMs will be an increasingly important part of the quantitative ecologist's toolbox (Zuur et al. 2009, 2012, 2013; Millar 2011).

APPENDIX

Probability distributions

Random variables are variables whose possible values are the outcomes of random processes. The probability distribution for a particular random variable describes the chance of occurrence of each possible value of that variable. We denote the random variable of interest as Y, and y is used to denote a specific value taken by that variable. Discrete random variables (e.g., count data) consist of values from a countable set (that is, a quantity that can be represented by integers). The probability that a discrete random variable takes on each of the set of possible values (e.g., a butterfly might lay 20, 21, 22 eggs on a leaf) is described by a probability mass function (PMF). Continuous random variables (e.g., lengths, rates, probabilities) consist of real values, and can take *any* value in some range (e.g., the mass of each butterfly egg can vary by quantities much smaller than we can ever measure). The probability density function (PDF) describes how the probability changes as the measurement changes. Consult basic statistics books for more detail.

Each PMF and PDF is described by one or more parameters that determine the shape of the distribution. The most common shape measurements are the mean, variance, and skewness, which describe the central tendency, the spread, and the asymmetry of the distribution, respectively. Distributions can be written in a number of mathematically equivalent ways, all with the same number of parameters. Why choose one of these as opposed to another? Some ways of writing these distributions are easier to interpret biologically, or to work with mathematically. Because the mean is such a useful descriptor, for all of the probability distributions presented in this appendix, we show how to calculate the mean, along with other useful parameters. These parameters are not always provided by software packages.

Because many statistical analyses, especially those involving likelihoods, require calculating the natural logarithm of a probability distribution (see chapter 3), in this appendix we present both the distribution/mass function and its natural logarithm (henceforth referred to as lnPDF or lnPMF), for each distribution.

This appendix provides only a summary of some of the more frequently used probability distributions mentioned throughout the book. As with the rest of the book, R code for the figures is available at our online site; you may find it useful to perform numerical experiments aimed at deepening your understanding of how the parameters affect the distributions. We have not included commonly encountered statistical PDFs that are used when performing null-hypothesis tests, such as the Student's t-distribution, the F-distribution, and the chi-squared distribution, because they are mainly used for hypothesis testing, while the distributions discussed below are often used for models of statistical populations.

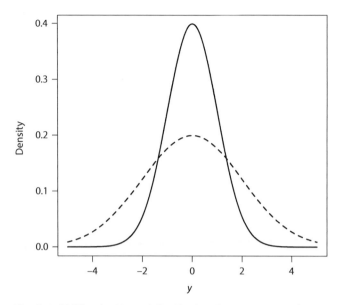

Fig. A.1 PDF for the Normal distribution, for mean $\mu = 0$; the solid and dashed lines are for $\sigma = (1, 2)$, respectively.

A.1 Continuous random variables

A.1.1 *Normal distribution (chapters 1, 3, 4, 5, 6, 7, 8, 9, 10, 11, 12, and 13)*

The Normal distribution is the most commonly used continuous probability distribution. It assumes that variation is symmetric about the mean μ. The typical form of this distribution is:

$$f_n(y|\mu, \sigma) = \frac{1}{\sqrt{2\pi\sigma^2}} \exp\left[-\frac{1}{2}\left(\frac{y-\mu}{\sigma}\right)^2\right],$$

where σ^2 is the variance of Y; see figure A.1 for some examples of the PDF. As discussed above, it is frequently more useful to work with the lnPDF:

$$\ln f_n(y|\mu, \sigma) = -\frac{1}{2}\ln(2\pi) - \ln(\sigma) - \frac{1}{2}\left(\frac{y-\mu}{\sigma}\right)^2.$$

Although this distribution takes on values from minus infinity to plus infinity, it is often used for variables that can only take on positive values. This misspecification is usually not problematic if the mean is sufficiently large and the standard deviation is small, so that negative outcomes are extremely unlikely.

A.1.2 *Lognormal distribution (chapters 3, 5, 6, and 13)*

If a continuous random variable is positive ($y > 0$), expected to have positive skew, and intermediate values are most likely, then the lognormal distribution may be appropriate. Some care is needed in using this distribution, because it describes the case in which the logarithm of y is normally distributed; this makes it essential to distinguish between y and its log, and between quantities like the mean or variance of y and the mean or variance of its log. If we set μ as the mean of $\ln(y)$ and σ as the standard deviation of $\ln(y)$, the typical form of this distribution is:

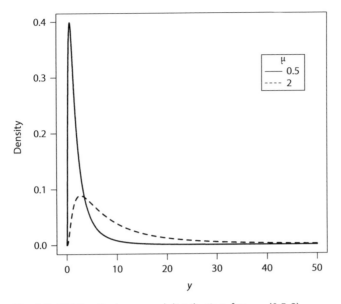

Fig. A.2 PDF for the lognormal distribution, for $\mu = (0.5, 2)$, $\sigma = 1$.

$$f_{\ln}(y|\mu, \sigma) = \frac{1}{y\sigma\sqrt{2\pi}} \exp\left\{-[\ln(y) - \mu]^2 / 2\sigma^2\right\}.$$

See figure A.2 for some examples of the PDF. While μ and σ^2 are the mean and variance of $\ln(y)$, they are not the mean and variance of the lognormal distribution; the mean is $E[Y] = e^{\mu + \sigma^2/2}$, the median is e^μ, and the variance is $\mathrm{Var}[Y] = \left(e^{\sigma^2} - 1\right)e^{2\mu + \sigma^2}$. It is sometimes convenient to rewrite the PDF with new parameters m and ϕ, so that the variance positively scales with ϕ (so ϕ is called a variance parameter); then the lnPDF for the lognormal distribution is:

$$\ln f_{\ln}(y|m, \phi) = -\frac{1}{2}\ln(2\pi) - \ln(\phi) - \ln(y) - \frac{1}{2}\left[\frac{\ln(y) - \ln(m) + \frac{\phi^2}{2}}{\phi}\right]^2.$$

In this case $E[Y] = m$ and $\mathrm{Var}[Y] = \left(e^{\phi^2} - 1\right)m^2$. Ecologists frequently use the lognormal distribution in studies of individual growth, time series analyses of populations, and species abundance distributions.

A.1.3 *Gamma distribution (chapters 3, 6, 12, and 13)*

If a continuous random variable is positive ($y > 0$) and expected to have positive skew, but there is the possibility of lower values always being more likely, then the gamma distribution may be a good model. The typical form of this distribution is:

$$f_g(y|a, b) = \frac{y^{a-1}e^{-y/b}}{\Gamma(a)\,b^a} \quad \text{for } a, b > 0.$$

See figure A.3 for some examples of the PDF. The mean of this distribution is $E[Y] = ab$, and $\mathrm{Var}[Y] = ab^2$. The lnPDF is given by:

$$\ln f_g(y|a, b) = (a-1)\ln(y) - \frac{y}{b} - a[\ln(b)] - \ln[\Gamma(a)],$$

where Γ is the complete gamma function.

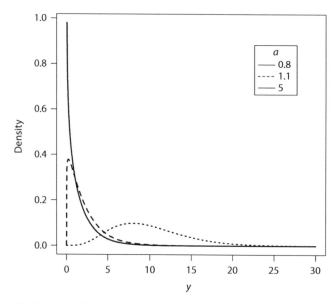

Fig. A.3 PDF for the gamma distribution, for $a = (0.8, 1.1, 5)$, $b = 2$.

The complete gamma function is essentially an extension of the factorial function, to include positive numbers that are not integers. (Recall that, for a positive integer n, the factorial $n!$ is the product of n and all smaller positive integers—e.g., $4! = 4 \times 3 \times 2 \times 1 = 24$.) The relationship between factorials and the complete gamma is $\Gamma(n) = (n-1)!$ for any positive integer n. The general formula (for any positive number q) is $\Gamma(q) = \int_0^\infty x^{q-1} \exp(-x) dx$. Factorials may seem more intuitive to you, but they are more problematic computationally, especially for large n; we recommend getting used to the complete gamma function!

There are several widely used alternative parameterizations of the gamma distribution, all of which are mathematically equivalent. For example, the gamma distribution can be formulated in terms of its mean, μ, and a positive variance parameter ϕ, by setting $a = \mu/\phi$ and $b = \phi$. In this case the variance is $\text{Var}[Y] = \mu\phi$. The gamma distribution arises naturally in event–time data, as the sum of time intervals that are each exponentially distributed, or as the waiting time until a given number of events has occurred. The gamma distribution is also an important component of mixture models; see discussion of the negative binomial distribution.

A.1.4 *Exponential distribution (chapters 5 and 6)*

If a stochastic event occurs at some fixed rate λ (the event rate), then the time between consecutive events is described by an exponential distribution. The typical form of this distribution is:

$$f_e(y|\lambda) = \lambda e^{-\lambda y}.$$

See figure A.4 for some examples of the PDF. The lnPDF is:

$$\ln f_e(y|\lambda) = \ln(\lambda) - \lambda y.$$

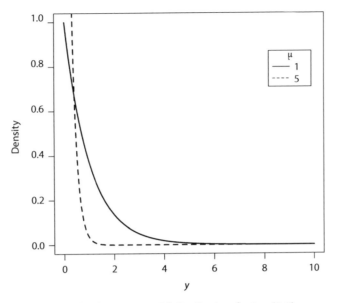

Fig. A.4 PDF for the exponential distribution, for $\lambda = (1, 5)$.

This distribution can only take non-negative values, and the event rate must also be positive. The mean of Y is $E[Y] = \mu = \lambda^{-1}$, and the variance of Y is $Var[Y] = \lambda^{-2}$. This distribution is most often used to describe event times, like time to death or flowering; exponentially distributed survival times mean a Type II survival curve. In ecology the exponential distribution is frequently taken as a null model for event times.

A.1.5 *Weibull distribution (chapter 5)*

If the rate at which events occur increases or decreases over time, then the time to the first event can be described by the Weibull distribution. Suppose the event occurs at rate $(a/b)t^{a-1}$ at time t. Then for $a > 1$ the rate increases over time, for $a = 1$ the rate is constant, and for $a < 1$ the rate decreases over time. The typical form of this distribution is:

$$f_W(y|a, b) = \frac{a}{b}\left(\frac{y}{b}\right)^{a-1} e^{-(y/b)^a}.$$

See figure A.5 for some examples of the PDF. The lnPDF is:

$$\ln f_W(y|a, b) = \ln\left(\frac{a}{b}\right) + (a - 1)\ln\left(\frac{y}{b}\right) - \left(\frac{y}{b}\right)^a.$$

The Weibull is a generalization of the exponential distribution; if $a = 1$, the expressions above are the same as for the exponential. For the time to the first event, the mean is $E[Y] = b\Gamma(1 + 1/a)$, and the variance is $Var[Y] = b^2[\Gamma(1 + 2/a) - (\Gamma(1 + 1/a))^2]$. This distribution can only model non-negative values. Ecologists frequently use this distribution when performing survival analyses, in which case the event rate corresponds to the age-dependent mortality rate. For each individual studied, the first event being modeled is death.

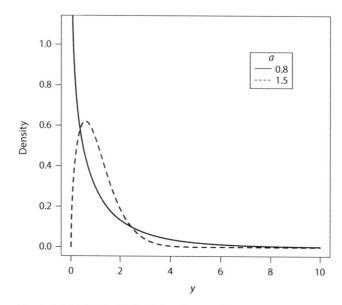

Fig. A.5 PDF for the Weibull distribution, for $a = (0.8, 1.5)$, $b = 1.2$.

A.1.6 *Beta distribution (chapters 3, 6, 12, and 13)*

If a continuous random variable is bounded by [0,1] (e.g., it describes a probability or a proportion), then the beta distribution can be a good choice. The typical form of this distribution is:

$$f_b (y|a, b) = y^{a-1}(1 - y)^{b-1} \frac{\Gamma(a + b)}{\Gamma(a)\Gamma(b)}.$$

See figure A.6 for some examples of the PDF. The lnPDF is:

$$\ln f_b(y \,|\, a, b) = (a - 1) \ln(y) + (b - 1) \ln (1 - y) + \ln [\Gamma (a + b)] - \ln [\Gamma(a)] - \ln [\Gamma(b)].$$

The mean is $E[Y] = \left(\frac{a}{a+b}\right)$ and the variance is $Var[Y] = \frac{ab}{(a+b)^2 (a+b+1)}$. This distribution is sometimes parameterized in terms of the mean, μ, and a dispersion parameter ϕ. In this case $a = \mu/\phi$ and $b = (1-\mu)/\phi$, which gives $E[Y] = \mu$ and $Var[Y] = \mu(1-\mu)\phi/(1+\phi)$. Ecologists have used the beta distribution to model quantities like the fraction of plant cover in a habitat, but recent use of this distribution is mainly to model random variation in the binomial parameter (overdispersion; see beta-binomial distribution, below).

A.2 Discrete random variables

A.2.1 *Poisson distribution (chapters 3, 4, 6, 8, 12, and 13)*

Count data in ecology are usually (in theory) unbounded and positively skewed ($y \geq 0$). If events occur at some constant rate, then the number of events counted during a fixed interval is given by the Poisson distribution. The typical form of the Poisson PMF is:

$$P_P (Y = y|\mu) = \frac{e^{-\mu} \mu^y}{y!}.$$

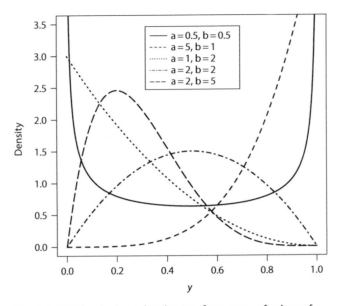

Fig. A.6 PDF for the beta distribution, for a range of values of *a* and *b*.

See figure A.7 for some examples of the PMF. The lnPMF is:

$$\ln P_{\mathrm{P}}\,(Y = y|\mu) = -\mu + y\,[\ln(\mu)] - \ln\,[\Gamma\,(y + 1)]\,.$$

We use the complete gamma function here, rather than a factorial, because it is much easier to compute. The expected number of counts is equal to the variance among counts, so $E[Y] = \mathrm{Var}[y] = \mu$. Alternatively, this distribution can be used to describe the number of

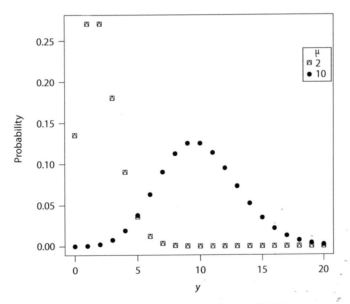

Fig. A.7 PMF for the Poisson distribution, for $\mu = (2, 10)$.

subjects counted in an area of given size when subjects are randomly distributed in space with given density. Ecologists frequently use this distribution as a simple model for the numbers of individuals counted in a unit area, transect, or similar measure.

A.2.2 *Binomial distribution (chapters 3, 4, 6, 7, 8, 12, and 13)*

Binomial distributions are used for data with two possible outcomes that occur with probability p and $1 - p$. Thus the binomial is quite different from the Poisson: the binomial models the number of successes or failures in a known number of trials (say, the number of seeds that germinate, of a known number planted), while the Poisson models the total number of units (say, the number of seeds found in a quadrat), with no fixed maximum. For n binomial trials, the PMF for the number of successes ($0 \le y \le n$) is:

$$P_{bn} (Y = y | n, p) = \binom{n}{y} p^y (1 - p)^{n-y} .$$

Here, $\binom{n}{y} = \dfrac{n!}{k! \, (n-k)!}$ is the binomial coefficient, often read as "n choose k."

See figure A.8 for some examples of the PMF. The lnPMF is:

$$\ln P_{bn}(Y = y | n, p) = \ln \left[\Gamma \, (n + 1) \right] - \ln \left[\Gamma \, (y + 1) \right] - \ln[\Gamma(n - y + 1)] + y \ln (p) + (n - y) \ln (1 - p) .$$

This distribution has mean $E[Y] = np$ and variance $\mathrm{Var}[Y] = np(1 - p)$ Ecologists frequently use this distribution when modeling outcomes like the number of seeds that germinate, or the number of individuals surviving a season.

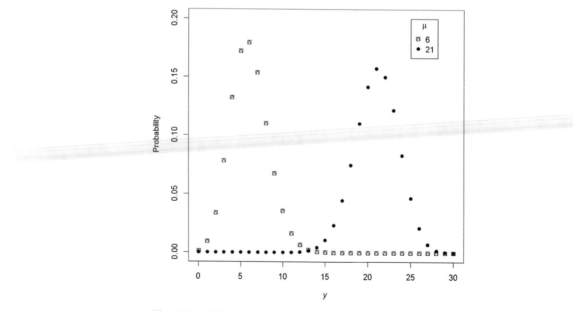

Fig. A.8 PMF for the binomial distribution, for $n = 30$, $p = (0.2, 0.7)$.

A.2.3 *Negative binomial distribution (chapters 3, 6, 8, 12, and 13)*

Often ecological count data are collected for which the variance is much greater than the mean. One discrete distribution for this case is the negative binomial; mathematically this distribution comes from allowing a Poisson distribution to have a parameter (the mean) that varies according to a gamma distribution. There are many different parameterizations of the negative binomial distribution. One that is especially useful for ecologists models the observed number of counts, y, given a mean, μ, and a positive clustering coefficient k:

$$P_{nb}(Y = y|\mu, k) = \left[\frac{\Gamma(k+y)}{\Gamma(k)\Gamma(y+1)}\right]\left(\frac{k}{k+\mu}\right)^k\left(\frac{\mu}{k+\mu}\right)^y;$$

See figure A.9 for some examples of the PMF. The lnPMF is:

$$\ln P_{nb}(Y = y|\mu, k) = \ln[\Gamma(k+y)] - \ln[\Gamma(k)] - \ln[\Gamma(y+1)] + k\ln(k) + y\ln(\mu) - (k+y)[\ln(k+\mu)].$$

This PMF has mean $E[Y] = \mu$, and variance $Var[Y] = \mu(1 + \mu/k)$. As k becomes large, the variance approaches the mean and the negative binomial approaches the Poisson distribution. In many ecological settings, $k < 1$.

Lindén and Mäntyniemi (2011) provide a derivation (and examples) of a useful generalization of the negative binomial that allows researchers more flexibility in modeling overdispersion. Their approach is to use a second parameter in the mean–variance relationship, so that $Var[Y] = \mu(1 + \phi_1 + \phi_2\mu)$; both ϕ_1 and ϕ_2 are non-negative. This approach allows models in which the mean–variance relationship is Poisson ($\phi_x = \phi_2 = 0$), linear ($\phi_2 = 0$), or quadratic. If we write $a = \mu/(\phi_1 + \phi_2\mu)$ and $b = 1/(\phi_1 + \phi_2\mu)$, the lnPMF is:

$$\ln P_{nb}(Y = y|\mu, \phi_1, \phi_2) = \ln[\Gamma(y+a)] - \ln[\Gamma(y+1)] - \ln[\Gamma(a)] + a\ln(b) - (a+y)\ln(1+b).$$

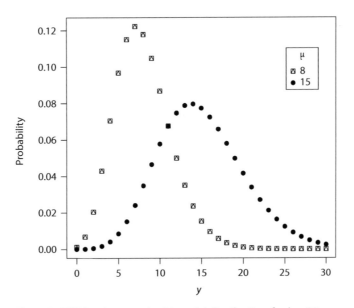

Fig. A.9 PMF for the negative binomial distribution, for $k = 20$, $\mu = (8, 15)$.

A.2.4 Beta-binomial distribution (chapters 3 and 12)

Suppose n binomial trials occur and the probability that each of the n trials is a success is drawn from a beta distribution with mean p and dispersion parameter θ; large values of θ mean that there is small overdispersion. If we are interested in the number of successes, then we have the beta-binomial distribution. The PMF describing the number of successes $(0 \le y \le n)$ is:

$$P_{bb}(Y = y | n, p, \theta) = \frac{\Gamma(\theta)}{\Gamma(p\theta)\Gamma[(1-p)\theta]} \binom{n}{y} \frac{\Gamma(y+p\theta)\Gamma[n-y+(1-p)\theta]}{\Gamma(n+\theta)}.$$

While this looks complicated, it reduces to the binomial for very large θ (that is, small overdispersion). For large overdispersion (as θ approaches 0), the mass is concentrated at zero (all failures) and n (all successes), with $P_{bb}(Y = 0 | n, p, 0) = 1 - p$ and $P_{bb}(Y = n | n, p, 0) = p$. See figure A.10 for some examples of the PMF. The lnPMF is:

$$\ln P_{bb}(Y = y | n, p, \theta) = \ln[\Gamma(\theta)] + \ln[\Gamma(n+1)] + \ln[\Gamma(y+p\theta)] + \ln\{\Gamma[n-y+(1-p)\theta]\}$$
$$- \ln[\Gamma(p\theta)] - \ln\{\Gamma[(1-p)\theta]\} - \ln[\Gamma(y+1)] - \ln[\Gamma(n-y+1)] - \ln[\Gamma(n+\theta)].$$

This distribution has a mean of $E[Y] = np$ and a variance of

$$\mathrm{Var}[Y] = np(1-p)\left(1 + \frac{n-1}{\theta+1}\right).$$

The last term in the variance is the variance inflation factor, relative to the binomial distribution; if θ is very large, the variance inflation factor approaches 0 because the denominator in the fraction becomes very large. There are many alternative parameterizations of the beta-binomial; the example presented in chapter 3 replaces θ with $1/\phi$. The beta-binomial is often used in ecology to model binary data (e.g., presence/absence, survived/died, success/failure of fertilization) when the probability of success varies among sample units.

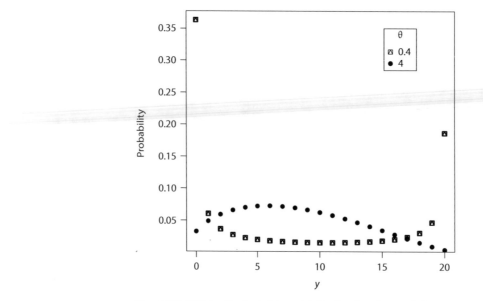

Fig. A.10 PMF for the beta-binomial distribution, for $p = 0.4$, $n = 20$, $\theta = (0.4, 4)$. This plot requires the emdbook library.

Glossary

Aggregation. In count data, high or low counts occurring more grouped in space than expected at random.

Akaike weight. A measurement of the *likelihood* that the model is the best, relative to all those being considered.

Akaike's Information Criterion (AIC). A quantity used to evaluate the support for a model, relative to other models. It incorporates the trade-off between goodness of fit and number of parameters. The classic AIC, derived from information theory, is $AIC = -2\ln(L_{max}) + 2k$, where L_{max} is the maximized log-likelihood for a model and k is the number of parameters it uses.

Anisotropic. Spatial pattern in which structures change with direction; for example, *patches* are elongated in one direction. Anisotropy implies a different degree and distance of *autocorrelation* or *aggregation* in different directions.

Areal data (or lattice). Consist of contiguous spatial units like grids or mosaics, where each cell has a single value of the response variable, and there is information about which cells are neighbors.

Autocorrelation. See *Spatial autocorrelation*.

Auxiliary variables. Variables that are included in an imputation process to increase the accuracy of imputed values; such variables are strongly correlated with variables with missing values.

Available case analysis. Analysis in which one only uses rows (cases) that do not have missing values as in *complete case analysis*, but adjusting the number of cases depending on which variables are used for a particular model. Stepwise regression often uses available case analysis.

Available variable analysis. Analysis in which one only uses variables that do not have any missing values.

Bias. In the context of study design, any tendency that prevents unprejudiced consideration of a particular question. In the context of statistical estimation, an estimator is biased if its expected value is different from the population parameter, given small sample size. Contrast with *consistency*.

Biased parameter estimates. Systematic deviation from true estimates of parameters including regression coefficients and variance components.

Biased uncertainty estimates. Systematic deviation from true estimates of uncertainty, for example standard errors or confidence (credible) intervals.

Bootstrapping. A *non-parametric* method for estimating statistics (often means or confidence intervals) by repeated resampling with replacement from the data.

Brownian motion evolution (BM). A stochastic model of evolution that assumes only drift acting to erode *phylogenetic correlations*.

Canonical links. A particular *link function* associated with a probability distribution that has useful statistical properties (e.g., minimal sufficiency).

Causal diagram. A graph indicating the *causal relationships* among a set of variables.

Causal hypothesis. A proposed *causal relationship*.

Causal network. The system of *causal relationships* among a set of variables.

Causal proposition. A *causal hypothesis* that can be tested with observational data.

Causal relationship. The relationship of X to Y is causal if induced variation in X can lead to changes in the value of Y.

Censored data. Data points for which inequalities (but not exact values) are known.

Censorship levels. The values at which data are censored.

Cholesky decomposition. A method for factoring a matrix into a lower and upper triangular matrix that (for matrices that are purely real numbers) are transposes of one another; useful in simulating correlated random numbers, and in solving linear equations.

Cognitive bias. A bias in thinking habits that can lead to systematic errors in reasoning.

Complete case analysis. Analysis where one only uses rows (cases) that do not have missing values.

Complete separation. A problem with estimation for binomial *generalized linear* and generalized linear mixed models that occurs when some linear combination of the predictor variables can separate the responses perfectly into a group of all zeros (failures) and a group of all positive values (successes). For example, in a logistic regression with a single continuous predictor, there might be a threshold value for which all observations below the threshold fail and all observations above the threshold succeed.

Conditional independence. Two variables in a model that are not causally linked once the effects of other variables are accounted for.

Conditional modes. The values predicted for deviations (of intercepts, slopes, or treatment effects) from the population average for each level of *random-effects* grouping variables; in *linear mixed models*, these are also called best linear unbiased predictors (BLUPs).

Confirmation bias. The tendency to confirm rather than deny a current hypothesis.

Confounding. Two measured variables with a cause–effect relationship that are both affected by another, unmeasured, variable.

Consistency. A consistent estimator converges asymptotically to the quantity being estimated as the sample size increases. A biased estimator may be consistent (e.g., the average sum of squares as an estimate of the population variance). While some inconsistent estimators simply do not converge to a single value, a more common inconsistency problem in ecological statistics is that many estimators converge to a value (i.e., the standard error goes to zero as sample size gets large), but this value differs from the quantity being estimated (e.g., the ordinary least squares estimate of the regression slope when there is measurement error in the predictor variable).

Contagious patch. *Autocorrelation* increases and decreases smoothly through space.

Continuous mixture model. An infinite mixture model in which the mixture weights are continuous.

Convergence. In many models, parameters are estimated through an iterative algorithm that assesses changes in the parameters. A model is determined to have converged when successive iterations of the algorithm no longer improve the fit, or improve it only a very small amount. Non-convergence can occur if a stable area of parameter space cannot be found.

Correlogram. Graph of Moran's I against distance classes. It represents the change of *autocorrelation* with distance between observations.

Countable mixture model. An infinite mixture model in which the mixture weights are discrete.

Covariate. A continuous variable in a statistical model that may predict the outcome (in older literature, an independent variable). Also called moderator in *meta-regression*.

Credible interval. The interval within which a parameter value from a Bayesian analysis is predicted to lie. The confidence of this prediction is given by a probability, usually expressed as a percentage.

Cumulative distribution function (CDF). A function giving the probability that a continuous *random variable* is less than or equal to a particular value.

Data augmentation. A process in which missing data imputation and data analysis are combined; these two steps feed back to one another.

Data constraints. The forces that determine the data set used to address a particular question.

Data gathering. The systematic acquisition of information in order to address a particular question.

Deviance. A quality-of-fit statistic that quantifies the difference between a specified model and the *saturated model* for a given set of data. Given a set of data $\mathbf{x}$, the deviance of a model defined by the parameters q, is $-2[LL(q^*|\mathbf{x})-LL(q_S^*|\mathbf{x})]$, where LL is the log-likelihood of the data, q^* are the maximum-likelihood parameter values of the specified model, and q_S^* are the maximum-likelihood values of the saturated model.

Deviance residuals. The contribution of each data point to total model *deviance*. Deviance residuals are a generalization of the residual sum of squares in linear models.

Directed relationship. A causal relationship in a *structural equation model*.

Disaggregation. Occurs when counts are evenly spaced.

Domain. See *extent*.

Doubly censored. Data sets in which both right- and left-censoring occurs. Contrast with *single censorship* and *multiple censorship*.

Dummy coding. Binary recoding of categorical variables; categorical variables with m levels are recorded into $m - 1$ binary variables.

Ecological inference problem. The situation in which population-level attributes are used incorrectly to assess individual attributes and vice versa.

Ecological spatial scales. The spatial scales (usually more than one) at which the structure of the ecological phenomenon occurs.

Effect size (in meta-analysis). An index that expresses the outcomes of all studies on the same scale.

Efficiency. Describes how well a statistical procedure is able to make use of data. An inefficient estimator requires a larger sample size than an efficient one.

Endogenous or **exogenous.** In structural equation modeling, a variable that has a causal variable within the model is endogenous. In regression modeling, a predictor variable that is correlated with the error term is endogenous; this leads to *inconsistent* parameter estimates. In spatial autoregressive modeling, spatial autocorrelation in a response variable related to organisms (e.g., population density) caused by interactions between individuals (e.g., competition). In all three contexts, variables that are not endogenous are called "exogenous."

Expected value. The most likely value of a random variable. Also called the expectation of a random variable, denoted by E[x]. It is the value of a random variable one would find if one could repeat the random sampling process an infinite number of times and take the average of the values obtained.

Extent. The total area (or distance in the case of transects) covered by a study.

Fail-safe number. A method proposed to address the effect of publication bias, by estimating how many studies would be required to nullify a significant meta-analytic mean; due to criticism of this method it is not currently used widely.

Finite mixture model. A mixture model that is constructed from a finite number of distributions.

Fixed effects. In linear and generalized linear mixed models, effects (parameters) in a statistical model that are estimated independently of each other, without any sharing of information among parameter estimates for different groups. In meta-analysis, the term refers to models that make the assumption that the differences in outcome among studies are due only to random sampling variation.

Forest plots. A graphical plot for meta-analysis results, showing effect sizes with their confidence intervals (CI). A vertical line of "no effect" and sometimes a vertical line for the grand mean are typically included; the size of the symbol for each mean effect size may be used to indicate study weight or sample size.

Fraction of missing information. A quantity reflecting the fraction of missing values as well as the information content of those missing values.

Full model. See *saturated model*.

Fully conditional specification. An imputation process in which each variable can be imputed separately when each variable is conditional on other values in the data.

Funnel plots. Scatter plots in which the effect sizes are plotted against study weights such as sample sizes or the inverse of sampling variance; a method of exploratory data analysis in meta-analysis.

Generalized least squares (GLS). A statistical technique for estimating regression parameters when assumptions of homoscedasticity and independence of samples are violated; GLS accounts for the correlation structure within the error term (*R-side effect*) but includes no random terms. When used with spatial data, often known as kriging regressions; the spatial correlation structure is incorporated by introducing the parameters of a variogram model (sill and range) in the covariance matrix.

Generalized linear model. A model composed of three components: a *linear predictor*, an error distribution, and a *link function*.

Geostatistical data. Include the precise coordinates of data points, with a value of the response variables for each location.

Global estimation. The process in structural equation modeling by which model fit is determined for the entire model with a single estimator.

Grain. Refers to the size of a single observation; for example a 2 m^2 sampling quadrat, 50 m radius observation station, or 5 m radius sampling point.

Grouping variable. A categorical predictor in a mixed model that defines the groups across which the random effects parameters vary.

G-side effects. Correlation in statistical models that is due to group membership.

Hauck–Donner effect. A problem with inference for generalized linear and generalized linear mixed models that occurs when modeling binomial data with large differences in the response across groups (see *complete separation*). In this case the commonly used *Wald* approximation substantially overestimates the standard errors of the parameter estimates, leading to overly wide confidence intervals and highly conservative *p*-values.

Heterogeneity. In many statistical contexts, samples that are not drawn from a random sample with the same parameters. In meta-analysis and meta-regression, variation in true effect sizes among the different study outcomes; modeled only in the framework of random-effects or mixed-effects models.

Identifiablity. The ability to be able to estimate a parameter from data. If one or more parameters are not estimable from data, then they are *unidentifiable*.

Ignorability. In missing data theory, missing data are ignorable if they are *missing at random* (MAR) or *missing completely at random* (MCAR), in the sense that the study design does not require a model of the way in which the data are missing, and which data points are missing or observed does not depend on the missing data, given the observed data.

Incorrect data points. Data points identified by exploratory graphical analysis with values that are too extreme to be believable, and are due to errors in measurement or data entry.

Indirect effect. A causal effect of one variable on another that occurs through a third variable.

Indirect impacts (also called spillovers). The spreading effect across space of an explanatory variable caused by the spatial autocorrelation in the response variable; a feature of SARlag and SARmix models, which allow the total effect of an explanatory variable in these models to be partitioned between its direct effect (β) and its indirect impacts.

Inefficiency. See *efficiency*.

Infinite mixture model. A mixture model that is constructed from an infinite number of distributions.

Informative censoring. Occurs when subjects are removed from the study due to their (sometimes anticipated) data value.

Instrumental variable. A variable correlated with a predictor variable in a regression, but having no direct causal effect on the response variable of the regression; used to eliminate *inconsistency* arising from any *endogeneity* in the predictor variable.

Interspecific data. Data from multiple species.

Interval censorship. Describes data whose values are not known, but are known to lie between two limits.

Interval. In spatial analyses, minimum distance between the *grains*.

Intrinsic hypothesis or **intrinsic assumption.** The variance of the difference between the observed values at pairs of locations depends only on the distance and the direction between them, but not on their specific location; it implies that (a) the mean of these differences is 0 and that (b) stationarity holds only within the range of a relatively small neighborhood.

Isotropic. Opposite of *anisotropic*.

Joint modeling. An imputation approach in which all (multivariate) missing values in a data set are imputed at once, assuming a multivariate-normal distribution.

Lag. A fixed distance or distance class.

Latent variables are inferred from a statistical or mathematical procedure, rather than observed directly. Examples are estimation of population growth rate from a model, and the true (but unobserved) value of a censored data point.

Left-censored data are points known only to be no larger than some value.

Likelihood. A function that is proportional to the probability of generating the data, given the model. By maximizing a likelihood function, one obtains the values of parameters most likely to have generated the data (given the model), called the maximum likelihood estimates (MLE).

Likelihood surface. The *likelihoods* of all possible sets of parameter values, as a function of the k parameters, form a surface with $k + 1$ dimensions. If there are two parameters, the surface can be just one peak (high likelihood values) in a large plane (low likelihood values) or it could be a series of hills of equal height indicating several sets of equally likely parameter values.

Limits of detection. Values beyond which data cannot be quantified (e.g., due to the sensitivity of measuring techniques or sampling designs).

Linear mixed models. A class of statistical models that extend linear models (i.e., the response variable is Normally distributed with a constant variance and mean equal to a linear combination of categorical and continuous predictors) by allowing random effects (i.e., the responses are grouped according to one or more categorical variables and parameters may vary across these groups).

Linear predictor. A simple linear model in which the response is predicted by a linear combination of continuous and/or categorical explanatory variables.

Link function. A mathematical function that links the linear predictor with the original response variable.

List-wise deletion. Deletion of rows (cases) that contain one or more missing values in a data set.

Local estimation. The process in structural equation modeling by which model fit is determined separately for each piece of the total model.

Location parameter. See *location-scale distribution*.

Location-scale distribution. A probability distribution for a random variable in which the CDF for y is some function of $(y - m)/s$, and does not depend on any other parameter. Here m is the *location parameter* and s the *scale parameter*. Use of these parameters translates the distribution to a new location, and rescales it, but the new distribution is in the same family. Familiar examples include the Normal, exponential, uniform, and Student's t distributions.

MAR. See *missing at random*.

Markov Chain Monte Carlo (MCMC). Methods that sample randomly from a Markov chain. When used in Bayesian analyses, the Markov chain is constructed so that its stationary (equilibrium) distribution is the posterior distribution of the analysis being conducted.

Maximum likelihood estimate (MLE). See *likelihood*.

MCAR. See *missing completely at random*.

Mean imputation. The use of the mean of a variable to fill in missing values for that variable.

Mediation. In structural equation models, intermediate steps by which one variable directly affects another. If X affects Y (i.e., $X \rightarrow Y$), mediation occurs if it operates through Z (i.e., $X \rightarrow Z \rightarrow Y$).

Meta-analysis. Quantitative synthesis of the results of different studies addressing equivalent questions with comparable methods, carried out in an unbiased and statistically defensible manner.

Meta-regression. A tool used in meta-analysis to examine the impact of *moderator* variables on study effect size using regression-based techniques.

Methodological underdetermination. The situation in which the relevant hypothesis is not tested because of an inadequate research design.

Missing at random (MAR). Missing values are distributed at random after controlling for other variables; compare with *Missing completely at random (MCAR)*.

Missing completely at random (MCAR). Missing values are distributed at random, and are not related to other variables; compare with *Missing at random (MAR)*.

Missing data mechanisms. The statistical relationship between observations and the probability of missing data.

Missing data pattern. A description of which values are missing in a data set.

Missing data theory. Theory on how missing values arise and how such missing values can be best treated for data analysis.

Missingness. A recoding of a data matrix into a binary matrix with missing values as 1 and non-missing values as 0.

Misspecified model. A model whose components are not supported by the data or the underlying process(es) that produced the data; e.g., a binomial distribution fit to Poisson-distributed errors.

Missing not at random (MNAR). Missing values are not distributed randomly, and they are related to unobserved data (i.e., missing values themselves and/or non-recorded variables).

Mixed models or **mixed-effect models.** Statistical models with both fixed and random predictor variables, in which the random variables describe the variation in effects among one or more categorical grouping variables. In meta-regression, mixed-effects models have the specific meaning that factors that categorize the studies (e.g., field or laboratory studies) are fixed, and there is random variation in true study effect sizes due to unspecified factors.

MNAR. See *missing not at random.*

Moderators. See *covariates.*

Modification index. An estimator in a structural equation model that indicates how much the fit of the model would be increased by adding a particular causal relationship.

Monotonic differentiable function. A monotonic function is one that does not change from increasing to decreasing or vice versa, and a differentiable function is one that can be differentiated; thus a monotonic differentiable function has a derivative of constant sign.

Monte Carlo experiment. A computational experiment that generates random data to explore the capabilities and limitations of statistical tests.

Multiple censorship refers to the case in which different samples are censored at different levels. Compares with *single censorship* and contrasts with *doubly censored.*

Multiple imputation. An imputation process that creates many copies of a data set with missing values replaced by imputed values.

Multivariate imputation by chained equation (MICE). An imputation process using *fully conditional specification.*

Non-detects. Left-censored data points in some fields (especially environmental science) are often called non-detects because recorded values are 0 but the true value lies between 0 and some *limits of detection.*

Non-parametric statistics. Statistical models that do not make specific assumptions about the probability distributions from which the data arose.

Non-stationary. Spatial processes in which both the mean and the autocovariance change with the location within the region studied.

Normal quantile plot. A plot in which the distribution of the standardized effect sizes is plotted against the standard Normal distribution, marked into quantiles. If some of the points fall outside of the confidence bands, it suggests that the data are not normally distributed.

Null model. A model that assumes variation in the data arises from sampling processes, and perhaps from ecological phenomena that are not the focus of the study/model. A model can thus be a null model with respect to the studied processes, but not to others.

Offset. A variable, or a function of a variable, that is added directly to a model equation (in contrast to a predictor variable, for which an associated coefficient is estimated). If x is a predictor variable in a model, the equation would be $Y = b_0 + b_1 x$; if it is an offset the equation would be just $Y = b_0 + x$. Offsets are most often used in GLMs or GLMMs to adjust for variable plot sizes or exposure times.

Order statistics. The nth order statistic of a sample is the nth smallest value in the sample.

Ordinary least squares (OLS). A statistical technique for estimating regression parameters, assuming homoscedasticity and independence of samples.

Ornstein–Uhlenbeck evolution (OU). A stochastic model of evolution that assumes that both drift and stabilizing selection act to erode phylogenetic correlations.

Overall spatial multiplier. In spatial autoregressive models, the average level by which the direct effect of an explanatory variable is multiplied to calculate the total effect of an explanatory variable, taking into account the spreading of *indirect impacts* in the system.

Overdispersion. Random variables that have greater variance than expected by the assumed distribution; thus, remarks regarding overdispersion must refer to a data set and a specified probability distribution.

Pair-wise deletion. In bivariate analysis (e.g., correlation analysis), deleting cases in which missing values occur in one or both variables, for that analysis only.

Parametric statistics. Statistical models that make specific assumptions about the probability distributions from which the data arose.

Parent–child relationship. A causal relationship in structural equation modeling, parents having causal effects on children.

Passive imputation. An imputation method used for missing values in derived variables such as interactions, quadratic terms, and transformed variables.

Patch. An area where values of a variable are more similar to each other (or individuals more aggregated in the case of point pattern or count data) than expected at random.

Pattern-mixture models. One of the common modeling approaches used when missing data are non-ignorable (i.e., *missing not at random*).

Phylogenetic comparative method. A statistical approach used to analyze interspecific (multi-species) data.

Phylogenetic correlations. A correlation that quantifies the relative phylogenetic similarity among related species. They are used in the *phylogenetic comparative method*.

Phylogenetic generalized least squares (PGLS). A statistical technique that uses phylogenetic correlations and models of evolution for estimating regression parameters from interspecific data. It is based on a *generalized least squares* model.

Phylogenetic tree. See *phylogeny*.

Phylogeny. A hypothesis on the shared evolutionary history of taxonomic groups and taxa.

Planned missing data design. A study design deliberately including missing values, for example, to reduce the cost of the study by measuring a low-cost variable in all individuals but a correlated high-cost variable only in some.

Point pattern data. Data that consist solely of locations; useful in studying spatial patterns of occurrence.

Posterior distribution or **posterior.** A probability distribution that is used to define uncertainty in a parameter of a Bayesian model after the data and the prior are considered.

Power. The probability that a test will reject a false null hypothesis.

Prediction interval. An estimate of an interval within which the prediction from a statistical model is expected to lie. The confidence with which this prediction is made is usually expressed as a percentage.

Preferred Reporting Items for Systematic Reviews and Meta-Analyses (PRISMA). Standards published by the International Prospective Register of Systematic Reviews based at the University of York, UK.

Prior distribution or **prior.** A probability distribution that is used to define uncertainty in a parameter of a Bayesian model before the data are considered. A prior should reflect what is previously known about that parameter.

PRISMA. See *Preferred Reporting Items for Systematic Reviews and Meta-Analyses*.

Probability density function (PDF). A function describing how the probability of a continuous random variable changes as the measurement changes; the derivative of the *cumulative distribution function*.

Probability mass function (PMF). Description of the probability that a discrete random variable takes on each of the set of possible values.

Publication bias. Statistical bias that exists when studies showing large and statistically significant effects are published preferentially.

Quasi-likelihood. An approximation to the true likelihood of a model that behaves like the true likelihood. Quasi-likelihoods are usually used to facilitate the finding of maximum-likelihood parameter estimates where the true likelihood is difficult to calculate.

Random censoring. Random censoring occurs when data points are censored not because of a limit of detection or because of the number of points already censored, but for some other reason extrinsic to the study.

Random effects. Effects (parameters) in a statistical model that are estimated assuming that the effects for different levels of a categorical grouping variable are random variables drawn from an underlying distribution; allows sharing of information among parameter estimates for different groups.

Random fields. *Random variables* associated with spatial location information.

Random spatial distribution. See *spatially independent*.

Random variable. A variable whose possible values are given by a probability distribution. Alternatively, the values are the outcomes of a random process.

Random-effects models. In meta-analysis, models that include both random sampling variation among studies and variation in the true effects among studies (due to unspecified causes). In generalized linear mixed models, models containing only random effects.

Regression imputation. The use of regression predictions to fill in missing values.

Regression on order statistics (ROS). A technique used for estimating the mean and variance of a sample that includes censored data. A linear regression is fit to the probability plot of the data; if the data are linear, the slope estimates the *scale parameter* of the distribution, and the intercept estimates the *location parameter*.

Relative efficiency. The number and importance of errors due to multiple imputation, relative to the minimum possible number of errors. This depends on the fraction of missing information and the number of sets of imputed data.

Research bias. Bias that exists when certain organisms, systems, etc., are well studied while others are not, so that research syntheses lack information on these overlooked organisms or systems.

Research synthesis. Application of the scientific method to reviewing and summarizing evidence about a scientific question, using methods that are unbiased, repeatable, and transparent.

Resolution. See *grain*.

Right-censored data are points known only to be at least as large as some value.

Robust ROS. A refinement of *regression on order statistics*. The slope and intercept from the regression are used to impute the censored values, and then the *location* and *scale parameters* are calculated directly from the data set including the imputed values.

Root. The ancestral divergence period of an entire lineage of taxa from a phylogenetic tree. See *phylogeny*.

R-side effects. Correlation in statistical models that is incorporated at the level of the residuals.

Saturated model. The most complex model that can be fit to a data set, often using the same number of parameters as there are observations.

Scale parameter. See *location-scale distribution*.

Second-order stationary (or **stationary**). Processes are stationary when they have constant mean and variance within a study area. The autocorrelation depends on the distance and direction between locations (and not on their absolute locations).

Selection models. One of the common modeling approaches used when missing data are non-ignorable (i.e., *missing not at random*).

Sequential regression imputation. See *fully conditional specification*.

Single censorship refers to the case where all censorship occurs at the same level. It compares with *multiple censorship*, where different samples are censored at different levels. Contrast with *doubly censored* data.

Single imputation. Any imputation process that creates only one copy of a data set, in which missing values are replaced by imputed values.

Spatial autocorrelation or **spatial covariance.** The degree of correlation of a variable with itself, which is also a function of the spatial positions. Near values are more similar than distant ones.

Spatial autoregressive models (SAM). Linear models that can model spatial patterns in ecological data as imposed by unknown factors; the three main approaches are conditional autoregressive (CAR), simultaneous autoregressive (SAR), and moving average (MA) models.

Spatial structure. See *patch*.

Spatially explicit question (hypothesis or problem). Statistical or scientific questions about the shape and scale of spatial patterns.

Spatially independent. Spatial autocorrelation that is not stronger than expected at random.

Step. Abrupt boundary of a patch.

Structural equation meta-model. A conceptual model of a theoretical set of causal relationships among a system of variables.

Structural equation model. A set of equations that model a network of cause–effect relationships.

Study spatial scale. The scale at which a spatial study is performed, entirely defined by the researcher. In *geostatistical data*, three dimensions define the study scale: grain, interval, and extent. In point pattern data, extent is the only dimension used. In areal data, grain and extent are used.

Systematic review. A complete, unbiased, reproducible literature search according to clearly specified search criteria and selection of evidence.

Topology. The branching pattern of a phylogenetic tree. See *phylogeny*.

Tree. See *phylogeny*.

Trend or **gradient.** A spatial pattern in which a measured variable changes monotonically over the entire area being studied.

Trim-and-fill. A method proposed to impute potentially missing data points and provide a corrected *meta-analytic* mean, based on *funnel-plot* asymmetry; best viewed as a sensitivity analysis to assess the robustness of the results to potential *publication bias*.

Truncated data sets are those for which observations that would fall in some range are not even known to exist. A common ecological example is assessment of the numbers of individuals in a population using remotely sensed data; typically, very small individuals are completely undetectable. Contrast with *censored data*.

Type I (time) censoring. Censorship occurs because of some limits to detection, including a planned date on which a study ends. The number of censored values is a random variable, but the censorship levels are fixed.

Type I error. A false positive outcome; it is the incorrect rejection of an existing null outcome.

Type II (failure) censoring. The number of censored values is planned, e.g., a design to study a number of individuals but stop after a particular number of events have occurred. The *censorship levels* are thus random.

Type II error. A false negative outcome; it is the failure to reject a false null outcome.

Ultrametric tree. A phylogenetic tree that has all terminal branches aligned contemporaneously.

Unidentifiable. A parameter that cannot be estimated from a given data set due to confounding or lack of information. For example, in a linear model $Y = a+b+cx$, the two intercept parameters a and b would be unidentifiable (or *jointly* unidentifiable), because any combination of values of a and b that added up to the value of Y when $x = 0$ would be equally good fits to the data.

Vote-counting. Counting the number of significant outcomes and weighing those against the number of non-significant outcomes as a research synthesis method. Vote-counts are biased and not meaningful ways to summarize the results of different studies and are not a currently accepted technique for quantitative research synthesis.

Wald. Standard errors and hypothesis tests (*p*-values) based on an assumption that the log-likelihood surface associated with a statistical model is quadratic.

Weighted least squares. Model fitting strategy that minimizes the weighted residual sum of squares, using, for example, the number of observations as weights.

Yule birth–death process. A model that simulates the evolution of taxa through time; used to model random phylogenies.

Zero-inflation occurs when data contain more zeroes than expected based on the probability distribution being used to describe the data. Zero-inflation is therefore always relative to a distribution.

References

Adams, D. C., Gurevitch, J., and Rosenberg, M. S. 1997. Resampling tests for meta-analysis of ecological data. *Ecology,* 78, 1277–83.

Adler, F. R. 2004. *Modeling the dynamics of life: calculus and probability for life scientists.* Pacific Grove, CA, Brooks/Cole.

Agresti, A. 2002. *Categorical data analysis.* Hoboken, NJ, Wiley.

Agresti, A. 2010. *Analysis of ordinal categorical data.* Hoboken, NJ, Wiley.

Akaike, H. 1983. Information measures and model selection. *International Statistical Institute,* 44, 277–91.

Akaike, H. 1973. Information theory as an extension of the maximum likelihood principle. *Proceedings of the Second International Symposium on Information Theory, Budapest,* 267–81.

Allen, T. F. H., and Starr, T. B. 1982. *Hierarchy: perspectives for ecological complexity.* Chicago. IL, University of Chicago Press.

Allignol, A. 2012. kmi: Kaplan-Meier multiple imputation for the analysis of cumulative incidence functions in the competing risks setting, R package version 0.4. http://cran.r-project.org/web/packages/kmi/index.html.

Allison, P.D. 1987. Estimation of linear models with incomplete data. *In* Clogg, C. C. (ed.), *Sociological Methodology,* pp. 71–103. San Francisco, Jossey-Bass.

Allison, P. D. 2002. *Missing data.* Thousand Oaks, CA, Sage.

Alsterberg, C., Eklöf, J. S., Gamfeldt, L., Havenhand, J. N., and Sundbäck, K. 2013. Consumers mediate the effects of experimental ocean acidification and warming on primary producers. *Proceedings of the National Academy of Sciences,* 110, 8603–08.

Anderson, D. R. 2008. *Model based inference in the life sciences: a primer on evidence.* New York, Springer.

Anderson, D. R., Burnham, K. P., and Thompson, W. L. 2000. Null hypothesis testing: problems, prevalence, and an alternative. *Journal of Wildlife Management,* 64, 912–23.

Anderson, D. R., Burnham, K. P., and White, G. C. 1994. AIC model selection in overdispersed capture-recapture data. *Ecology,* 75, 1780–93.

Anderson, T. M., Shaw, J., and Olff, H. 2011. Ecology's cruel dilemma, phylogenetic train evolution and the assembly of Serengeti plant communities. *Journal of Ecology,* 99, 797–806.

Angrist, J. D. 2010. Multiple endogenous variables—now what?! *Mostly Harmless Economics Blog.* http://www.mostlyharmlesseconometrics.com/2010/02/multiple-endogenous-variables-what-now/

Angrist, J. D., and Pischke, J.-S. 2009. *Mostly harmless econometrics.* Princeton, NJ, Princeton University Press.

Arhonditsis, G., Stow, C., Steinberg, L., Kenney, M., Lathrop, R., McBride, S., and Reckhow, K. 2006. Exploring ecological patterns with structural equation modeling and Bayesian analysis. *Ecological Modelling,* 192, 385–409.

Armsworth, P. R., Gaston, K. J., Hanley, N. D., and Ruffell, R. J. 2009. Contrasting approaches to statistical regression in ecology and economics. *Journal of Applied Ecology,* 46, 265–68.

Bagshaw, S. M., and Ghali, W. A. 2005. Theophylline for prevention of contrast-induced nephropathy - A systematic review and meta-analysis. *Archives of Internal Medicine,* 165, 1087–93.

Balvanera, P., Pfisterer, A. B., Buchmann, N., He, J. S., Nakashizuka, T., Raffaelli, D., and Schmid, B. 2006. Quantifying the evidence for biodiversity effects on ecosystem functioning and services. *Ecology Letters,* 9, 1146–56.

Baraldi, A. N., and Enders, C. K. 2010. An introduction to modern missing data analyses. *Journal of School Psychology,* 48, 5–37.

Baron, J. 2007. *Thinking and deciding.* New York, Cambridge University Press.

Barr, D. J., Levy, R., Scheepers, C., and Tily, H. J. 2013. Random effects structure for confirmatory hypothesis testing: Keep it maximal. *Journal of Memory and Language,* 68, 255–78.

Barrio, I. C., Hik, D. S., Bueno, C. G., and Cahill, J. F. 2013. Extending the stress-gradient hypothesis - is competition among animals less common in harsh environments? *Oikos,* 122, 516–23.

Barry, S., and Elith, J. 2006. Error and uncertainty in habitat models. *Journal of Applied Ecology,* 43, 413–23.

Barto, E. K., and Rillig, M. C. 2012. Dissemination biases in ecology: effect sizes matter more than quality. *Oikos,* 121, 228–35.

Bayes, T. R. 1763. An essay towards solving a problem in the doctrine of chances. *Philosophical Transactions,* 53, 370–418.

Beaujean, A. A. 2012. BaylorEdPsych: R package for Baylor University Educational Psychology Quantitative Courses, R package version 0.5. http://cran.r-project.org/web/packages/BaylorEdPsych/index.html.

Becker, B. J. 2005. Fail safe *N* or file-drawer number. *In:* Rothstein, H., Sutton, A. J. and Borenstein, M. (eds.) *Publication bias in meta-analysis: prevention, assessment and adjustments,* pp. 111–25. Chichester, UK, Wiley.

Beckerman, A. P., Benton, T. G., Lapsley, C. T., and Koesters, N. 2006. How effective are maternal effects at having effects? *Proceedings of the Royal Society B: Biological Sciences,* 273, 485–93.

Begg, C. B., and Mazumdar, M. 1994. Operating characteristics of a rank correlation test for publication bias. *Biometrics,* 50, 1088–1101.

Begon, M., and Mortimer, M. 1986. *Population ecology: a unified study of animals and plants.* Sunderland, MA, Sinauer.

Beguin, J., Pothier, D., and Cote, S. D. 2011. Deer browsing and soil disturbance induce cascading effects on plant communities: a multilevel path analysis. *Ecological Applications,* 21, 439–451.

Bellio, R., and Brazzale, A. R. 2011. Restricted likelihood inference for generalized linear mixed models. *Statistics and Computing,* 21, 173–83.

Belovsky, G. E., Stephens, D., Perschon, C., et al. 2011. The Great Salt Lake ecosystem (Utah, USA): long term data and a structural equation approach. *Ecosphere,* 2, art33.

Belshe, E. F., Schuur, E. a. G., and Bolker, B. M. 2013. Tundra ecosystems observed to be CO_2 sources due to differential amplification of the carbon cycle. *Ecology Letters,* 16, 1307–15.

Berger, J. O. 1985. *Statistical decision theory and Bayesian analysis.* New York, Springer-Verlag.

Bernstein, C., Auger, P., and Poggiale, J. C. 1999. Predator migration decisions, the ideal free distribution, and predator-prey dynamics. *American Naturalist,* 153, 267–81.

Biro, P. A., and Dingemanse, N. J. 2009. Sampling bias resulting from animal personality. *Trends in Ecology & Evolution,* 24, 66–7.

Bivand, R. S., Pebesma, E., and Gómez-Rubio, V. 2013. *Applied spatial data analysis with R.* New York, Springer.

Blomberg, S. P., and Garland, T. 2002. Tempo and mode in evolution: phylogenetic inertia, adaptation and comparative methods. *Journal of Evolutionary Biology,* 15, 899–910.

Blomberg, S. P., Lefevre J. G., Wells J. A., and Waterhouse, M. 2012. Independent contrasts and PGLS regression estimators are equivalent. *Systematic Biology,* 61, 382–391.

Bodner, T. E. 2006. Missing data: Prevalence and reporting practices. *Psychological Reports,* 99, 675–80.

Bolker, B. M. 2008. *Ecological models and data in R.* Princeton, New Jersey, USA, Princeton University Press.

Bolker, B. M., Brooks, M. E., Clark, C. J., Geange, S. W., Poulsen, J. R., Stevens, M. H. H., and White, J. S. S. 2009. Generalized linear mixed models: a practical guide for ecology and evolution. *Trends in Ecology & Evolution,* 24, 127–35.

Bolker, B. M., Gardner, B., Maunder, M., Berg, C. W., et al. 2013. Strategies for fitting nonlinear ecological models in R, AD Model Builder, and BUGS. *Methods in Ecology and Evolution,* 4, 501–12.

Bollen, K. A. 1989. *Structural equations with latent variables.* New York, Wiley.

Bonds, M. H., Dobson, A. P., and Keenan, D. C. 2012. Disease ecology, biodiversity, and the latitudinal gradient in income. *PLOS Biology,* 10, e1001456.

Borcard, D., Gillet, F., and Legendre, P. 2011. *Numerical ecology with R.* New York, Springer.

Borcard, D., Legendre, P., Avois-Jacquet, C., and Tuomisto, H. 2004. Dissecting the spatial structure of ecological data at multiple scales. *Ecology,* 85, 1826–32.

Borenstein, M. 2009. Effect sizes for continuous data. *In:* Cooper, H., Hedges, L. V. and Valentine, J. C. (eds.) *The handbook of research synthesis and meta-analysis,* pp. 221–36. NY, Russell Sage Foundation.

Borenstein, M., Hedges, L. V., Higgins, J. P. I., and Rothstein, H. R. 2009. *Introduction to meta-analysis.* Chichester, UK, Wiley.

Bowker, M. A., Maestre, F. T., and Escolar, C. 2010. Biological crusts as a model system for examining the biodiversity–ecosystem function relationship in soils. *Soil Biology and Biochemistry* 42, 405–417.

Breen, R. 1996. *Regression models: censored, sample selected, or truncated data.* Thousand Oaks, CA, Sage Publications.

Breslow, N. E. 2004. Whither PQL? *In:* Lin, D. Y., and Heagerty, P. J. (eds.). *Analysis of correlated data.* Proceedings of the Second Seattle Symposium in Biostatistics, pp. 1–22. New York, Springer.

Brockway, D. G., and Outcalt, K. W. 1998. Gap-phase regeneration in longleaf pine wiregrass ecosystems. *Forest Ecology and Management,* 106, 125–39.

Bruggeman, J., Heringa, J., and Brandt, B. W. 2009. PhyloPars: estimation of missing parameter values using phylogeny. *Nucleic Acids Research,* 37, W179–84.

Buckland, S. T., Anderson, K. P., and Augustin, N. H. 1997. Model selection: an integral part of inference. *Biometrics,* 53, 603–18.

Buckland, S. T., Newman, K. B., Thomas, L., and Koesters, N. B. 2004. State-space models for the dynamics of wild animal populations. *Ecological Modelling,* 171, 157–75.

Buonaccorsi, J. P. 2010. *Measurement error: models, methods, and applications.* Boca Raton, FL, Chapman and Hall/CRC.

Burgman, M. 2005. *Risks and decisions for conservation and environmental management.* Cambridge, Cambridge University Press.

Burnham, K. P., and Anderson, D. R. 2002. *Model selection and multimodel inference: a practical information-theoretic approach.* New York, Springer-Verlag.

Burnham, K. P., Anderson, D. R., and White, G. C. 1994. Evaluation of the Kullback–Leibler discrepancy for model selection in open population capture–recapture models. *Biometrical Journal,* 36, 299–315.

Burrough, P. A. 1987. Spatial aspects of ecological data. *In:* R. H. G. Jongman, C. J. F. T. Braak, and O. F. R. van Tongeren, eds., *Data analysis in community and landscape ecology,* pp 213–51. Cambridge, Cambridge University Press.

Butler, M. A., and King, A. A. 2004. Phylogenetic comparative analysis: A modeling approach for adaptive evolution. *American Naturalist,* 164, 683–95.

Calabrese, J. M., Brunner, J. L., and Ostfeld, R. S. 2011. Partitioning the aggregation of parasites on hosts into intrinsic and extrinsic components via an extended Poisson-gamma mixture model. *PLOS One,* 6.

Caldwell, D. M., Ades, A. E., and Higgins, J. P. T. 2005. Simultaneous comparison of multiple treatments: combining direct and indirect evidence. *British Medical Journal,* 331, 897–900.

Caplat, P., Nathan, R., and Buckley, Y. M. 2012. Seed terminal velocity, wind turbulence, and demography drive the spread of an invasive tree in an analytical model. *Ecology,* 93, 368–77.

Carnicer, J., Brotons, L., Sol, D., and de Caceres, M. 2008. Random sampling, abundance–extinction dynamics and niche-filtering immigration constraints explain the generation of species richness gradients. *Global Ecology and Biogeography,* 17, 352–362.

Chamberlain, S. A., Hovick, S. M., Dibble, C. J., Rasmussen, N. L., Van Allen, B. G., Maitner, B. S., Ahern, J. R., Bell-Dereske, L. P., Roy, C. L., Meza-Lopez, M., Carrillo, J., Siemann, E., Lajeunesse, M. J., and Whitney, K. D. 2012. Does phylogeny matter? Assessing the impact of phylogenetic information in ecological meta-analysis. *Ecology Letters,* 15, 627–36.

Charlier, T. D., Underhill, C., Hammond, G. L., and Soma, K. K. 2009. Effects of aggressive encounters on plasma corticosteroid-binding globulin and its ligands in white-crowned sparrows. *Hormones and Behavior,* 56, 339–47.

Chung, Y., Rabe-Hesketh, S., Dorie, V., Gelman, A., and Liu, J. 2013. A nondegenerate penalized likelihood estimator for variance parameters in multilevel models. *Psychometrika,* 1–25.

Claeskens, G., and Hjort, N. L. 2008. *Model selection and model averaging.* Cambridge, Cambridge University Press.

Clark, J. S. 2005. Why environmental scientists are becoming Bayesians. *Ecology Letters,* 8, 2.

Clark, J. S., Silman, M., Kern, R., Macklin, E., and HilleRisLambers, J. 1999. Seed dispersal near and far: Patterns across temperate and tropical forests. *Ecology*, 80, 1475–94.

Clark, R. G., and Allingham, S. 2011. Robust resampling confidence intervals for empirical variograms. *Mathematical Geosciences*, 43, 243–59.

Cleasby, I. R., Burke, T., Schroeder, J., and Nakagawa, S. 2011. Food supplements increase adult tarsus length, but not growth rate, in an island population of house sparrows (*Passer domesticus*). *BMC Research Notes*, 4, 431.

Cleasby, I. R., and Nakagawa, S. 2012. The influence of male age on within-pair and extra-pair paternity in passerines. *Ibis*, 154, 318–24.

Cohen, J. 1988. *Statistical Power Analysis for the Behavioral Sciences*. Hillsdale, NJ, Lawrence Erlbaum Associates.

Cohen, J. E. 1971. Mathematics as metaphor. *Science*, 172, 674–5.

Colwell, R. K., Mao, C. X., and Chang, J. 2004. Interpolating, extrapolating, and comparing incidence-based species accumulation curves. *Ecology*, 85, 2717–27.

Congdon, P. D. 2010. *Applied Bayesian hierarchical methods*. Boca Raton, FL, CRC.

Connor, E. F., and McCoy, E. D. 1979. The statistics and biology of the species-area relationship. *American Naturalist*, 113, 791–833.

Connor, E. F., and Simberloff, D. 1979. The assembly of species communities: chance or competition? *Ecology*, 60, 1132–40.

Conover, W. J. 1998. *Practical nonparametric statistics*. New York, Wiley.

Cooper, H., Hedges, L. V., and Valentine, J. C. 2009. *The handbook of research synthesis and meta-analysis*. New York, Russell Sage Foundation.

Cooper, J. K., Li, J., and Montagnes, D. J. S. 2012. Intermediate fragmentation per se provides stable predator-prey metapopulation dynamics. *Ecology Letters*, 15, 856–63.

Correa, A., Gurevitch, J., Martins-Loucao, M. A., and Cruz, C. 2012. C allocation to the fungus is not a cost to the plant in ectomycorrhizae. *Oikos*, 121, 449–63.

Côté, I. M., Curtis, P. S., Rothstein, H. R., and Stewart, G. B. 2013. Gathering data: searching literature and selection criteria. *In:* Koricheva, J., Gurevitch, J. and Mengersen, K. (eds.) *The handbook of meta-analysis in ecology and evolution*, pp. 37–51. Princeton, NJ, Princeton University Press.

Coutts, S. R., Caplat, P., Cousins, K., Ledgard, N., and Buckley, Y. M. 2012. Reproductive ecology of *Pinus nigra* in an invasive population: individual- and population-level variation in seed production and timing of seed release. *Annals of Forest Science*, 69, 467–76.

Cox, D. R., and Donnelly, C. A. 2011. *Principles of applied statistics*. Cambridge, Cambridge University Press.

Crawford, E. D., Blumenstein, B., and Thompson, I. 1998. Type III statistical error. *Urology*, 51, 675.

Crawley, M. J. 2002. Statistical computing: an introduction to data analysis using S-Plus. New York, Wiley.

Crawley, M. J. 2007. *The R book*. Chichester, UK, Wiley.

Creel, S., and Creel, M. 2009. Density dependence and climate effects in Rocky Mountain elk: an application of regression with instrumental variables for population time series with sampling error. *Journal of Animal Ecology*, 78, 1291–7.

Cressie, N. A. C. 1993. *Statistics for spatial data*. New York, Wiley.

Crome, F. H. J. 1997. Researching Tropical Forest Fragmentation: Shall We Keep on Doing What We're Doing? *In:* Laurance, W. F. R. and Bierregard, O. (eds.) *Tropical forest remnants: ecology, management, and conservation of fragmented communities*, pp. 485-501. Chicago, IL.: University of Chicago Press.

Csada, R. D., James, P. C., and Espie, R. H. M. 1996. The "file drawer problem" of nonsignificant results Does it apply to biological research? *Oikos*, 76, 591–3.

Cumming, G. 2011. *Understanding the new statistics: effect sizes, confidence intervals, and meta-analysis*. New York, Routledge.

Cumming, G., and Finch, S. 2001. A primer on the understanding, use, and calculation of confidence intervals that are based on central and noncentral distributions. *Educational and Psychological Measurement*, 61, 532–84.

Cunningham, R. B., and Lindenmayer, D. B. 2005. Modeling count data of rare species: some statistical issues. *Ecology*, 86, 1135–42.

Curtis, P. S. 1996. A meta-analysis of leaf gas exchange and nitrogen in trees grown under elevated carbon dioxide. *Plant Cell and Environment,* 19, 127–37.

Curtis, P. S., and Wang, X. Z. 1998. A meta-analysis of elevated CO_2 effects on woody plant mass, form, and physiology. *Oecologia,* 113, 299–313.

Daniels, M. J., and Hogan, J. W. 2008. *Missing data in longitudinal studies: strategies for Bayesian modeling and sensitivity analysis.* Boca Raton, Chapman & Hall/CRC.

de Jong, G., and van Noordwijk, A. J. 1992. Acquisition and allocation of resources: genetic (co)variances, selection, and life histories. *American Naturalist,* 139, 749–70.

de Knegt, H. J., van Langevelde, F., Coughenour, M. B., Skidmore, A. K., de Boer, W. F., Heitkonig, I. M., Knox, A. N. M., Slotow, R., van der Waal, C., and Prins, H. H. T. 2010. Spatial autocorrelation and the scaling of species-environment relationships. *Ecology,* 91, 2455–65.

de Winter, J., and Happee, R. 2013. Why selective publication of statistically significant results can be effective. *PLOS One,* 8, e66463.

Dean, C. B. 1992. Testing for overdispersion in Poisson and binomial regression models. *Journal of the American Statistical Association,* 87, 451–7.

Del Re, A. 2012. Compute.es: Compute effect sizes. R package version 0.2.1. http://cran.r-project.org/web/packages/compute.es

Demirtas, H., and Schafer, J. L. 2003. On the performance of random-coefficient pattern-mixture models for non-ignorable drop-out. *Statistics in Medicine,* 22, 2553–75.

Dempster, A. P., Laird, N. M., and Rubin, D. B. 1977. Maximum likelihood from incomplete data via EM algorithm. *Journal of the Royal Statistical Society Series B-Methodological,* 39, 1–38.

Dennis, B. 1996. Discussion: should ecologists become Bayesians? *Ecological Applications,* 6, 1095–1103.

DerSimonian, R., and Laird, N. 1986. Meta-analysis in clinical trials. *Controlled Clinical Trials,* 7, 177–88.

DeVellis, R. F. 2011. *Scale development: theory and applications.* Los Angeles, CA, Sage Publications.

Diaz-Uriarte, R., and Garland, T. 1996. Testing hypotheses of correlated evolution using phylogenetically independent contrasts: Sensitivity to deviations from Brownian motion. *Systematic Biology,* 45, 27–47.

DiCiccio, T. J., and Efron, B. 1996. Bootstrap confidence intervals (with Discussion). *Statistical Science,* 11, 189–228.

Diggle, P., and Ribeiro, P. J. 2007. *Model-based geostatistics.* New York, Springer.

Diggle, P. J. 1990. *Time series: a biostatistical introduction.* Oxford, Oxford University Press.

Diggle, P. J. 2013. *Statistical analysis of spatial and spatio-temporal point patterns,* 3rd ed. Boca Raton, CRC Press.

Di Stefano, J. 2001. Power analysis and sustainable forest management. *Forest Ecology and Management* 154, 141–153.

Doak, D. F., Marino, P. C., and Kareiva, P. M. 1992. Spatial scale mediates the influence of habitat fragmentation on dispersal success: implications for conservation. *Theoretical Population Biology,* 41, 315–36.

Dobson, A. J. 2002. *An introduction to generalized linear models.* Boca Raton, FL, Chapman and Hall/CRC.

Dobzhansky, T. 1973. Nothing in biology makes sense except in light of evolution. *American Biology Teacher,* 35, 125–9.

Donoghue, M. J., and Ackerly, D. D. 1996. Phylogenetic uncertainties and sensitivity analyses in comparative biology. *Philosophical Transactions of the Royal Society B,* 351, 1241–9.

Dormann, C. F., McPherson, J. M., Araújo, B. M., Bivand, R., Bolliger, J., Carl, G. , Davies, R. G. , Hirzel, A., Jetz, W., Daniel Kissling, W., Kühn, I. , Ohlemüller, R., Peres-Neto, P. R., Reineking, B., Schröder, B., Schurr, F. M., and Wilson, R. 2007. Methods to account for spatial autocorrelation in the analysis of species distributional data: a review. *Ecography,* 30, 609–28.

Dray, S., Legendre, P., and Peres-Neto, P. R. 2006. Spatial modelling: a comprehensive framework for principal coordinate analysis of neighbour matrices (PCNM). *Ecological Modelling,* 196, 483–93.

Drew, G. S. 1994. The scientific method revisited. *Conservation Biology,* 8, 596–7.

Dungan, J. L., Perry, J. N., Dale, M. R. T., Legendre, P., Citron-Pousty, S., Fortin, M. J. , Jakomulska, A., Miriti, M., and Rosenberg, M. S. 2002. A balanced view of scale in spatial statistical analysis. *Ecography,* 25, 626–40.

Dungan, R. J., Duncan, R. P., and Whitehead, D. 2003. Investigating leaf lifespans with interval-censored failure time analysis. *New Phytologist,* 158, 593–600.

Durrett, R., and Levin, S. A. 1996. Spatial models for species-area curves. *Journal of Theoretical Biology,* 179, 119–27.

Duval, S. 2005. The trim and fill method. *In:* Rothstein, H. R., Sutton, A. J., and Borenstein, M. (eds.) *Publication bias in meta-analysis: prevention, assessment and adjustments,* pp. 127–44. Chichester, UK, Wiley.

Duval, S., and Tweedie, R. 2000a. A nonparametric "trim and fill" method of accounting for publication bias in meta-analysis. *Journal of the American Statistical Association,* 95, 89–98.

Duval, S., and Tweedie, R. 2000b. Trim and fill: A simple funnel-plot-based method of testing and adjusting for publication bias in meta-analysis. *Biometrics,* 56, 455–63.

Dwyer, J., Fensham, R., Fairfax, R., and Buckley, Y. 2010. Neighbourhood effects influence drought-induced mortality of savanna trees in Australia. *Journal of Vegetation Science,* 21, 573–85.

Edwards, A. W. F., Cavalli-Sforza, L. L., and Heywood, V. H. 1963. Phenetic and phylogenetic classification. *Systematics Association Publication,* 6, 67–76.

Edwards, M., C., Wirth, R. J., Houts, C. R., and Xi, N. 2012. Categorical data in the structural equation modeling framework. *In:* Hoyle, R. H. (ed.) *Handbook of Structural Equation Modeling,* pp. 195–208. New York, Guilford Press.

Egger, M., Smith, G. D., Schneider, M., and Minder, C. 1997. Bias in meta-analysis detected by a simple, graphical test. *British Medical Journal,* 315, 629–34.

Ellison, A. M. 1993. Exploratory data analysis and graphic display. *In:* Scheiner, S. M., and Gurevitch, J. (eds.) *Design and analysis of ecological experiments,* pp. 14–41. New York, Chapman and Hall.

Elston, D. A., Moss, R., Boulinier, T., Arrowsmith, C., and Lambin, X. 2001. Analysis of aggregation, a worked example: numbers of ticks on red grouse chicks. *Parasitology,* 122, 563–9.

Enders, C. K. 2010. *Applied missing data analysis.* New York, Guilford Press.

Enders, C. K., and Gottschall, A. C. 2011. Multiple Imputation Strategies for Multiple Group Structural Equation Models. *Structural Equation Modeling,* 18, 35–54.

Erman, D. C., and Pister, E. P. 1989. Ethics and the environmental biologist. *Fisheries,* 14, 4–7.

Fagerström, T. 1987. On theory, data and mathematics in ecology. *Oikos,* 50, 258–61.

Felsenstein, J. 1985. Phylogenies and the comparative method. *American Naturalist,* 125, 1–15.

Felsenstein, J. 2008. Comparative methods with sampling error and within-species variation: Contrasts revisited and revised. *American Naturalist,* 171, 713–25.

Felsenstein, J. 2004. *Inferring phylogenies.* Sunderland, MA, Sinauer.

Fidler, F., Burgman, M. A., Cumming, G., Buttrose, R., and Thomason, N. 2006. Impact of criticism of null-hypothesis significance testing on statistical reporting practices in conservation biology. *Conservation Biology,* 20, 1539–44.

Field, C. A., and Welsh, A. H. 2007. Bootstrapping clustered data. *Journal of the Royal Statistical Society: Series B (Statistical Methodology),* 69, 369–90.

Fisher, D. O., Blomberg, S. P., and Owens, I. P. F. 2003. Extrinsic versus intrinsic factors in the decline and extinction of Australian marsupials. *Proceedings of the Royal Society of London Series B-Biological Sciences,* 270, 1801–08.

Fleiss, J. L., and Berlin, J. A. 2009. Effect sizes for dichotomous data. *In:* Cooper, H., Hedges, L. V. and Valentine, J. C. (eds.) *The handbook of research synthesis and meta-analysis,* pp. 237–54. New York, Russell Sage Foundation.

Folmer, E. O., Olff, H., and Piersma, T. 2012. The spatial distribution of flocking foragers: disentangling the effects of food availability, interference and conspecific attraction by means of spatial autoregressive modeling. *Oikos,* 121, 551–61.

Folmer, E. O., and Piersma, T. 2012. The contributions of resource availability and social forces to foraging distributions: a spatial lag modelling approach. *Animal Behaviour,* 84, 1371–80.

Forber, P. 2009. Spandrels and a pervasive problem of evidence. *Biology and Philosophy,* 24, 247–66.

Ford, C. R., Minor, E. S., and Fox, G. A. 2010. Long-term effects of fire and fire-return interval on population structure and growth of longleaf pine (*Pinus palustris*). *Canadian Journal of Forest Research-Revue Canadienne De Recherche Forestiere,* 40, 1410–20.

Forister, M. L., Fordyce, J. A., McCall, A. C., and Shapiro, A. M. 2011. A complete record from colonization to extinction reveals density dependence and the importance of winter conditions for a population of the silvery blue, *Glaucopsyche lygdamus*. *Journal of Insect Science,* 11, art130.

Forstmeier, W., and Schielzeth, H. 2011. Cryptic multiple hypotheses testing in linear models: overestimated effect sizes and the winner's curse. *Behavioral Ecology and Sociobiology,* 65, 47–55.

Fortin, M.-J., James, P. M. A., MacKenzie, A., Melles, S. J., and Rayfield, B. 2012. Spatial statistics, spatial regression, and graph theory in ecology. *Spatial Statistics,* 1, 100–9.

Fortin, M. J., and Dale, M. R. T. 2005. *Spatial analysis: a guide for ecologists.* Cambridge, Cambridge University Press.

Fox, G. A. 2001. Failure time analysis: studying times-to-events and rates at which events occur. *In:* Scheiner, S. M., and Gurevitch, J. (eds.) *Design and analysis of ecological experiments,* 2nd ed, pp. 253–89. Oxford, Oxford University Press.

Fox, G. A., Kendall, B. E., Fitzpatrick, J., and Woolfenden, G. 2006. Consequences of heterogeneity in survival in a population of Florida scrub-jays. *Journal of Animal Ecology,* 75, 921–7.

Fox, J., Nie, Z., and Byrnes, J. 2013. sem: Structural equation models (Version 3.1-3). http://CRAN.R-project.org/package=sem

Freckleton, R. P. 2009. The seven deadly sins of comparative analysis. *Journal of Evolutionary Biology,* 22, 1367–75.

Freckleton, R. P. 2011. Dealing with collinearity in behavioural and ecological data: model averaging and the problems of measurement error. *Behavioral Ecology and Sociobiology,* 65, 91–101.

Freckleton, R. P., Cooper, N., and Jetz, W. 2011. Comparative methods as a statistical fix: the dangers of ignoring an evolutionary model. *American Naturalist,* 178, E10–E17.

Freckleton, R. P., Harvey, P. H., and Pagel, M. 2002. Phylogenetic analysis and comparative data: A test and review of evidence. *American Naturalist,* 160, 712–26.

Freckleton, R. P., Harvey, P. H., and Pagel, M. 2003. Bergmann's rule and body size in mammals. *American Naturalist,* 161, 821–5.

Garamszegi, L. Z., and Møller, A. P. 2011. Nonrandom variation in within-species sample size and missing data in phylogenetic comparative studies. *Systematic Biology,* 60, 876–80.

Garland, T., Bennett, A. F., and Rezende, E. L. 2005. Phylogenetic approaches in comparative physiology. *Journal of Experimental Biology,* 208, 3015–35.

Garland, T., and Ives, A. R. 2000. Using the past to predict the present: Confidence intervals for regression equations in phylogenetic comparative methods. *American Naturalist,* 155, 346–64.

Garland, T., Midford, P. E., and Ives, A. R. 1999. An introduction to phylogenetically based statistical methods, with a new method for confidence intervals on ancestral values. *American Zoologist,* 39, 374–88.

Gauch, H. G. 2003. *Scientific method in practice.* Cambridge, Cambridge University Press.

Gelfand, A. E. 2012. Hierarchical modeling for spatial data problems. *Spatial Statistics,* 1, 30–9.

Gelfand, A. E., Mallick, B. K., and Polasek, W. 1997. Broken biological size relationships: A truncated semiparametric regression approach with measurement error. *Journal of the American Statistical Association,* 92, 836–45.

Gelman, A. 2005. Analysis of variance: why it is more important than ever. *Annals of Statistics,* 33, 1–53.

Gelman, A., and Hill, J. 2007. *Data analysis using regression and multilevel/hierarchical models.* Cambridge, Cambridge University Press.

Gelman, A., Pasarica, C., and Dodhia, R. 2002. Let's practice what we preach: turning tables into graphs. *American Statistician,* 56, 121–30.

Gelman, A., and Tuerlinckx, F. 2000. Type S error rates for classical and Bayesian single and multiple comparison procedures. *Computational Statistics,* 15, 373–90.

George, T. L., and Zack, S. 2001. Spatial and temporal considerations in restoring habitat for wildlife. *Restoration Ecology,* 9, 272–9.

Gimenez, O., Anker-Nilssen, T., and Grosbois, V. 2012. Exploring causal pathways in demographic parameter variation: path analysis of mark–recapture data. *Methods in Ecology and Evolution,* 3, 427–32.

González-Suárez, M., Lucas, P. M., and Revilla, E. 2012. Biases in comparative analyses of extinction risk: mind the gap. *Journal of Animal Ecology,* 81, 1211–22.

Gotelli, N. J., and Ellison, A. M. 2004. *A Primer of Ecological Statistics.* Sunderland, MA, Sinauer.

Gotelli, N. J., and Graves, G. R. 1996. *Null Models in Ecology.* Washington, D.C., Smithsonian Press.

Gough, L., and Grace, J. B. 1999. Predicting effects of environmental change on plant species density: experimental evaluations in a coastal wetland. *Ecology,* 80, 882–90.

Gould, S. J., Raup, D. M., Sepkoski, J., and Simberloff, D. S. 1977. The shape of evolution: a comparison of real and random clades. *Paleobiology,* 3, 23–40.

Grace, J. B. 2006. Structural equation modeling and natural systems. Cambridge, Cambridge University Press.

Grace, J. B., Anderson, T. M., Olff, H., and Scheiner, S. M. 2010. On the specification of structural equation models for ecological systems. *Ecological Monographs,* 80, 67–87.

Grace, J. B., and Bollen, K. A. 2005. Interpreting the results from multiple regression and structural equation models. *Bulletin of the Ecological Society of America,* 86, 283–95.

Grace, J. B., and Bollen, K. A. 2008. Representing general theoretical concepts in structural equation models: the role of composite variables. *Environmental and Ecological Statistics,* 15, 191–213.

Grace, J. B., Harrison, S., and Damschen, E. I. 2011. Local richness along gradients in the Siskiyou herb flora: R. H. Whittaker revisited. *Ecology* 92, 108–20.

Grace, J. B., Schoolmaster Jr., D. R., Guntenspergen, G. R., Little, A. M., Mitchell, B. R., Miller, K. M., and Schweiger, E. W. 2012. Guidelines for a graph-theoretic implementation of structural equation modeling. *Ecosphere* 3:art73. http://dx.doi.org/10.1890/ES12-00048.1

Grace, J. B., Youngblood, A., and Scheiner, S. M. 2009. Structural equation modeling and ecological experiments. *In:* Miao, S., Carstenn, S., and Nungesser, M. (eds.) *Real World Ecology,* pp. 19–45. New York, Springer-Verlag.

Grace, S. L., and Platt, W. J. 1995. Neighborhood effects on juveniles in an old-growth stand of longleaf pine, *Pinus palustris. Oikos,* 72, 99–105.

Grafen, A. 1989. The phylogenetic regression *Philosophical Transactions of the Royal Society of London Series B-Biological Sciences,* 326, 119–57.

Grafen, A., and Hails, R. 2002. *Modern statistics for the life sciences: learn to analyse your own data.* Oxford, Oxford University Press.

Graham, J. W. 2009. Missing data analysis: making it work in the real world. *Annual Review of Psychology,* 60, 549–76.

Graham, J. W. 2012. *Missing data: analysis and design.* New York, Springer.

Graham, J. W., and Coffman, D. L. 2012. Structural equation modeling with missing data. *In:* Hoyle, R. H. (ed.) *Handbook of Structural Equation Modeling,* pp. 277–94. New York, Guilford Press.

Graham, J. W., Olchowski, A. E., and Gilreath, T. D. 2007. How many imputations are really needed? Some practical clarifications of multiple imputation theory. *Prevention Science, 8,* 206–13.

Graham, J. W., and Schafer, J. L. 1999. On the performance of multiple imputation for multivariate data with small sample size. *In:* Hoyle, R. (ed.) *Statistical strategies for small sample research,* pp. 1–29. Thousand Oaks, CA: Sage.

Graham, J. W., Taylor, B. J., Olchowski, A. E., and Cumsille, P. E. 2006. Planned missing data designs in psychological research. *Psychological Methods,* 11, 323–43.

Greene, W. 2005. Censored data and truncated distributions. *In:* Mills, T., and Patterson, K. (eds.) *Handbook of econometrics,.* London, Palgrave.

Greene, W. H. 2008. *Econometric analysis.* Upper Saddle River, NJ, Pearson/Prentice Hall.

Greenland, S. 2004. Ecologic inference problems in the analysis of surveillance data. *In:* Brookmayer, R., and Stroup, D. F. (eds.) *Monitoring the health of populations,* pp. 315–40. New York, Oxford University Press.

Greenland, S., and O'Rourke, K. 2001. On the bias produced by quality scores in meta-analysis, and a hierarchical view of proposed solutions. *Biostatistics,* 2, 463–71.

Greenland, S., Pearl, J., and Robins, J. M. 1999. Causal diagrams for epidemiologic research. *Epidemiology,* 10, 37–48.

Greven, S., and Kneib, T. 2010. On the behaviour of marginal and conditional Akaike information criteria in linear mixed models. *Biometrika,* 97, 773–89.

Griffith, D. A. 2012. Spatial statistics: a quantitative geographer's perspective. *Spatial Statistics,* 1, 3–15.

Griffith, D. A., and Peres-Neto, P. R. 2006. Spatial modeling in ecology: the flexibility of eigenfunction spatial analyses. Ecology, 87, 2603–13.

Griffith, S. C., Owens, I. P. F., and Thuman, K. A. 2002. Extra pair paternity in birds: a review of interspecific variation and adaptive function. *Molecular Ecology,* 11, 2195–212.

Guillera-Arroita, G., and Lahoz-Monfort, J. J. 2012. Designing studies to detect changes in species occupancy: power analysis under imperfect detection. *Methods in Ecology & Evolution,* 3, 860–9.

Gunton, R. M., and Kunin, W. E. 2009. Density-dependence at multiple scales in experimental and natural plant populations. *Journal of Ecology,* 97, 567–80.

Gurevitch, J., Curtis, P. S., and Jones, M. H. 2001. Meta-analysis in ecology. *Advances in Ecological Research, 32,* 199–247.

Gurevitch, J., and Hedges, L. V. 1999. Statistical issues in ecological meta-analyses. *Ecology,* 80, 1142–9.

Gurevitch, J., and Koricheva, J. 2013. Conclusions: past, present, and future of meta-analysis in ecology. *In:* Koricheva, J., Gurevitch, J., and Mengersen, K. (eds.) *The handbook of meta-analysis in ecology and evolution,* pp. 426–31. Princeton, NJ, Princeton University Press.

Gusset, M., Stewart, G. B., Bowler, D. E., and Pullin, A. S. 2010. Wild dog reintroductions in South Africa: A systematic review and cross-validation of an endangered species recovery programme. *Journal for Nature Conservation,* 18, 230–4.

Hadfield, J. D. 2008. Estimating evolutionary parameters when viability selection is operating. *Proceedings of the Royal Society B-Biological Sciences,* 275, 723–34.

Hadfield, J. D. 2010. MCMC methods for multi-response generalised linear mixed models: the MCMCglmm R package. *Journal of Statistical Software,* 33, 1–22.

Hadfield, J. D., Krasnov, B. R., Poulin, R., and Nakagawa, S. 2014. A tale of two phylogenies: comparative analyses of ecological interactions. *American Naturalist,* 183, 174–87.

Hadfield, J. D., and Nakagawa, S. 2010. General quantitative genetic methods for comparative biology: phylogenies, taxonomies and multi-trait models for continuous and categorical characters. *Journal of Evolutionary Biology,* 23, 494–508.

Haila, Y., and Kouki, J. 1994. The phenomenon of biodiversity in conservation biology. *Annales Zoologici Fennici,* 31, 5–18.

Haining, R. 1990. *Spatial data analysis in the social and environmental sciences.* Cambridge, Cambridge University Press.

Haining, R., Law, J., and Griffith, D. 2009. Modelling small area counts in the presence of overdispersion and spatial autocorrelation. *Computational Statistics & Data Analysis,* 53, 2923–37.

Halekoh, U., and Højsgaard, S. 2013. pbkrtest: Parametric bootstrap and Kenward Roger based methods for mixed model comparison. R package version 0.3-8. http://CRAN.R-project.org/package=pbkrtest

Hall, D. 2000. Zero-inflated Poisson and binomial regression with random-effects: a case study. *Biometrics,* 56, 1030–9.

Hanley, N., Davies, A., Angelopoulos, K., Hamilton, A., Ross, A., Tinch, D., and Watson, F. 2008. Economic determinants of biodiversity change over a 400-year period in the Scottish uplands. *Journal of Applied Ecology,* 45, 1557–65.

Hansen, T. F. 1997. Stabilizing selection and the comparative analysis of adaptation. *Evolution,* 51, 1341–51.

Hansen, T. F., and Bartoszek, K. 2012. Interpreting the Evolutionary Regression: The Interplay Between Observational and Biological Errors in Phylogenetic Comparative Studies. *Systematic Biology,* 61, 413–25.

Hanson, N. R. 1958. *Patterns of discovery.* Cambridge, Cambridge University Press.

Harmon, L. J., Weir, J. T., Brock, C. D., Glor, R. E., and Challenger, W. 2008. GEIGER: investigating evolutionary radiations. *Bioinformatics,* 24, 129–31.

Haroldson, M. A., Schwartz, C. C., and White, G. C. 2006. Survival of independent grizzly bears in the greater Yellowstone ecosystem, 1983–2001. *Wildlife Monographs,* 161, 33–43.

Harvey, P. H., and Pagel, M. D. 1991. *The comparative method in evolutionary biology.* Oxford, Oxford University Press.

Harvey, P. H., and Rambaut, A. 1998. Phylogenetic extinction rates and comparative methodology. *Proceedings of the Royal Society B,* 265, 1691–6.

Hastie, T., Tibshiran, R., and Friedman, J. 2009. *The elements of statistical learning: data mining, inference, and prediction.* New York, Springer.

Hastings, A. 1977. Spatial heterogeneity and the stability of predator-prey systems. *Theoretical Population Biology,* 12, 37–48.

Hedges, L. V., and Vevea, J. L. 1998. Fixed- and random-effects models in meta-analysis. *Psychological Methods,* 3, 486–504.

Hedges, L. V., and Vevea, J. L. 2005. Selection method approaches. *In:* Rothstein, H., Sutton, A. J., and Borenstein, M. (eds.) *Publication bias in meta-analysis: prevention, assessment and adjustments,* pp. 145–74. Chichester, UK, Wiley.

Hedges, S. B., Dudley, J., and Kumar, S. 2006. TimeTree: a public knowledge-base of divergence times among organisms. *Bioinformatics,* 22, 2971–2.

Heider, F. 1958. *The psychology of interpersonal relations.* New York, Wiley.

Helsel, D. R. 2005. *Nondetects and data analysis: statistics for censored environmental data.* Hoboken, NJ, Wiley.

Helsel, D. R. 2012. *Statistics for censored environmental data using Minitab and R.* New York, Wiley.

Henningsen, A., and Hamann, J. D. 2007. systemfit: A package for estimating systems of simultaneous equations in R. *Journal of Statistical Software,* 23, 4.

Higgins, J. P., and Green, S. (eds.) 2011. *Cochrane Handbook for Systematic Reviews of Interventions.* Available from www.cochrane-handbook.org.

Higgins, J. P. T., and Thompson, S. G. 2002. Quantifying heterogeneity in a meta-analysis. *Statistics in Medicine,* 21, 1539–58.

Higgins, J. P. T., and Thompson, S. G. 2004. Controlling the risk of spurious findings from meta-regression. *Statistics in Medicine,* 23, 1663–82.

Higgins, J. P. T., Thompson, S. G., Deeks, J. J., and Altman, D. G. 2003. Measuring inconsistency in meta-analyses. *British Medical Journal,* 327, 557–60.

Hilborn, R., and Mangel, M. 1997. *The ecological detective: confronting models with data.* Princeton, NJ, Princeton University Press.

Hilborn, R., and Stearns, S. C. 1982. On inference in ecology and evolutionary biology: the problem of multiple causes. *Acta Biotheoretica,* 31, 145–64.

Hillebrand, H., and Gurevitch, J. 2014. Meta-analysis results are unlikely to be biased by differences in variance and replication between ecological lab and field studies. *Oikos* 123: 794–9.

Hinde, J., and Demetrio, C. G. B. 1998. Overdispersion: models and estimation. *Computational Statistics & Data Analysis,* 27, 151–70.

Hobbs, N. T., and Hilborn, R. 2006. Alternatives to statistical hypothesis testing in ecology: a guide to self teaching. *Ecological Applications,* 16, 5–19.

Hoeksema, J. D., Chaudhary, V. B., Gehring, C. A., Johnson, N. C., Karst, J., Koide, R. T., Pringle, A., Zabinski, C., Bever, J. D., Moore, J. C., Wilson, G. W. T., Klironomos, J. N., and Umbanhowar, J. 2010. A meta-analysis of context-dependency in plant response to inoculation with mycorrhizal fungi. *Ecology Letters,* 13, 394–407.

Hoeting, J. A., Davis, R. A., Merton, A. A. , and Thompson, S. E. 2006. Model selection for geostatistical models. *Ecological Applications,* 16, 87–98.

Holyoak, M., Leibold, M. A., and Holt, R. D. 2005. *Metacommunities: spatial dynamics and ecological communities.* Chicago, University of Chicago Press.

Hom, C. L., and Cochrane, M. E. 1991. Ecological experiments: assumptions, approaches, and prospects. *Herpetologica,* 47, 460–73.

Honaker, J., and King, G. 2010. What to do about missing data values in time series cross-section data. *American Journal of Political Science,* 54, 561–81.

Honaker, J., King, G., and Blackwell, M. 2011. Amelia II: a program for missing data. *Journal of Statistical Software,* 45, 1–47.

Horton, N. J., and Kleinman, K. P. 2007. Much ado about nothing: A comparison of missing data methods and software to fit incomplete data regression models. *American Statistician,* 61, 79–90.

Houle, D., Pélabon, C., Wagner, G. P., and Hansen, T. F. 2011. Measurement and meaning in biology. *The Quarterly Review of Biology,* 86, 3–34.

Hoyle, R. H. 2012a. *Handbook of Structural Equation Modeling.* New York, Guilford Press.

Hoyle, R. H. 2012b. Model specification in structural equation modeling. *In:* Hoyle, R. H. (ed.) *Handbook of Structural Equation Modeling,* pp. 126–44. New York, Guilford Press.

Hozo, S. P., Djulbegovic, B., and Hozo, I. 2005. Estimating the mean and variance from the median, range, and the size of a sample. *BMC Med Res Methodol,* 5, 13.

Huffaker, C. B. 1958. Experimental studies on predation: dispersion factors and predator-prey oscillations. *Hilgardia,* 27, 343–83.

Huffaker, C. B., Shea, K. P., and Herman, S. G. 1963. Experimental studies on predation: complex dispersion and levels of food in an acarine predator-prey interaction. *Hilgardia,* 34, 305–29.

Hull, D. 1988. *Science as a process,* Chicago, University of Chicago Press.

Hurlbert, S. H. 1984. Pseudoreplication and the design of ecological field experiments. *Ecological Monographs,* 187–211.

Hurlbert, S. H. 1994. Old shibboleths and new syntheses (book review). *Trends in Ecology and Evolution,* 9, 495–6.

Hurlbert, S. H. 1997. Functional importance vs keystoneness: Reformulating some questions in theoretical biocoenology. *Australian Journal of Ecology,* 22, 369–82.

Hurlbert, S. H., and Lombardi, C. M. 2009. Final collapse of the Neyman-Pearson decision theoretic framework and rise of the neoFisherian. *Annales Zoologici Fennici,* 46, 311–49.

Hurvich, C. M., and Tsai, C. 1989. Regression and time series model selection in small samples. *Biometrika,* 76, 297–307.

Huston, M. A. 1997. Hidden treatments in ecological experiments: re-evaluating the ecosystem function of biodiversity. *Oecologia,* 110, 449–60.

Hyatt, L. A., Rosenberg, M. S., Howard, T. G., Bole, G., Fang, W., Anastasia, J., Brown, K., Grella, R., Hinman, K., Kurdziel, J. P., and Gurevitch, J. 2003. The distance dependence prediction of the Janzen-Connell hypothesis: a meta-analysis. *Oikos,* 103, 590–602.

Ibañez, I. 2002. Effects of litter, soil surface conditions, and microhabitat on *Cercocarpus ledifolius* Nutt. Seedling emergence and establishment. *Journal of Arid Environments,* 52, 209–21.

Isaaks, E. H., and Srivastava, R. M. 1989. *Applied geostatistics.* Oxford, Oxford University Press.

Ives, A. R., Midford, P. E., and Garland, T. 2007. Within-species variation and measurement error in phylogenetic comparative methods. *Systematic Biology,* 56, 252–70.

Ives, A. R., and Zhu, J. 2006. Statistics for correlated data: phylogenies, space, and time. *Ecological Applications,* 16, 20–32.

Jackson, C. I. 1986. *Honor in science.* New Haven, CT, Sigma Xi.

Jansen, J. P., and Naci, H. 2013. Is network meta-analysis as valid as standard pairwise meta-analysis? It all depends on the distribution of effect modifiers. *Bmc Medicine,* 11, 159.

Janssen, M. H., Arcese, P., Kyser, T. K., Bertram, D. F., and Norris, D. R. 2011. Stable isotopes reveal strategic allocation of resources during juvenile development in a cryptic and threatened seabird, the Marbled Murrelet (*Brachyramphus marmoratus*). *Canadian Journal of Zoology,* 89, 856–868.

Janssen, V. 2012. Indirect tracking of drop bears using GNSS technology. *Australian Geographer,* 43, 445–52.

Jaynes, E. T. 2003. *Probability theory: the logic of science.* Cambridge, Cambridge University Press.

Jennions, M. D., Lorite, C. J., and Koricheva, J. 2013a. Role of meta-analysis in interpreting the scientific literature. *In:* Koricheva, J., Gurevitch, J., and Mengersen, K. (eds.) *The handbook of meta-analysis in ecology and evolution,* pp. 364–80. Princeton, NJ, Princeton University Press.

Jennions, M. D., Lorite, C. J., Rosenberg, M. S., and Rothstein, H. R. 2013b. Publication and related biases. *In:* Koricheva, J., Gurevitch, J., and Mengersen, K. (eds.) *The handbook of meta-analysis in ecology and evolution,* pp. 207–36. Princeton, NJ, Princeton University Press.

Jennions, M. D., and Møller, A. P. 2002. Publication bias in ecology and evolution: an empirical assessment using the 'trim and fill' method. *Biological Reviews,* 77, 211–22.

Jennions, M. D., and Møller, A. P. 2003. A survey of the statistical power of research in behavioral ecology and animal behavior. *Behavioural Ecology,* 14, 438–45.

Jiménez, L., Negrete-Yankelevich, S., and Macías-Ordóñez, R. 2012. Spatial association between floral resources and hummingbird activity in a Mexican tropical montane cloud forest. *Journal of Tropical Ecology,* 28, 497–506.

Johnson, D. H. 2002. The importance of replication in wildlife research. *Journal of Wildlife Management,* 66, 919–32.

Johnson, N. L., Kemp, A. W., and Kotz, S. 2005. *Univariate Discrete Distributions.* Hoboken, NJ, Wiley.

Jonsson, M., and Wardle, D. A. 2010. Structural equation modeling reveals plant-community drivers of carbon storage in boreal forest ecosystems. *Biology Letters,* 6, 116–19.

Jöreskog, K. G. 1973. A general method for estimating a linear structural equation system. *In:* Goldberger, A. S., and Duncan, O. D. (eds.) *Structural Equation Models in the Social Sciences,* pp. 85–112. New York, Seminar Press.

Kabacoff, R. I. 2011. *R in action: data analysis and graphics with R.* New York, Manning.

Kahl, S., Manski, D., Flora, M., and Houtman, N. 2000. Water resources management plan. US National Park Service, Acadia National Park, Mount Desert Island, ME, USA.

Kalbfleisch, J. D., and Prentice, R. L. 2002. *The statistical analysis of failure time data.* New York, Wiley.

Karban, R., and Huntzinger, M. 2006. *How to do Ecology.* Princeton, Princeton University Press.

Kareiva, P. 1987. Habitat fragmentation and the stability of predator-prey interactions. *Nature,* 326, 388–90.

Kareiva, P., and Odell, G. 1987. Swarms of predators exhibit "preytaxis" if individual predators use area-restricted search. *American Naturalist,* 130, 233–70.

Karst, J., Marczak, L., Jones, M. D., and Turkington, R. 2008. The mutualism-parasitism continuum in ectomycorrhizas: A quantitative assessment using meta-analysis. *Ecology,* 89, 1032–42.

Kass, R. E., and Raftery, A. E. 1995. Bayes factors. *Journal of the American Statistical Association,* 90, 773–95.

Keeley, J. E., Brennan, T., and Pfaff, A. H. 2008. Fire severity and ecosystem responses following crown fires in California shrublands. *Ecological Applications,* 18, 1530–46.

Keith, D. A. 2002. Population dynamics of an endangered heathland shrub, *Epacris stuartii* (Epacridaceae): Recruitment, establishment and survival. *Austral Ecology,* 27, 67.

Keller, D. R., and Golley, F. B. 2000. *The philosophy of ecology: from science to synthesis.* Athens, GA, University of Georgia Press.

Kendall, B. E., and Wittmann, M. E. 2010. A stochastic model for annual reproductive success. *American Naturalist,* 175, 461–8.

Kendall, W. L., and White, G. C. 2009. A cautionary note on substituting spatial subunits for repeated temporal sampling in studies of site occupancy. *Journal of Applied Ecology,* 46, 1182–8.

Kenward, M. G., and Roger, J. H. 1997. Small sample inference for fixed effects from restricted maximum likelihood. *Biometrics,* 53, 983–97.

Kéry, M. 2002. Inferring the absence of a species: a case study of snakes. *Journal of Wildlife Management,* 66, 330–8.

Kéry, M. 2010. *Introduction to WinBUGS for ecologists: Bayesian approach to regression, ANOVA, mixed models and related analyses.* Boston, Elsevier.

Kimball, A. W. 1957. Errors of the third kind in statistical consulting. *Journal of the American Statistical Association,* 52, 133–42.

King, G. 1997. *A solution to the ecological inference problem.* Princeton, NJ, Princeton University Press.

Kleiber, C., and Zeileis, A. 2008. *Applied econometrics with R.* New York, Springer-Verlag.

Klein, J. P., and Moeschberger, M. L. 2003. *Survival analysis: techniques for censored and truncated data.* New York, Springer.

Kleinbaum, D. G., and Kupper, L. L. 1978. *Applied regression analysis and other multivariable methods.* North Scituate, RI, USA, Duxbury Press.

Kline, R. B. 2012. Assumptions in structural equation modeling. *In:* Hoyle, R. H. (ed.) *Handbook of Structural Equation Modeling,* pp. 111–25. New York, Guilford Press.

Kodre, A. R., and Perme, M. P. 2013. Informative censoring in relative survival. *Statistics in Medicine,* 32, 4791–802.

Kodric-Brown, A., and Brown, J. H. 1993. Incomplete data sets in community ecology and biogeography: a cautionary tale. *Ecological Applications,* 3, 736–42.

Kooijman, S. A. L. M. 2010. *Dynamic energy budget theory for metabolic organisation.* Cambridge, Cambridge University Press.

Koopman, J. S. 1977. Causal models and sources of interaction. *American Journal of Epidemiology,* 106, 439–44.

Koricheva, J., and Gurevitch, J. 2013. Place of meta-analysis among other methods of research synthesis. *In:* Koricheva, J., Gurevitch, J., and Mengersen, K. (eds.), *The handbook of meta-analysis in ecology and evolution,* pp. 1–13. Princeton, NJ, Princeton University Press.

Koricheva, J., and Gurevitch, J. 2014. Uses and misuses of meta-analysis in plant ecology. *Journal of Ecology* 102, 828–844.

Koricheva, J., Gurevitch, J., and Mengersen, K. 2013a. *The handbook of meta-analysis in ecology and evolution*. Princeton, Princeton University Press.

Koricheva, J., Jennions, M. D., and Lau, J. 2013b. Temporal trends in effect sizes: causes, detection and implications. *In:* Koricheva, J., Gurevitch, J., and Mengersen, K. (eds.), *The handbook of meta-analysis in ecology and evolution,* pp. 237–54 Princeton, NJ, Princeton University Press.

Kruess, A., and Tscharntke, T. 1994. Habitat fragmentation, species loss, and biological control. *Science* 264, 1581–4..

Kuhn, T. 1970. *The structure of scientific revolutions*. Chicago, University of Chicago Press.

Kuhn, T. S., Mooers, A. O., and Thomas, G. H. 2011. A simple polytomy resolver for dated phylogenies. *Methods in Ecology & Evolution,* 2, 427–36.

Kullback, S. 1959. *Information theory and statistics*. New York, Wiley.

Lajeunesse, M. J. 2009. Meta-analysis and the comparative phylogenetic method. *American Naturalist,* 174, 369–81.

Lajeunesse, M. J. 2011. phyloMeta: a program for phylogenetic comparative analyses with meta-analysis. *Bioinformatics,* 27, 2603–04.

Lajeunesse, M. J. 2013. Recovering missing or partial data from studies: a survey of conversions and imputations for meta-analysis. *In:* Koricheva, J., Gurevitch, J., and Mengersen, K. (eds.), *The handbook of meta-analysis in ecology and evolution,* pp. 195–206. Princeton, NJ, Princeton University Press.

Lajeunesse, M. J., Rosenberg, M. S., and Jennions, M. D. 2013. Phylogenetic nonindependence and meta-analysis. *In:* Koricheva, J., Gurevitch, J., and Mengersen, K. (eds.), *The handbook of meta-analysis in ecology and evolution,* pp. 284–99. Princeton, NJ, Princeton University Press.

Laliberte, E., and Tylianakis, J. M. 2010. Deforestation homogenizes tropical parasitoid-host networks. *Ecology,* 91, 1740–7.

Lamb, E. G., and Cahill Jr., J. F. 2008. When competition does not matter: Grassland diversity and community composition. *American Naturalist,* 171, 777–87.

Lambert, D. 1992. Zero-inflated Poisson regression, with an application to defects in manufacturing. *Technometrics,* 34, 1–14.

Lande, R. 1976. Natural-selection and random genetic drift in phenotypic evolution. *Evolution,* 30, 314–34.

Landwehr, J. M., Pregibon, D., and Shoemaker, A. C. 1984. Graphical methods for assessing logistic regression models. *Journal of the American Statistical Association,* 79, 61–71.

Larson, D. L., and Grace, J. B. 2004. Temporal dynamics of leafy spurge (*Euphorbia esula*) and two species of flea beetles (*Aphthona* spp.) used as biological control agents. *Biological Control,* 29, 207–14.

Lau, J., Rothstein, H. R., and Stewart, G. B. 2013. History and progress of meta-analysis. *In:* Koricheva, J., Gurevitch, J., and Mengersen, K. (eds.), *The handbook of meta-analysis in ecology and evolution,* pp. 407–19. Princeton, NJ, Princeton University Press.

Lau, J. A., McCall, A. C., Davies, K. F., McKay, J. K., and Wright, J. W. 2008. Herbivores and edaphic factors constrain the realized niche of a native plant. *Ecology,* 89, 754–62.

Laughlin, D. C. 2011. Nitrification is linked to dominant leaf traits rather than functional diversity. *Journal of Ecology,* 99, 1091–9.

Laughlin, D. C., Abella, S. R., Covington, W. F., P, and Grace, J. B. 2007. A structural equation modeling analysis of plant species richness and soil properties in *Pinus ponderosa* forests ecosystem. *Journal of Vegetation Science,* 18, 231–42.

Law, R., Weatherby, A. J., and Warren, P. H. 2000. On the invisibility of persistent protist communities. *Oikos,* 88, 319–26.

Lebreton, J. D., Burnham, K. P., Clobert, J., and Anderson, D. R. 1992. Modelling survival and testing biological hypotheses using marked animals: a unified approach with case studies. *Ecological Monographs,* 62, 67–118.

Lee, S.-Y. 2007. *Structural Equation Modeling: A Bayesian Approach*. New York, Wiley.

Lee, S.-Y., and Song, X.-Y. 2004. Evaluation of the Bayesian and maximum likelihood approaches in analyzing structural equation models with small sample sizes. *Multivariate Behavioral Research,* 39, 653–86.

Legendre, P., and Fortin, M. J. 1989. Spatial pattern and ecological analysis. *Vegetatio*, 80, 107–38.

Legendre, P., and Legendre, L. 1998. *Numerical ecology.* New York, Elsevier.

Lei, P.-W., and Wu, Q. 2012. Estimation in structural equation modeling. *In:* Hoyle, R. H. (ed.) *Handbook of Structural Equation Modeling,* pp. 164–80. New York, Guilford Press.

Leibold, M. A., and McPeek, M. A. 2006. Coexistence of the niche and neutral perspectives in community ecology. *Ecology*, 87, 1399–1410.

LeSage, J., and Pace, R. K. 2009. *Introduction to spatial econometrics.* Boca Raton, CRC Press.

Leventhal, L., and Huynh, C. L. 1996. Directional decisions for two-tailed tests: Power, error rates, and sample size. *Psychological Methods,* 1, 278–92.

Levin, S. A. 1992. The problem of pattern and scale in ecology. *Ecology*, 73, 1943–67.

Levins, R., and Lewontin, R. 1985. *The dialectical biologist.* Cambridge, MA, Harvard University Press.

Li, K. H., Raghunathan, T. E., and Rubin, D. B. 1991. Large-sample significance levels from multiply imputed data using moment-based statistics and an F-reference distribution. *Journal of the American Statistical Association,* 86, 1065–73.

Lindén, A., and Mäntyniemi, S. 2011. Using the negative binomial distribution to model overdispersion in ecological count data. *Ecology,* 92, 1414–21.

Lindsey, J. K. 2004. *Introduction to Applied Statistics a Modeling Approach.* Oxford, Oxford University Press.

Lipsey, M. W., and Wilson, D. B. 2001. *Practical meta-analysis.* Beverly Hills, CA, Sage.

Little, A. M., Guntenspergen, G. R., and Allen, T. F. 2010. Conceptual hierarchical modeling to describe wetland plant community organization. *Wetlands,* 30, 55–65.

Little, R. J. A. 1988. A test of missing completely at random for multivariate data with missing values. *Journal of the American Statistical Association,* 83, 1198–1202.

Little, R. J. A. 1992. Regression with missing Xs: a review. *Journal of the American Statistical Association,* 87, 1227–37.

Little, R. J. A. 1995. Modeling the drop-out mechanism in repeated-measures studies. *Journal of the American Statistical Association,* 90, 1112–21.

Little, R. J. A., and Rubin, D. B. 1987. *Statistical analysis with missing data.* New York, Wiley.

Little, R. J. A., and Rubin, D. B. 2002. *Statistical analysis with missing data,* 2nd ed. Hoboken, NJ, Wiley.

Loehle, C. 1987. Hypothesis testing in ecology: psychological aspects and the importance of theory maturation. *The Quarterly Review of Biology,* 62, 397–409.

Loehle, C. 2011. Complexity and the problem of ill-posed questions in ecology. *Ecological Complexity,* 8, 60–7.

Long, J. S. 1997. *Regression models for categorical and limited dependent variables.* Thousand Oaks, CA, Sage Publications.

Longino, H. 1990. *Science as social knowledge.* Princeton, N. J., Princeton University Press.

Lowry, E., Rollinson, E. J., Laybourn, A. J., Scott, T. E., Aiello-Lammens, M. E., Gray, S. M., Mickley, J., and Gurevitch, J. 2012. Biological invasions: a field synopsis, systematic review, and database of the literature. *Ecology and Evolution,* 3, 182–96.

Lynch, M. 1991. Methods for the analysis of comparative data in evolutionary biology. *Evolution,* 45, 1065–80.

Lyons, I. M., and Beilock, S. L. 2012. When math hurts: math anxiety predicts pain network activation in anticipation of doing math. *PLOS One,* 7, e48076.

Mackenzie, D. I., Nichols, J. D., Lachman, G. B., Droege, S., Royle, J. A., and Langtimm, C. A. 2002. Estimating site occupancy rates when detection probabilities are less than one. *Ecology,* 83, 2248–55.

Mackenzie, D. I., Nichols, J. D., Royle, J. A., Pollock, K. H., Bailey, L. L., and Hines, J. E. 2006. *Occupancy Estimation and Modeling.* Burlington, MA, Academic Press.

Mackinnon, D. P. 2008. *Introduction to statistical mediation analysis.* New York, Lawrence Erlbaum Associates.

Maddison, W. P. 1989. Reconstructing character evolution on polytomous cladograms. *Cladistics,* 5, 365–77.

Malone, P. S., and Lubansky, J. B. 2012. Preparing data for structural equation modeling: doing your homework. *In:* Hoyle, R. H. (ed.) *Handbook of Structural Equation Modeling,* pp. 263–76. New York, Guilford Press.

Mangel, M. 2006. *The theoretical biologist's toolbox: quantitative methods for ecology and evolutionary biology.* Cambridge, Cambridge University Press.

Marascuilo, L. A., and Levin, J. R. 1970. Appropriate post-hoc comparisons for interaction and nested hypotheses in analysis of variance designs: the elimination of errors. *American Educational Research Journal,* 7, 397–421.

Marra, P. P., and Holmes, R. T. 1997. Avian removal experiments: do they test for habitat saturation or female availability? *Ecology,* 78, 947–52.

Martin, T. G., Kuhnert, P. M., Mengersen, K., and Possingham, H. P. 2005a. The power of expert opinion in ecological models using Bayesian methods: impact of grazing on birds. *Ecological Applications,* 15, 266–80.

Martin, T. G., Wintle, B. A., Rhodes, J. R., Kuhnert, P. M., Field, S. A., Low Choy, S. J., Tyre, A. J., and Possingham, H. P. 2005b. Zero tolerance ecology: improving ecological inference by modelling the source of zero observations. *Ecology Letters,* 8, 1235–46.

Martins, E. P. 1996. *Phylogenies and the comparative method in animal behavior.* Oxford, Oxford University Press.

Martins, E. P. 2000. Adaptation and the comparative method. *Trends in Ecology & Evolution,* 15, 296–9.

Martins, E. P., Diniz, J. a. F., and Housworth, E. A. 2002. Adaptive constraints and the phylogenetic comparative method: A computer simulation test. *Evolution,* 56, 1–13.

Martins, E. P., and Garland, T. 1991. Phylogenetic analyses of the correlated evolution of continuous characters - A simulation study. *Evolution,* 45, 534–57.

Martins, E. P., and Hansen, T. F. 1997. Phylogenies and the comparative method: A general approach to incorporating phylogenetic information into the analysis of interspecific data. *American Naturalist,* 149, 646–67.

Martins, E. P., and Housworth, E. A. 2002. Phylogeny shape and the phylogenetic comparative method. *Systematic Biology,* 51, 873–80.

Marzolin, G. 1988. Polygynie du Cincle pongeur (*Cinclus cinclus*) dans les côtes de Lorraine. *L'Oiseau et la Revue Francaise d'Ornithologie,* 58, 277–86.

Masicampo, E. J., and Lalande, D. R. 2012. A peculiar prevalence of p values just below .05. *Quarterly Journal of Experimental Psychology,* 65, 2271–9.

Mayerhofer, M. S., Kernaghan, G., and Harper, K. A. 2013. The effects of fungal root endophytes on plant growth: a meta-analysis. *Mycorrhiza,* 23, 119–28.

Mayo, D. G. 1996. *Error and the growth of experimental knowledge.* Chicago, University of Chicago Press.

McAlpine, C. A., Rhodes, J. R., Bowen, M. E., Lunney, D., Callaghan, J. G., Mitchell, D. L., and Possingham, H. P. 2008. Can multi-scale models of species' distribution be generalised from region to region? A case study of the koala. *Journal of Applied Ecology,* 45, 558567.

McAuliffe, J. R. 1984. Competition for space, disturbance, and the structure of a benthic stream community. *Ecology,* 65, 894–908.

McCarthy, M. A. 2007. *Bayesian methods for ecology.* Cambridge, Cambridge University Press.

McCarthy, M. A. 2011. Breathing some air into the single-species vacuum: multi-species responses to environmental change. *Journal of Animal Ecology,* 80, 1–3.

McCarthy, M. A., Citroen, R., and McCall, S. C. 2008. Allometric scaling and Bayesian priors for annual survival of birds and mammals. *American Naturalist,* 172, 216–22.

McCarthy, M. A., and Masters, P. 2005. Profiting from prior information in Bayesian analyses of ecological data. *Journal of Applied Ecology,* 42, 1012–19.

McCoy, E. D. 1984. Colonization by herbivores of *Heliconia* spp. plants (Zingiberales: Heliconiaceae). *Biotropica,* 16, 10–13.

McCoy, E. D. 1985. Interactions among leaf-top herbivores of *Heliconia imbricata* (Zingiberales: Heliconiaceae). *Biotropica,* 17, 326–9.

McCoy, E. D. 1990. The distribution of insect associations along elevational gradients. *Oikos,* 58, 313–22.

McCoy, E. D. 1996. Advocacy as part of conservation biology. *Conservation Biology,* 10, 919–20.

McCoy, E. D. 2002. The "veiled gradients" problem in ecology. *Oikos,* 99, 189–192.

McCoy, E. D. 2008. A data aggregation problem in studies of upper respiratory tract disease in the gopher tortoise. *Herpetological Review,* 39, 419–22.

McCoy, E. D., and Connor, E. F. 1980. Latitudinal gradients in the species-diversity of North America mammals. *Evolution,* 34, 193–203.

McCoy, E. D., and Shrader-Frechette, K. S. 1992. *Community ecology, scale, and the instability of the stability concept.* East Lansing, MI, Philosophy of Science Association.

McCoy, E. D., and Shrader-Frechette, K. S. 1995. Natural landscapes, natural communities, and natural ecosystems. *Forest and Conservation History,* 39, 138–42.

McCullagh, P., and Nelder, J. A. 1989. *Generalized Linear Models,* 2nd ed. New York, Chapman & Hall.

McCune, B., and Grace, J. B. 2002. *Analysis of ecological communities.* Gleneden Beach, OR, MjM Software.

McDonald, R. I., Peet, R. K., and Urban, D. L. 2003. Spatial pattern of *Quercus* regeneration limitation and Acer rubrum invasion in a piedmont forest. *Journal of Vegetation Science,* 14, 441–50.

McKeon, C. S., Stier, A., McIlroy, S., and Bolker, B. 2012. Multiple defender effects: synergistic coral defense by mutualist crustaceans. *Oecologia,* 169, 1095–1103.

McKnight, P. E., McKnight, K. M., Sidani, S., and Figueredo, A. J. 2007. *Missing data: a gentle introduction.* New York, Guilford Press.

McLachlan, G., and Peel, D. 2000. *Finite Mixture Models.* New York, Wiley.

McMahon, S. M., and Diez, J. M. 2007. Scales of association: hierarchical linear models and the measurement of ecological systems. *Ecology Letters,* 10, 437–52.

Mengersen, K., and Gurevitch, J. 2013. Using other metrics of effect size in meta-analysis. *In:* Koricheva, J., Gurevitch, J., and Mengersen, K. (eds.) *The handbook of meta-analysis in ecology and evolution,* pp. 72–85. Princeton, NJ, Princeton University Press.

Mengersen, K., Jennions, M. D., and Schmid, C. H. 2013a. Statistical models for the meta-analysis of nonindependent data. *In:* Koricheva, J., Gurevitch, J., and Mengersen, K. (eds.) *The handbook of meta-analysis in ecology and evolution,* pp. 255–83. Princeton, NJ, Princeton University Press.

Mengersen, K., and Schmid, C. H. 2013. Maximum likelihood approaches to meta-analysis. *In:* Koricheva, J., Gurevitch, J., and Mengersen, K. (eds.) *The handbook of meta-analysis in ecology and evolution,* pp. 125–44. Princeton, NJ, Princeton University Press.

Mengersen, K., Schmid, C. H., Jennions, M. D., and Gurevitch, J. 2013b. Statistical models and approaches to inference. *In:* Koricheva, J., Gurevitch, J., and Mengersen, K. (eds.) *The handbook of meta-analysis in ecology and evolution,* pp. 89–107. Princeton, NJ, Princeton University Press.

Mengersen, K. L., Robert, C. P., and Titterington, D. M. (eds.) 2011. *Mixtures: Estimation and Applications.* Chichester, UK, Wiley.

Merow, C., Dahlgren, J. P., Metcalf, C. J. E., Childs, D. Z., Evans, M. E. K., Jongejans, E., Record, S., Rees, M., Salguero-Gómez, R., and McMahon, S. M. 2014. Advancing population ecology with integral projection models: a practical guide. *Methods in Ecology & Evolution,* 5, 99–110.

Millar, R. B. 2011. *Maximum likelihood estimation and inference: with examples in R, SAS and ADMB.* Chichester, UK, Wiley.

Miller, A. 1993. The role of analytical science in natural resource decision making. *Environmental Management,* 17, 563–74.

Mitchell, R. J. 1992. Testing evolutionary and ecological hypotheses using path analysis and structural equation modeling. *Functional Ecology,* 6, 123–9.

Mittelhammer R. C., Judge, G. C., and Miller, D. J. 2000. *Econometric foundations.* Cambridge, Cambridge University Press.

Moher, D., Liberati, A., Tetzlaff, J., Altman, D. G., and PRISMA Group. 2009. Preferred reporting items for systematic reviews and meta-analyses: The PRISMA Statement. *PLOS Medicine,* 6, e1000097.

Molenberghs, G., and Kenward, M. G. 2007. *Missing data in clinical studies.* Chichester, UK, Wiley.

Møller, A. P., and Jennions, M. D. 2001. Testing and adjusting for publication bias. *Trends in Ecology & Evolution,* 16, 580–6.

Moore, J. A. 1993. *Science as a way of knowing.* Cambridge, MA, Harvard University Press.

Morgan, M. S. 1990. *The history of econometric ideas.* Cambridge, Cambridge University Press.

Morgan, S. L., and Winship, C. 2007. *Counterfactuals and Causal Inference: Methods and Principles for Social Research.* Cambridge, Cambridge University Press.

Moritz, M. A. 2003. Spatiotemporal analysis of controls on shrubland fire regimes: age dependency and fire hazard. *Ecology*, 84, 351–61.

Motulsky, H. J., and Chistopoulus, A. 2003. *Fitting models to biological data using linear and nonlinear regression. A practical guide to curve fitting.* San Diego CA, GraphPad Software Inc.,

Müller, S., Scealy, J. L., and Welsh, A. H. 2013. Model selection in linear mixed models. *Statistical Science*, 28, 135–67.

Mumby, P. J., Vitolo, R., and Stephenson, D. B. 2011. Temporal clustering of tropical cyclones and its ecosystem impacts. *Proceedings of the National Academy of Sciences of the United States of America*, 108, 17626–30.

Murtaugh, P. A. 2002. Journal quality, effect size, and publication bias in meta-analysis. *Ecology*, 83, 1162–6.

Murtaugh, P. A. 2007. Simplicity and complexity in ecological data analysis. *Ecology*, 88, 56–62.

Nachappa, P., Margolies, D. C., Nechols, J. R., and Campbell, J. F. 2011. Variation in predator foraging behaviour changes predator-prey spatio-temporal dynamics. *Functional Ecology*, 25, 1309–17.

Naeem, S. 1998. Species redundancy and ecosystem reliability. *Conservation Biology*, 12, 39–44.

Nagelkerke, N. J. D. 1991. A note on a general definition of the coefficient of determination. *Biometrika*, 78, 691–2.

Nagy, J. A., and Haroldson, M. A. 1990. Comparisons of some home range and population parameters among four grizzly bear populations in Canada. *In:* Darling, L. M., and Archibald, W. R. (eds.), *Bears: Their Biology and Management, Vol. 8, A Selection of Papers from the Eighth International Conference on Bear Research and Management, Victoria, British Columbia, Canada, February 1989*, pp. 227–35. International Association for Bear Research and Management.

Nakagawa, S., and Cuthill, I. C. 2007. Effect size, confidence interval and statistical significance: a practical guide for biologists. *Biological Reviews*, 82, 591–605.

Nakagawa, S., and Freckleton, R. P. 2008. Missing inaction: The dangers of ignoring missing data. *Trends in Ecology & Evolution*, 23, 592–6.

Nakagawa, S., and Freckleton, R. 2011. Model averaging, missing data and multiple imputation: A case study for behavioural ecology. *Behavioral Ecology and Sociobiology*, 65, 103–16.

Nakagawa, S., and Hauber, M. E. 2011. Great challenges with few subjects: Statistical strategies for neuroscientists. *Neuroscience and Biobehavioral Reviews*, 35, 462–73.

Nakagawa, S., Ockendon, N., Gillespie, D., Hatchwell, B., and Burke, T. 2007a. Does the badge of status influence parental care and investment in house sparrows? An experimental test. *Oecologia*, 153, 749–60.

Nakagawa, S., Ockendon, N., Gillespie, D. O. S., Hatchwell, B. J., and Burke, T. 2007b. Assessing the function of house sparrows' bib size using a flexible meta-analysis method. *Behavioral Ecology*, 18, 831–840.

Nakagawa, S., and Santos, E. S. A. 2012. Methodological issues and advances in biological meta-analysis. *Evolutionary Ecology*, 26, 1253–74.

Negrete-Yankelevich, S., Maldonado-Mendoza, I. E., Lázaro-Castellanos, J. O., Sangabriel-Conde, W., and Martínez-Álvarez, J. C. 2013. Arbuscular mycorrhizal root colonization and soil P availability are positively related to agrodiversity in Mexican maize polycultures. *Biology and Fertility of Soils*, 49, 201–12.

Neuhauser, C. 2010. *Calculus for biology and medicine.* Upper Saddle River, NJ, Pearson.

Nunn, C. L. 2011. *The comparative approach in evolutionary anthropology and biology.* Chicago, University of Chicago Press.

Nur, N., Holmes, A. L., and Geupel, G. R. 2004. Use of survival time analysis to analyze nesting success in birds: an example using loggerhead shrikes. *Condor*, 106, 457.

O'Hara, R. B., and Kotze, D. J. 2010. Do not log-transform count data. *Methods in Ecology & Evolution*, 1, 118–22.

O'Hara, B. 2007. *Focus on DIC.* Deep thoughts and silliness. http://deepthoughtsandsilliness.blogspot.com/2007/12/focus-on-dic.html, 2007 (retrieved 6 April 2014).

O'Hara, R. B. 2009. How to make models add up - a primer on GLMMs. *Annales Zoologici Fennici*, 46, 124–37.

O'Meara, B. C. 2012. Evolutionary Inferences from Phylogenies: A Review of Methods. *Annual Review of Ecology, Evolution, and Systematics,* 43, 267–85.

O'Meara, B. C., Ané, C., Sanderson, M. J., and Wainwright, P. C. 2006. Testing for different rates of continuous trait evolution using likelihood. *Evolution,* 60, 922–33.

Osenberg, C. W., Sarnelle, O., and Cooper, S. D. 1997. Effect size in ecological experiments: The application of biological models in meta-analysis. *American Naturalist,* 150, 798–812.

Osenberg, C. W., and St. Mary, C. M. 1998. Meta-analysis: synthesis or statistical subjugation? *Integrative Biology Issues News and Reviews,* 1, 37–41.

Ottaviani, D., Lasinio, G. J., and Boitani, L. 2004. Two statistical techniques to validate habitat suitability models using presence-only data. *Ecological Modelling,* 179, 417–43.

Ozgul, A., Oli, M. K., Bolker, B. M., and Perez-Heydrich, C. 2009. Upper respiratory tract disease, force of infection, and effects on survival of gopher tortoises. *Ecological Applications,* 19, 786–98.

Pagel, M. 1993. Seeking the evolutionary regression coefficient -An analysis of what comparative methods measure. *Journal of Theoretical Biology,* 164, 191–205.

Pagel, M. 1997. Inferring evolutionary processes from phylogenies. *Zoologica Scripta,* 26, 331–48.

Pagel, M. 1999. Inferring the historical patterns of biological evolution. *Nature,* 401, 877–84.

Palmer, A. R. 2000. Quasireplication and the contract of error: Lessons from sex ratios, heritabilities, and fluctuating asymmetry. *Annual Review of Ecology and Systematics,* 31, 441–80.

Paradis, E. 2011. *Analysis of Phylogenetics and Evolution with R,* 2nd ed. New York, Springer.

Paradis, E., Claude, J., and Strimmer, K. 2004. APE: Analyses of Phylogenetics and Evolution in R language. *Bioinformatics,* 20, 289–90.

Parris, K. M. 2006. Urban amphibian assemblages as metacommunities. *Journal of Animal Ecology,* 75, 757–64.

Parris, K. M., Norton, T. W., and Cunningham, R. B. 1999. A comparison of techniques for sampling amphibians in the forests of south-east Queensland, Australia. *Herpetologica,* 55, 271–83.

Patterson, T. A., Thomas, L., Wilcox, C., Ovaskainen, O., and Matthiopoulos, J. 2008. State-space models of individual animal movement. *Trends in Ecology & Evolution,* 23, 87–94.

Pawitan, Y. 2001. *In all likelihood: statistical modelling and inference using likelihood.* Oxford, Oxford University Press.

Pearl, J. 1988. *Probabilistic Reasoning in Intelligent Systems: Networks of Plausible Inference.* San Mateo, CA, Morgan Kaufmann.

Pearl, J. 2009. *Causality.* Cambridge, Cambridge University Press.

Pearl, J. 2012. The causal foundations of structural equation modeling. *In:* Hoyle, R. H. (ed.) *Handbook of Structural Equation Modeling,* pp. 68–91. New York, Guilford Press.

Peng, H., and Lu, Y. 2012. Model selection in linear mixed effect models. *Journal of Multivariate Analysis,* 109, 109–29.

Perry, J. N., Bell, E. D., Smith, R. H., and Woiwod, I. P. 1996. SADIE: software to measure and model spatial patterns. *Aspects of Applied Biology,* 46, 95–102.

Perry, J. N., Liebhold, A. M., Rosenberg, M. S., Dungan, J., Miriti, M., Jakomulska, A., and Citron-Pousty, S. 2002. Illustrations and guidelines for selecting statistical methods for quantifying spatial pattern in ecological data. *Ecography,* 25, 578–600.

Peters, R. H. 1977. The unpredictable problems of tropho-dynamics. *Environmental Biology of Fishes,* 2, 97–101.

Peugh, J. L., and Enders, C. K. 2004. Missing data in educational research: A review of reporting practices and suggestions for improvement. *Review of Educational Research,* 74, 525–56.

Pfeiffer, D. U., Robinson, T. P., Stevenson, M., Stevens, K. B., Rogers, D. J., and Clements, A. C. A. 2008. *Spatial analysis in epidemiology.* Oxford, Oxford University Press.

Pigott, T. D. 2009. Handling missing data. *In:* Cooper, H., Hedges, L. V., and Valentine, J. C. (eds.) *The handbook of research synthesis and meta-analysis,* pp. 399–416. New York, Russell Sage Foundation.

Pigott, T. D. 2012. *Advances in meta-analysis.* New York, Springer.

Pinheiro, J. C., and Bates, D. M. 2000. *Mixed-effects models in S and S-Plus.* New York, Springer-Verlag.

Plant, R. E. 2012. *Spatial data analysis in ecology and agriculture using R.* Boca Raton, CRC Press.

Platt, J. R. 1964. Strong inference: certain systematic methods of scientific thinking may produce much more rapid progress than others. *Science,* 146, 347–53.

Platt, W. J., Evans, G. W., and Rathbun, S. L. 1988. The population dynamics of a long-lived conifer (*Pinus palustris*). *American Naturalist,* 131, 491–525.

Pledger, S., and Schwarz, C. J. 2002. Modelling heterogeneity of survival in band-recovery data using mixtures. *Journal of Applied Statistics,* 29, 315–27.

Polakow, D. A., and Dunne, T. T. 1999. Modelling fire-return interval T: stochasticity and censoring in the two-parameter Weibull model. *Ecological Modelling,* 121, 79–102.

Polanyi, M. 1964. *Personal knowledge.* New York, Harper and Row.

Polanyi, M. 1959. *The study of man.* Chicago, University of Chicago Press.

Pollock, K. H., Nichols, J. D., Brownie, C., and Hines, J. E. 1990. Statistical inference for capture-recapture experiments. *Wildlife Society Monographs,* 107, 3–97.

Ponciano, J. M., Taper, M. L., Dennis, B., and Lele, S. R. 2009. Hierarchical models in ecology: confidence intervals, hypothesis testing, and model selection using data cloning. *Ecology,* 90, 356–62.

Popper, K. R. 1959. *The logic of scientific discovery.* New York, Basic Books.

Popper, K. R. 1963. *Conjectures and refutations: the growth of scientific knowledge.* London, Routledge.

Powell, J. L. 1986. Censored regression quantiles. *Journal of Econometrics,* 32, 143–55.

Power, M. E., Tilman, D., Estes, J. A., Menge, B. A., Bond, W. J., Mills, L. S., Daily, G., Castilla, J. C., Lubchenco, J., and Paine, R. T. 1996. Challenges in the quest for keystones. *Bioscience,* 46, 609–20.

Preisser, E. L., Bolnick, D. I., and Benard, M. F. 2005. Scared to death? The effects of intimidation and consumption in predator-prey interactions. *Ecology,* 86, 501–9.

Preston, F. W. 1948. The commonness, and rarity, of species. *Ecology,* 29, 254–83.

Price, T. 1997. Correlated evolution and independent contrasts. *Philosophical Transactions of the Royal Society of London B, Biological Sciences,* 352, 519–29.

Prugh, L. R., and Brashares, J. S. 2012. Partitioning the effects of an ecosystem engineer: Kangaroo rats control community structure via multiple pathways. *Journal of Animal Ecology,* 81, 667–78.

Pullin, A. S., and Stewart, G. B. 2006. Guidelines for systematic review in conservation and environmental management. *Conservation Biology,* 20, 1647–56.

Purvis, A., and Garland, T. 1993. Polytomies in comparative analyses of continuous characters. *Systematic Biology,* 42, 569–75.

Quinn, G. P., and Keough, M. J. 2002. *Experimental Design and Data Analysis for Biologists.* Cambridge, Cambridge University Press.

Quintero, H. E., Abebe, A., and Davis, D. A. 2007. Zero-inflated discrete statistical models for fecundity data analysis in channel catfish, *Ictalurus punctatus. Journal of the World Aquaculture Society,* 38, 175–87.

R Core Team. 2014. R: A language and environment for statistical computing. R Foundation for Statistical Computing, Vienna, Austria. URL http://www.R-project.org/.

Raghunathan, T. E. 2004. What do we do with missing data? Some options for analysis of incomplete data. *Annual Review of Public Health,* 25, 99–117.

Raudenbush, S. W., and Bryk, A. S. 2002. *Hierarchical linear models: applications and data analysis methods.* Thousand Oaks, Sage Publications.

Raykov, T., and Marcoulides, G. A. 2010. *Introduction to Psychometric Theory.* New York, Taylor & Francis.

Reich, P. B., Frelich, L. E., Voldseth, R. A., Bakken, P., and Adair, E. C. 2012. Understorey diversity in southern boreal forests is regulated by productivity and its indirect impacts on resource availability and heterogeneity. *Journal of Ecology,* 100, 539–45.

Revell, L. J. 2009. Size-correction and principal components for interspecific comparative studies *Evolution,* 63, 3258–68.

Revell, L. J. 2010. Phylogenetic signal and linear regression on species data. *Methods in Ecology and Evolution,* 1, 319–29.

Revelle, W. 2012. psych: Procedures for Psychological, Psychometric, and Personality Research, R package version 1.2.12. http://personality-project.org/r.

Rey, J. R., and McCoy, E. D. 1979. Application of island-biogeographic theory to pests of cultivated crops. *Environmental Entomology*, 8, 577–82.

Rhemtulla, M., and Little, T. D. 2012. Planned Missing Data Designs for Research in Cognitive Development. *Journal of Cognition and Development*, 13, 425–38.

Rhodes, J., Lunney, D., Moon, C., Matthews, A., and McAlpine, C. A. 2011. The consequences of using indirect signs that decay to determine species' occupancy. *Ecography*, 34, 141–50.

Rhodes, J. R., Callaghan, J. G., McAlpine, C. A., de Jong, C., Bowen, M. E., Mitchell, D. L., Lunney, D., and Possingham, H. P. 2008a. Regional variation in habitat-occupancy thresholds: a warning for conservation planning. *Journal of Applied Ecology*, 45, 549–57.

Rhodes, J. R., Grist, E. P. M., Kwok, K. W. H., and Leung, K. M. Y. 2008b. A Bayesian mixture model for estimating intergeneration chronic toxicity. *Environmental Science & Technology*, 42, 8108–14.

Richards, S. A. 2005. Testing ecological theory using the information-theoretic approach: examples and cautionary results. *Ecology*, 86, 2805–14.

Richards, S. A. 2008. Dealing with overdispersed count data in applied ecology. *Journal of Applied Ecology*, 45, 218–27.

Richards, S. A., Whittingham, M. J., and Stephens, P. A. 2011. Model selection and model averaging in behavioural ecology: the utility of the IT-AIC framework. *Behavioural Ecology and Sociobiology*, 65, 77–89.

Riginos, C., and Grace, J. B. 2008. Savanna tree density, herbivores, and the herbaceous community: bottom-up vs. top-down effects. *Ecology*, 89, 2228–38.

Riseng, C. M., Wiley, M. J., Seelbach, P. W., and Stevenson, R. J. 2010. An ecological assessment of Great Lakes tributaries in the Michigan peninsulas. *Journal of Great Lakes Research* 36, 505–19.

Rohlf, F. J. 2001. Comparative methods for the analysis of continuous variables: geometric interpretations. *Evolution*, 55, 2143–60.

Rose, G. A. 1992. *The strategy of preventative medicine.* Oxford, Oxford University Press.

Rosenberg, M. S. 2005. The file-drawer problem revisited: A general weighted method for calculating fail-safe numbers in meta-analysis. *Evolution*, 59, 464–8.

Rosenberg, M. S. 2013. Moment and least-squares based approaches to meta-analytic inference. *In:* Koricheva, J., Gurevitch, J., and Mengersen, K. (eds.) *The handbook of meta-analysis in ecology and evolution*, pp. 108–24. Princeton, NJ, Princeton University Press.

Rosenberg, M. S., Adams, D. C., and Gurevitch, J. 2000. *MetaWin: statistical software for meta-analysis.* 2nd ed. Sunderland, MA, Sinauer.

Rosenberg, M. S., Rothstein, H. R., and Gurevitch, J. 2013. Effect sizes: conventional choices and calculations. *In:* Koricheva, J., Gurevitch, J., and Mengersen, K. (eds.) *The handbook of meta-analysis in ecology and evolution*, pp. 61–71. Princeton, NJ, Princeton University Press.

Rosenthal, R. 1979. The "file drawer problem" and tolerance for null results. *Psychological Bulletin*, 86, 638–41.

Rosseel, Y. 2012. lavaan: An R package for structural equation modeling. *Journal of Statistical Software*, 48, 1–36.

Rosseel, Y., Oberski, D., Byrnes, J., Vanbrabant, L., Savalei, V., Merkle, E., Hallquist, M., Rhemtulla, M., Katsikatsou, M., and Barendse, M. 2014. Package 'lavaan'. http://cran.r-project.org/web/packages/lavaan/lavaan.pdf

Rossi, R. E., Mulla, D. J., Journel, A. G., and Franz, E. H. 1992. Geostatistical tools for modeling and interpreting spatial dependence. *Ecological Monographs*, 62, 277–314.

Rota, C. T., Fletcher, R. J., Dorazio, R. M., and Betts, M. G. 2009. Occupancy estimation and the closure assumption. *Journal of Applied Ecology*, 46, 1173–81.

Rothstein, H., Sutton, A. J., and Borenstein, M. (eds.) 2005. *Publication bias in meta-analysis: prevention, assessment and adjustments.* Chichester, UK, Wiley.

Rothstein, H. R., Lorite, C. J., Stewart, G. B., Koricheva, J., and Gurevitch, J. 2013. Quality standards for research syntheses. *In:* Koricheva, J., Gurevitch, J., and Mengersen, K. (eds.) *The handbook of meta-analysis in ecology and evolution*, pp. 323–38. Princeton, NJ, Princeton University Press.

Royle, J. A. 2004. N-mixture models for estimating population size from spatially replicated counts. *Biometrics*, 60, 108–15.

Royle, J. A., and Dorazio, R. M. 2008. *Hierarchical Modeling and Inference in Ecology.* London, Academic Press.

Rubin, D. B. 1976. Inference and missing data. *Biometrika,* 63, 581–90.

Rubin, D. B. 1987. *Multiple imputation for nonresponse in surveys.* New York, Wiley.

Rubin, D. B. 1996. Multiple imputation after 18+ years. *Journal of the American Statistical Association,* 91, 473–89.

Rubinstein, R. Y., and Kroese, D. P. 2011. *Simulation and the Monte Carlo method.* New York, Wiley.

Rufibach, K. 2011. selectMeta: Estimation of weight functions in meta-analysis. R package version 1.0.4. ed. http://CRAN.R-project.org/package=selectMeta

Saino, N., Romano, M., Ambrosini, R., Rubolini, D., Boncoraglio, G., Caprioli, M., and Romano, A. 2012. Longevity and lifetime reproductive success of barn swallow offspring are predicted by their hatching date and phenotypic quality. *Journal of Animal Ecology,* 1004–12.

Schafer, J. L. 1997. *Analysis of incomplete multivariate data.* London, Chapman & Hall.

Schafer, J. L. 1999. Multiple imputation: a primer. *Statistical Methods in Medical Research,* 8, 3–15.

Schafer, J. L. 2001. Multiple imputation with PAN. *In* Sayer, A. J., and Collins, L. M. (eds.) *New Methods for the Analysis of Change,* pp. 355–77. Washington, DC, American Psychological Association.

Schafer, J. L. 2003. Multiple imputation in multivariate problems when the imputation and analysis models differ. *Statistica Neerlandica,* 57, 19–35.

Schafer, J. L., and Graham, J. W. 2002. Missing data: our view of the state of the art. *Psychological Methods,* 7, 147–77.

Schafer, J. L., and Yucel, R. M. 2002. Computational strategies for multivariate linear mixed-effects models with missing values. *Journal of Computational and Graphical Statistics,* 11, 437–57.

Scheiner, S. M., Cox, S. B., Willig, M., Mittelbach, G. C., Osenberg, C., and Kaspari, M. 2000. Species richness, species-area curves and Simpson's paradox. *Evolutionary Ecology Research,* 2, 791–802.

Scheiner, S. M., and Gurevitch, J. 2001. *Design and Analysis of Ecological Experiments,* 2nd ed. New York, Oxford University Press.

Scheipl, F., Greven, S., and Kuechenhoff, H. 2008. Size and power of tests for a zero random effect variance or polynomial regression in additive and linear mixed models. *Computational Statistics & Data Analysis,* 52, 3283–99.

Schermelleh-Engel, K., Moosbrugger, H., and Müller, H. 2003. Evaluating the fit of structural equation models: Tests of significance and descriptive goodness-of-fit measures. *Methods of psychological research online,* 8, 23–74.

Schielzeth, H. 2010. Simple means to improve the interpretability of regression coefficients. *Methods in Ecology & Evolution,* 1, 103–13.

Schielzeth, H., and Forstmeier, W. 2009. Conclusions beyond support: overconfident estimates in mixed models. *Behavioral Ecology,* 20, 416–20.

Schluter, D. 2000. *The ecology of adaptive radiation.* Oxford, Oxford University Press.

Schmid, B., Polasek, W., Weiner, J., Krause, A., and Stoll, P. 1994. Modeling of discontinuous relationships in biology with censored regression. *American Naturalist,* 143, 494–507.

Schmid, C. H., and Mengersen, K. 2013. Bayesian meta-analysis. *In:* Koricheva, J., Gurevitch, J., and Mengersen, K. (eds.) *The handbook of meta-analysis in ecology and evolution,* pp. 145–73. Princeton, NJ, Princeton University Press.

Schmid, C. H., Stewart, G. B., Rothstein, H. R., Lajeunesse, M. J., and Gurevitch, J. 2013. Software for statistical meta-analysis. *In:* Koricheva, J., Gurevitch, J., and Mengersen, K. (eds.) *The handbook of meta-analysis in ecology and evolution,* pp. 174–91. Princeton, NJ, Princeton University Press.

Schoolmaster Jr., D. R., Grace, J. B., and Schweiger, E. W. 2012. A general theory of multimetric indices and their properties. *Methods in Ecology & Evolution,* 3, 773–81.

Schoolmaster Jr., D. R., Grace, J. B., Schweiger, E. W., Mitchell, B. R., and Guntenspergen, G. R. 2013. A causal examination of the effects of confounding factors on multimetric indices. *Ecological Indicators,* 29, 411–19.

Schroeder, J., Burke, T., Mannarelli, M. E., Dawson, D. A., and Nakagawa, S. 2012. Maternal effects and heritability of annual productivity. *Journal of Evolutionary Biology,* 25, 149–56.

Schwartz, C. C., Haroldson, M. A., White, G. C., Harris, R. B., Cherry, S., Keating, K. A., Moody, D., and Servheen, C. 2006. Temporal, spatial, and environmental influences on the demographics of grizzly bears in the Greater Yellowstone Ecosystem. *Wildlife Monographs,* 1–68.

Schwartz, S., and Carpenter, K. M. 1999. The right answer for the wrong question: consequences of type III error for public health research. *American Journal of Public Health,* 89, 1175–80.

Schwarzer, G., Carpenter, J., and Rucker, G. 2010. Empirical evaluation suggests Copas selection model preferable to trim-and-fill method for selection bias in meta-analysis. *Journal of Clinical Epidemiology,* 63, 282–8.

Schwenk, C. R. 1990. Effects of devil's advocacy and dialectical inquiry on decision making: a meta-analysis. *Organizational Behavior and Human Decision Processes,* 47, 161–76.

Seabloom, E. W., Williams, J. W., Slayack, D., Stoms, D. M., Viers, J. H., and Dobson, A. P. 2006. Human impacts, plant invasion, and imperiled plant species in California. *Ecological Applications,* 16, 1338–50.

Shadish, W. R., Cook, T. D., Campbell, D. T., et al. 2002. *Experimental and quasi-experimental designs for generalized causal inference.* Boston, MA, Houghton & Mifflin.

Shang, J., and Cavanaugh, J. E. 2008. Bootstrap variants of the Akaike information criterion for mixed model selection. *Computational Statistics & Data Analysis,* 52, 2004–21.

Shipley, B. 2000. *Cause and Correlation in Biology.* Cambridge, Cambridge University Press.

Shipley, B. 2013. The AIC model selection method applied to path analytic models compared using a d-separation test. *Ecology,* 94, 560–4.

Shrader-Frechette, K. S. 1994. *Ethics of scientific research.* Savage, MD, Rowman & Littlefield

Shrader-Frechette, K. S., and McCoy, E. D. 1992. Statistics, costs and rationality in ecological inference. *Trends in Ecology & Evolution,* 7, 96–9.

Shrader-Frechette, K. S., and McCoy, E. D. 1993. *Method in ecology.* Cambridge, Cambridge University Press.

Siegelman, S. S. 1991. Assassins and zealots: variations in peer review. Special report. *Radiology,* 178, 637–42.

Silvertown, J., and Bullock, J. M. 2003. Do seedlings in gaps interact? A field test of assumptions in ESS seed size models. *Oikos,* 101, 499–504.

Sim, I. M. W., Rebecca, G. W., Ludwig, S. C., Grant, M. C., and Reid, J. M. 2011. Characterizing demographic variation and contributions to population growth rate in a declining population. *Journal of Animal Ecology,* 80, 159–70.

Siqueira, T., Bini, L. M., Roque, F. O., Marques Couceiro, S. R., Trivinho-Strixino, S., and Cottenie, K. 2012. Common and rare species respond to similar niche processes in macroinvertebrate metacommunities. *Ecography,* 35, 183–92.

Skidmore, A. K., Franklin, J., Dawson, T. P., and Pilesjö, P. 2011. Geospatial tools address emerging issues in spatial ecology: a review and commentary on the Special Issue. *International Journal of Geographical Information Science,* 25, 337–65.

Sokal, R. R., and Rohlf, F. J. 1995. *Biometry: the principles and practice of statistics in biological research,* 3rd ed. New York, Freeman.

Solow, A. R. 1998. On fitting a population model in the presence of observation error. *Ecology,* 79, 1463–6.

Sólymos, P. 2010. dclone: data cloning in R. *The R Journal,* 2, 29–37.

Sommers, K. P., M. Elswick, G. I. Herrick, and G. A. Fox. 2011. Inferring microhabitat preferences of *Lilium catesbaei* (Liliaceae). *American Journal of Botany,* 98, 819–28.

Song, X.-Y., and Lee, S.-Y. 2012. *Basic and Advanced Bayesian Structural Equation Modeling: With Applications in the Medical and Behavioral Sciences.* West Sussex, UK, Wiley.

Spiegelhalter, D. J., Best, N., Carlin, B. P., and der Linde, A. V. 2002. Bayesian measures of model complexity and fit. *Journal of the Royal Statistical Society B,* 64, 583–640.

Spiller, D. A., and Schoener, T. W. 1995. Long-term variation in the effect of lizards on spider density is linked to rainfall. *Oecologia,* 103, 133–9.

Spiller, D. A., and Schoener, T. W. 1998. Lizards reduce spider species richness by excluding rare species. *Ecology,* 79, 503–16.

Steidl, R. J., and Thomas, L. 2001. Power analysis and experimental design. *In:* Scheiner, S. M. and Gurevitch, J. (eds.) *Design and analysis of ecological experiments,* 2nd ed, pp. 14–36. Oxford, Oxford University Press.

Stein, M. L. 1999. *Interpolation of spatial data: some theory for kriging.* New York, Springer.

Stephens, P. A., Buskirk, S. W., Hayward, G. D., and Del Rio, C. 2005. Information theory and hypothesis testing: a call for pluralism. *Journal of Animal Ecology,* 42, 4–12.

Sterne, J. A., and Egger, M. 2001. Funnel plots for detecting bias in meta-analysis: guidelines on choice of axis. *Journal of Clinical Epidemiology,* 54, 1046–55.

Sterne, J. A. C., Becker, B. J., and Egger, M. 2005. The funnel plot. *In:* Rothstein, H., Sutton, A. J., and Borenstein, M. (eds.) *Publication bias in meta-analysis: prevention, assessment and adjustments,* pp. 75–98. Chichester, UK, Wiley.

Sterne, J. a. C., White, I. R., Carlin, J. B., Spratt, M., Royston, P., Kenward, M. G., Wood, A. M., and Carpenter, J. R. 2009. Multiple imputation for missing data in epidemiological and clinical research: potential and pitfalls. *British Medical Journal,* 339, 157–60.

Stewart, A. J. A., and Lees, D. R. 1996. The colour/pattern polymorphism of *Philaenus spumarius* (L.) (Homoptera: Cercopidae) in England and Wales. *Philosophical Transactions of the Royal Society B,* 351, 69–89.

Stewart, G. B., Coles, C. F., and Pullin, A. S. 2005. Applying evidence-based practice in conservation management: Lessons from the first systematic review and dissemination projects. *Biological Conservation,* 126, 270–8.

Stock, J. H., and Trebbi, F. 2003. Who invented instrumental variable regression? *Journal of Economic Perspectives,* 17, 177–94.

Streiner, D. L. 2002. Breaking up is hard to do: the heartbreak of dichotomizing continuous data. *Canadian journal of psychiatry / Revue canadienne de psychiatrie,* 47, 262–6.

Strong, D. R. 1983. Natural variability and the manifold mechanisms of ecological communities. *American Naturalist,* 122, 636–60.

Stroup, D. F., Berlin, J. A., Morton, S. C., Olkin, I., Williamson, G. D., Rennie, D., Moher, D., Becker, B. J., Sipe, T. A., and Thacker, S. B. 2000. Meta-analysis of observational studies in epidemiology - A proposal for reporting. *JAMA: Journal of the American Medical Association,* 283, 2008–12.

Stroup, W. W. 2014. Rethinking the Analysis of Non-Normal Data in Plant and Soil Science. *Agronomy Journal,* 106, 1–17.

Stuart, A., and Ord, J. K. 1994. *Kendall's advanced theory of statistics. Vol. I. Distribution theory.* London, Griffin.

Su, Y. S., Gelman, A., Hill, J., and Yajima, M. 2011. Multiple Imputation with Diagnostics (mi) in R: Opening Windows into the Black Box. *Journal of Statistical Software,* 45, 1–31.

Sutton, A. J. 2009. Publication bias. *In:* Cooper, H., Hedges, L. V., and Valentine, J. C. (eds.) *The handbook of research synthesis and meta-analysis,* pp. 435–52. New York, Russell Sage Foundation.

Tanner, M. A., and Wong, W. H. 1987. The calculation of posterior distributions by data augmentation. *Journal of the American Statistical Association,* 82, 528–40.

Taper, M. L., and Lele, S. R. 2004. *The nature of scientific evidence: Statistical, philosophical, and empirical considerations.* Chicago, University of Chicago Press.

Taylor, M. K., Laake, J., McLoughlin, P. D., Born, E. W., Cluff, H. D., Ferguson, S. H., Rosing-Asvid, A., Schweinsburg, R., and Messier, F. 2005. Demography and viability of a hunted population of polar bears. *Arctic,* 58, 203–14.

Therneau, T. M., and Grambsch, P. M. 2000. *Modeling survival data: extending the Cox model.* New York, Springer-Verlag.

Thomas, D. L., Johnson, D., and Griffith, B. 2006. A Bayesian random effects discrete-choice model for resource selection: population-level selection inference. *Journal of Wildlife Management,* 70, 404–12.

Thompson, B. 2002. What future quantitative social science research could look like: confidence intervals for effect sizes. *Educational Researcher,* 31, 25–32.

Tilman, D., and Kareiva, P. M. 1997. *Spatial ecology: the role of space in population dynamics and interspecific interactions.* Princeton, Princeton University Press.

Tischendorf, L. 2001. Can landscape indices predict ecological processes consistently? *Landscape Ecology,* 16, 235–54.

Tobin, J. 1958. Estimation of relationships for limited dependent variables. *Econometrica,* 26, 24–36.

Tonsor, S. J., and Scheiner, S. M. 2007. Plastic trait integration across a CO_2 gradient in *Arabidopsis thaliana. American Naturalist,* 169, E119–40.

Tool, M. C. 1979. *The discretionary economy: a normative theory of political economy.* Santa Monica, CA, Goodyear.

Toulmin, S. 1961. *Foresight and understanding*. New York, Harper & Row.

Trikalinos, T. A., and Ioannidis, J. P. 2005. Assessing the evolution of effect sizes over time. *In:* Rothstein, H., Sutton, A. J., and Borenstein, M. (eds.) *Publication bias in meta-analysis: prevention, assessment and adjustments,* pp. 241–59. Chichester, UK, Wiley.

Tukey, J. W. 1977. *Exploratory data analysis*. Reading, MA, Addison-Wesley.

Tyre, A. J., Tenhumberg, B., Field, S. A., Niejalke, D., Parris, K., and Possingham, H. P. 2003. Improving precision and reducing bias in biological surveys: estimating false-negative error rates. *Ecological Applications,* 13, 1790–1801.

Uhlenbeck, G. E., and Ornstein, L. S. 1930. On the theory of the Brownian motion. *Physical Review,* 36, 0823–41.

Underwood, A. J. 1997. *Experiments in ecology: their logical design and interpretation using analysis of variance*. Cambridge, Cambridge University Press.

Underwood, A. J. 1999. Publication of so-called "negative" results in marine ecology. *Marine Ecology Progress Series,* 191, 307–09.

Vaida, F., and Blanchard, S. 2005. Conditional Akaike information for mixed-effects models. *Biometrika,* 92, 351–70.

van Buuren, S. 2011. Multiple imputation of multilevel data. *In* Hox J, J., and Roberts, J. K. *(eds.) The handbook of advanced multilevel analysis* (10), pp. 173–96. Milton Park, UK, Routledge.

van Buuren, S. 2012. *Flexible imputation of missing data*. Boca Raton, FL, CRC Press.

van Buuren, S., Brand, J. P. L., Groothuis-Oudshoorn, C. G. M., and Rubin, D. B. 2006. Fully conditional specification in multivariate imputation. *Journal of Statistical Computation and Simulation,* 76, 1049–64.

van Buuren, S., and Groothuis–Oudshoorn, K. 2011. mice: multivariate imputation by chained equations in R. *Journal of Statistical Software,* 45, 1–67.

Van Dyke, F. 2008. *Conservation biology: Foundations, concepts, applications*. London, Springer.

Vandermeer, J. 1981. *Elementary mathematical ecology*. New York, Wiley.

Venables, W. N., and Ripley, B. D. 2002. *Modern Applied Statistics with S*. London, Springer.

Venzon, D. J., and Moolgavkar, S. H. 1988. A method for computing profile likelihood-based confidence intervals. *Applied Statistics,* 37, 89–94.

Ver Hoef, J. M., and Boveng, P. L. 2007. Quasi-Poisson vs. negative binomial regression: how should we model overdispersed count data? *Ecology,* 88, 2766–72.

Vetter, D., Rucker, G., and Storch, I. 2013. Meta-analysis: A need for well-defined usage in ecology and conservation biology. *Ecosphere,* 4, 74.

Viechtbauer, W. 2010. Conducting meta-analyses in R with the metafor package. *Journal of Statistical Software,* 36, 1–48.

Vile, D., Shipley, B., and Garnier, E. 2006. A structural equation model to integrate changes in functional strategies during old-field succession. *Ecology,* 87, 504–17.

Viswanathan, M. 2005. *Measurement Error and Research Design*. Thousand Oaks, CA, Sage Publications.

von Hippel, P. T. 2009. How to impute interactions, squares and other transformed variables. *Sociological Methodology,* 39, 265–91.

Wagenmakers, E.-J. 2007. A practical solution to the pervasive problems of p values. *Psychonomic Bulletin & Review,* 14, 779–804.

Wainer, H. 2007. The most dangerous equation. *American Scientist,* 95, 249–56.

Wall, M. M. 2004. A close look at the spatial structure implied by the CAR and SAR models. *Journal of Statistical Planning and Inference,* 121, 311–24.

Wang, J.-F., Stein, A., Gao, B.-B., and Ge, Y. 2012. A review of spatial sampling. *Spatial Statistics,* 2, 1–14.

Wang, M. C., and Bushman, B. J. 1998. Using the normal quantile plot to explore meta-analytic data sets. *Psychological Methods,* 3, 46–54.

Wardle, D. A. 2002. *Communities and ecosystems: linking the aboveground and belowground components*. Princeton, Princeton University Press.

Warton, D. I., and Hui, F. K. C. 2010. The arcsine is asinine: the analysis of proportions in ecology. *Ecology,* 92, 3–10.

Webb, C. O., and Donoghue, M. J. 2005. Phylomatic: tree assembly for applied phylogenetics. *Molecular Ecology Notes,* 5, 181–3.

Webster, R., and Oliver, M. A. 1990. *Statistical methods in soil and land resource survey.* Oxford, Oxford University Press.

Webster, R., and Oliver, M. A. 1992. Sample adequately to estimate variograms of soil properties. *Journal of Soil Science,* 43, 177–92.

Weiher, E. 2003. Species richness along multiple gradients: testing a general multivariate model in oak savannas. *Oikos,* 101, 311–16.

Wenger, S. J., and Freeman, M. C. 2008. Estimating species occurrence, abundance, and detection probability using zero-inflated distribution. *Ecology,* 89, 2953–9.

West, G. B., Brown, J. H., and Enquist, B. J. 1997. A general model for the origin of allometric scaling laws in biology. *Science,* 276, 122–6.

West, S. G., Taylor, A. B., and Wu, W. 2012. Model fit and model selection in structural equation modeling. *In:* Hoyle, R. H. (ed.) *Handbook of Structural Equation Modeling,* pp. 209–32. New York, Guilford Press.

Whittingham, M. J., Stephens, P. A., Bradbury, R. B., and Freckleton, R. P. 2006. Why do we still use stepwise modelling in ecology and behaviour? *Journal of Animal Ecology,* 75, 1182–9.

Wiens, J. A. 1981. On skepticism and criticism in ornithology. *Auk,* 98, 848–9.

Wintle, B. A., Walshe, T. V., Parris, K. M., and McCarthy, M. A. 2012. Designing occupancy surveys and interpreting non-detection when observations are imperfect. *Diversity and Distributions,* 18, 417–24.

Wood, A. M., White, I. R., and Thompson, S. G. 2004. Are missing outcome data adequately handled? A review of published randomized controlled trials in major medical journals. *Clinical Trials,* 1, 368–76.

Wooldridge, J. M. 2010. *Econometric analysis of cross section and panel data.* Cambridge, Mass., MIT Press.

Woolfenden, G. E., and Fitzpatrick, J. W. 1984. *The Florida scrub jay: demography of a cooperative-breeding bird.* Princeton, NJ, Princeton University Press.

Worm, B., Lotze, H. K., Hillebrand, H., and Sommer, U. 2002. Consumer versus resource control of species diversity and ecosystem functioning. *Nature,* 417, 848–51.

Wright, S. 1921. Correlation and causation. *Journal of Agricultural Research,* 10, 557–85.

Youngblood, A., Grace, J. B., and McIver, J. D. 2009. Delayed conifer mortality after fuel reduction treatments: Interactive effects of fuel, fire intensity, and bark beetles. *Ecological Applications,* 19, 321–37.

Yuan, K.-H., and Zhang, Z. 2012. rsem: Robust Structural Equation Modeling with Missing Data and Auxiliary Variables, R package version 0.4.4. http://cran.r-project.org/web/packages/rsem/index.html.

Yucel, R. M. 2011. State of the Multiple Imputation Software. *Journal of Statistical Software,* 45, 1–7.

Zar, J. H. 2010. *Biostatistical Analysis.* Upper Saddle River, NJ, Prentice-Hall/Pearson.

Zhang, Z., and Wang, L. 2012. bmem: Mediation analysis with missing data using bootstrap, R package version 1.3. http://cran.r-project.org/web/packages/bmem/index.html.

Zuur, A. F., Hilbe, J. M., and Ieno, E. N. 2013. *A beginner's guide to GLM and GLMM with R: a frequentist and Bayesian perspective for ecologists.* Newburgh, UK, Highland Statistics Ltd.

Zuur, A. F., Ieno, E. N., and Smith, G. M. 2007. *Analysing Ecological Data.* New York, Springer.

Zuur, A. F., Ieno, E. N., Walker, N., Saveliev, A. A., and Smith, G. M. 2009. *Mixed effects models and extensions in ecology with R.* New York, Springer.

Zuur, A. F., Saveliev, A. A., and Ieno, E. N. 2012. *Zero inflated models and generalized linear mixed models with R.* Newburgh, UK, Highland Statistics Ltd.

Index

Printed and bound by CPI Group (UK) Ltd, Croydon, CR0 4YY